MECHANICS OF POROELASTIC MEDIA

SOLID MECHANICS AND ITS APPLICATIONS
Volume 35

Series Editor: G.M.L. GLADWELL
Solid Mechanics Division, Faculty of Engineering
University of Waterloo
Waterloo, Ontario, Canada N2L 3G1

Aims and Scope of the Series

The fundamental questions arising in mechanics are: *Why?*, *How?*, and *How much?* The aim of this series is to provide lucid accounts written by authoritative researchers giving vision and insight in answering these questions on the subject of mechanics as it relates to solids.

The scope of the series covers the entire spectrum of solid mechanics. Thus it includes the foundation of mechanics; variational formulations; computational mechanics; statics, kinematics and dynamics of rigid and elastic bodies; vibrations of solids and structures; dynamical systems and chaos; the theories of elasticity, plasticity and viscoelasticity; composite materials; rods, beams, shells and membranes; structural control and stability; soils, rocks and geomechanics; fracture; tribology; experimental mechanics; biomechanics and machine design.

The median level of presentation is the first year graduate student. Some texts are monographs defining the current state of the field; others are accessible to final year undergraduates; but essentially the emphasis is on readability and clarity.

For a list of related mechanics titles, see final pages.

Mechanics of Poroelastic Media

Edited by

A. P. S. SELVADURAI

Department of Civil Engineering and Applied Mechanics,
McGill University,
Montreal, Canada

SPRINGER-SCIENCE+BUSINESS MEDIA, B.V.

Library of Congress Cataloging-in-Publication Data

Mechanics of poroelastic media / edited by A.P.S. Selvadurai.
p. cm. -- (Solid mechanics and its applications ; v. 41)
Includes indexes.
ISBN 978-90-481-4513-3 ISBN 978-94-015-8698-6 (eBook)
DOI 10.1007/978-94-015-8698-6
1. Porous materials. 2. Elasticity. I. Selvadurai, A. P. S.
II. Series.
TA418.9.P6M45 1995
620.1'1692--dc20 95-24707

ISBN 978-90-481-4513-3

Printed on acid-free paper

Table of Contents

PREFACE

The mathematical theory of poroelasticity deals with the mechanical behaviour of an elastic porous medium which is either completely or partly filled with a pore fluid. The elastic porous phase can be either an assemblage of solid particles, a solid with a network of interconnected pores or a system of isolated pores. The pore fluids generally include air, water, oil, gas and biological fluids. During thermal, hydraulic, mechanical or electrochemical loading of a poroelastic medium, the various components of the multiphase system experience stresses and strains. The mutual interaction between the various phases of the poroelastic material will ultimately govern its thermo-hydraulic-mechanical behaviour.

The earliest recognition of the importance of the multiphase nature of porous geomaterials can be attributed to the pioneering studies by K.Terzaghi, published in 1925. In his development of the "*theory of effective stress*" for a geomaterial, Terzaghi postulated that when a saturated soil is subjected to external loading, this loading is carried partly by the porous soil skeleton and partly by the pore fluid. The ability of the pore fluid to be stressed during the application of an external loading is an important development in the understanding of the mechanical behaviour of saturated porous geomaterials. The second aspect of the influence of multiphase behaviour of geomaterials in their *time-dependent* behaviour was demonstrated by Terzaghi in the development of the classical theory of "*consolidation of soils*". This theory, which can be regarded as the origin of modern theories of poroelasticity, postulates that for soils with relatively low permeability, the load is initially borne by the pore fluid. With progress of time, the pore fluid pressures will dissipate and, at the termination of the consolidation process, the external loadings are borne entirely by the porous soil skeleton. The resulting time-dependent consolidation process results in deformations which are conventionally referred to as consolidation settlements.

The original developments by Terzaghi were primarily restricted to the examination of the one-dimensional consolidation behaviour of a saturated soil. These concepts were extended by M.A. Biot in a series of articlcs published between 1941 and 1956 to include three dimensional elastic behaviour of the soil skeleton, anisotropy of elastic behaviour and viscoelastic response of the soil skeleton. The theory of poroelasticity developed by Biot has, over the past five decades, received considerable attention within the geomechanics community. The theory has been used to develop elegant analytical solutions and computational schemes including finite element and boundary element solutions to problems in geomechanics associated with soil consolidation, offshore geotechnique, hydraulic fracturing for energy resources exploration and estimation of ground subsidence due to fluid withdrawal or heave due to fluid injection.

In recent years, the theory of poroelasticity has found extensive application in other areas, such as biomechanics of soft tissues, mechanics of bone, transport of multiphase fluids in porous media with special reference to applications in environmental geomechanics and energy resources recovery, geophysical applications with reference to the study of earthquake phenomena and in the study of advanced materials such as saturated micro-cellular foams and polymer composites. The renewed interest has resulted in the modification of Biot's classical theory to include more complex phenomena in the description of both the porous skeleton and the interstitial fluid.

The present volume contains a collection of papers which record the recent advances in the application of the theories of poroelasticity to problems of diverse interest. The contributions are grouped into primary topic areas covering constitutive modelling and analytical aspects, numerical modelling, applications in geomechanics, applications in biomechanics and dynamic effects. The contributions within these groups, however, do have several themes of common interest. The volume presents complete articles which document the versatility of theories of poroelasticity in addressing technologically important problems with significant contents in mathematics and mechanics. It is hoped that the volume will serve as a useful reference to students and researchers in geomechanics, biomechanics, applied mechanics and material science in illustrating the versatility of the classical theories of poroelasticity and the prospects for their further development.

A.P.S. Selvadurai
McGill University
Montréal, Canada

ACKNOWLEDGEMENTS

The papers contained in this volume were largely derived from the Mini-Symposium on "*Recent Developments in Poroelasticity*" organized by the Elasticity Committee of the Applied Mechanics Division of the American Society of Mechanical Engineers. The Symposium was a part of the Technical Program of the 1994 International Mechanical Engineering Congress held in Chicago, Illinois, U.S.A. As the organizer of the Symposium, the Editor would like to express his sincere thanks to Professor Millard F. Beatty (University of Nebraska), Chairman of the Elasticity Committee of the Applied Mechanics Division of the ASME, for his very generous support and encouragement in all aspects related to its scheduling within the activities of ASME. Thanks are also due to Professors Steve C. Cowin (The City College of CUNY) and Alex H.-D. Cheng (University of Delaware) for providing the names of additional participants which considerably enhanced the scope of both the Symposium and the Volume. The willingness of the authors to attend the Symposium and to contribute to the Volume is greatly appreciated both by the Editor and the Elasticity Committee of the Applied Mechanics Division of ASME. The authors are thanked for promptly attending to the various revisions associated with the contributions. The Editor would like to express his appreciation to Professor Graham M.L. Gladwell (University of Waterloo), Series Editor, for recommending the project as a Volume in the Kluwer Academic Publishers series '*Solid Mechanics and its Applications*'. The assistance provided by Mrs. Ann Bless, of the Department of Civil Engineering and Applied Mechanics at McGill University, in coordinating the correspondence is gratefully acknowledged.

Finally, the preparation of this Volume was greatly facilitated by the editorial assistance provided by Mrs. Sally Selvadurai, who completed the copy editing and compiled the Author and Keyword (Subject) Index.

SECTION 1

POROELASTICITY: CONSTITUTIVE MODELS AND ANALYTICAL ASPECTS

Moving and Stationary Dislocations in Poroelastic Solids and Applications to Aseismic Slip in the Earth's Crust

J. W. RUDNICKI
Departments of Civil Engineering,
Mechanical Engineering and Geological Sciences
Northwestern University
Evanston, IL 60208-3109

Abstract Results for the pore pressure induced by a plane strain shear dislocation that starts from rest, moves a finite distance at constant speed and stops demonstrate that coupling between deformation and diffusion causes a complex response even though the spatial distribution of slip is simple. A summary of recent solutions for stationary, instantaneous plane strain shear and opening dislocations and steadily moving shear dislocations demonstrates that coupling between deformation and diffusion is significant for locations near the dislocation edge and for short times. In addition, the response depends strongly on whether the plane of the dislocation is permeable or impermeable. Applications of these solutions to interpret water well level changes caused by aseismic slip (creep) in the Earth's crust are discussed.

1. Introduction

In portions of the earth's shallow crust that are infiltrated with ground water, the coupling of deformation with diffusion can significantly affect the mechanical response. In particular, this coupling has been identified as a factor in the propagation of hydraulic fractures (Ruina, 1978; Cleary, 1978; Detournay and Cheng, 1991a; Detournay *et al.*, 1990; Atkinson and Craster, 1991), time-dependent deformation following earthquake slip (Nur and Booker, 1972; Booker, 1974; Rice and Cleary, 1976; Rudnicki, 1986), propagating aseismic slip (creep) events (Rice and Cleary, 1976; Cleary, 1978; Rice and Simons, 1976; Simons, 1979; Rudnicki and Koutsibelas, 1991; Rudnicki, 1991). Detournay and Cheng (1991b) have recently summarized the effects of the coupling on a variety of rock mechanics problems, including borehole deformation and hydraulic fracture.

Solutions for dislocations in elastic solids provide elemental models for shear and tensile fractures. Even when the approximation of ideally elastic behavior outside of the

A. P. S. Selvadurai (ed.), Mechanics of Poroelastic Media, 3–22.

fracture is extreme, the solutions can provide insight into the magnitude and time scale of deformation. Furthermore, in applications to the earth's crust, material properties are often not known in sufficient detail to warrant more elaborate modeling. When more elaborate modeling is justified, dislocation solutions can provide the basis for efficient numerical modeling.

This paper reviews solutions for plane strain dislocations in poroelastic solids. The coupling of diffusion with deformation introduces a time dependence into the response in addition to any associated with imposition of the dislocation. The solutions reviewed include those for stationary shear and opening dislocations introduced instantaneously and for shear dislocations propagating steadily. In all cases, the plane containing the dislocation may be permeable or impermeable. In addition, we derive the solution for the pore pressure induced by a shear dislocation that moves non-steadily on a permeable plane. In particular, the dislocation starts from rest, moves at constant speed and stops.

The paper briefly summarizes application of the solutions to interpretation of water well level changes due to episodic creep events in the earth's crust. Creep events are slip episodes that occur too slowly to generate seismic waves. On some portions of large fault systems, like the San Andreas, creep can occur steadily at a rate comparable to the relative motion between tectonic plates (a few centimeters per year on the San Andreas). At other places, creep occurs intermittently in discrete events that appear to propagate laterally (parallel to the surface), toward the surface from depth or a combination of both. Because creep events occur relatively slowly, at least by comparison to propagation speeds of seismic wave or rupture velocities of earthquakes, the time scale of the associated deformation is comparable to that for ground water diffusion and the effects of coupling can be significant.

2. Governing Equations

In this section, we briefly summarize the governing equations for an isotropic poroelastic solid, beginning with the constitutive realtions. The presentation of the constitutive relations follows that of Rice and Cleary (1976). They improved upon Biot's (1941) formulation by exploiting the observation that in the limit of undrained response (no change in fluid mass content of material elements), the strains of the solid matrix are given by the usual elastic relation but with a different value of the Poisson's ratio.

For general deformations the strain tensor of the solid matrix ϵ_{ij} and the alteration of the fluid mass content m from some reference value m_0 depend on the total stress tensor σ_{ij} and the pore pressure p (measured from some ambient value) as follows:

$$\epsilon_{ij} = \frac{1}{2G}\left[\sigma_{ij} - \frac{\nu}{1+\nu}\sigma_{kk}\right] + p\delta_{ij}\,\frac{3(\nu_u - \nu)}{2GB(1+\nu)(1+\nu_u)} \tag{1}$$

$$m - m_0 = \frac{3\rho_0(\nu_u - \nu)}{2GB(1+\nu)(1+\nu_u)}\left[\sigma_{kk} + \frac{p}{3B}\right] \tag{2}$$

where ρ_0 is the density of the pore fluid (in the reference state), δ_{ij} is the Kronecker delta, the indicies (i, j) range over (1,2,3) and a repeated index denotes summation. In the limit of drained deformation, any alterations of pore fluid pressure are eliminated by fluid mass diffusion and, consequently, p=0. Thus, ν is identified as the drained Poisson's ratio and G is the shear modulus (which turns out to be the same for drained and undrained response). In the alternative limit of undrained deformation, the fluid mass in material elements remains constant, $m = m_0$, and from (2), p=-$B\ \sigma_{kk}/3$, where B is Skempton's coefficient. If this expression for the pore pressure is substituted into (1), the equation can be arranged in the form of the usual elasticity relation with ν_u replacing ν. Thus, ν_u is the undrained Poisson's ratio.

Rice and Cleary (1976) give the following expressions for B and ν_u:

$$B = \frac{1 - K/K_s'}{1 - K/K_s' + v_0 K/K_f - v_0 K/K_s''} \tag{3}$$

$$\nu_u = \frac{3\nu + B(1-2\nu)(1-K/K_s')}{3 - B(1-2\nu)(1-K/K_s')} \tag{4}$$

where K (=$2G(1+\nu)/3(1\text{-}2\nu)$) is the drained bulk modulus, K_f is the bulk modulus of the pore fluid, v_0 is porosity (in the reference state) and K_s'' and K_s' are bulk moduli that can be identified with the bulk modulus of the solid constituents if the pore space is interconnected and the solid constituent is locally isotropic and homogeneous (Nur and Byerlee, 1971; Rice and Cleary, 1976). From (3) and (4), it can be deduced that ν_u and B satisfy the following limits: $\nu \leq \nu_u \leq 1/2$ and $0 \leq B \leq 1$. In each case the upper bound occurs for separately incompressible solid and fluid constituents and the lower for a highly compressible pore fluid.

The final constitutive relation is Darcy's law which states that the fluid mass flux per unit area q_i is proportional to the gradient of the pore fluid pressure:

$$q_i = -\rho_0 \kappa \partial p/\partial x_i \tag{5}$$

where κ is a permeability more commonly expressed as k/μ where k has units of area and μ is the viscosity of the pore fluid.

The governing equations are completed by the strain-displacement relations

$$\epsilon_{ij} = \frac{1}{2}\left(\frac{\partial u_i}{\partial x_j} + \frac{\partial u_j}{\partial x_i}\right) \tag{6}$$

and equations expressing equilibrium of total stresses (in the absence of body forces) and fluid mass conservation

$$\frac{\partial \sigma_{ij}}{\partial x_i} = 0 \tag{7}$$

$$\frac{\partial q_i}{\partial x_i} + \frac{\partial m}{\partial t} = 0 \tag{8}$$

Rice and Simons (1976) showed that for plane strain deformation, the governing equations could be reduced to a set of four for the three in-plane stress components and pore pressure. Rudnicki (1987) and Rudnicki and Roeloffs (1990) followed this approach to obtain the dislocations solutions. Here, we illustrate an alternative approach and write the equations entirely in terms of the displacements and pore pressure. To this end, we invert the constitutive relations (1) and (2). The results are

$$\sigma_{ij} = 2G\epsilon_{ij} + (K - 2G/3)\epsilon_{kk}\delta_{ij} - \zeta p \delta_{ij} \tag{9}$$

$$m - m_0 = \rho_0 \zeta [\epsilon_{kk} + p/K_u B] \tag{10}$$

where $\zeta = 1 - K/K_s'$ and K_u is the undrained bulk modulus, $2G(1+\nu_u)/3(1-2\nu_u)$, which can also be expressed as

$$K_u = K/(1 - \zeta B) \tag{11}$$

Using the strain-displacement relations (6) in (9) and (10) and then substituting into equilibrium (7) and fluid mass conservation (8) yields

$$(K + G/3)\frac{\partial^2 u_k}{\partial x_k \partial x_j} + G\frac{\partial^2 u_j}{\partial x_k \partial x_k} - \zeta\frac{\partial p}{\partial x_j} = 0 \tag{12}$$

$$-\kappa\frac{\partial^2 p}{\partial x_k \partial x_k} + \zeta\frac{\partial}{\partial t}\left[\frac{\partial u_k}{\partial x_k} + \frac{p}{K_u B}\right] = 0 \tag{13}$$

A useful result is obtained by taking the divergence of (12):

$$\nabla^2[(K + 4G/3)e - \zeta p] = 0 \tag{14}$$

where $e = \partial u_k/\partial x_k$ and $\Delta^2(\ldots) = \partial^2(\ldots)/\partial x_k \partial x_k$ is the Laplacian. Using the result makes it possible to rewrite (13) as

$$c\nabla^2\left[e + \zeta p/K_u B\right] = \frac{\partial}{\partial t}\left[e + \zeta p/K_u B\right] \tag{15}$$

where c is the diffusivity:

$$c = \kappa\frac{K_u B(K + 4G/3)}{\zeta^2 K_u B + \zeta(K + 4G/3)} \tag{16}$$

or, in terms of the quantities used by Rice and Cleary (1976),

$$c = \kappa \frac{2GB^2(1+\nu_u)^2(1-\nu)}{9(\nu_u-\nu)(1-\nu_u)} \quad \textbf{(17)}$$

Comparison of (15) with (10) reveals that, as emphasized by Rice and Cleary (1976), the fluid mass content satisfies the homogeneous diffusion equation.

3. Formulation of Dislocation Problem

Consider plane strain deformation in the *xy* plane. As already mentioned, to solve the dislocation problem, Rudnicki (1987) and Rudnicki and Roeloffs (1990) followed the procedure of Rice and Simons (1976) in writing the governing equations in terms of the three in-plane stress components and the pore pressure. The conditions resulting from imposition of the dislocation were then expressed in terms of conditions on the stress and pore pressure on $y = 0$ and the resulting boundary value problem was solved in $y \geq 0$. Here, the equations (12-15) are in terms of displacements (where, for plane strain, the indicies range over (1,2) and $(x, y) = (x_1, x_2)$). Consequently, the boundary conditions can also be left in terms of displacements.

An instantaneous, plane strain shear dislocation at the origin in the *xy* plane corresponds to cutting the entire negative *x* axis, displacing the upper half ($y = 0^+$) to the right an amount $b/2$, the bottom ($y = 0^-$) to the left $b/2$ and then rejoining the two together. If this is done instantaneously at $t=0$, the condition on the *x*-direction displacement as $y=0$ is approached through positive values is

$$u_x(x,y=0^+) = (b/2)H(-x)H(t) \quad \textbf{(18)}$$

where H is the unit step function. Because u_y is anti-symmetric about $y = 0$ and is continuous, u_y vanishes on $y=0$:

$$u_y(x,y=0) = 0 \quad \textbf{(19)}$$

The pore pressure is also anti-symmetric about $y=0$. For a permeable fault the pore pressure is continous and, hence, must vanish on $y= 0$:

$$p(x,y=0) = 0 \quad \textbf{(20)}$$

The gradient $\partial p/\partial y$ will not equal zero on $y = 0$. According to Darcy's law (5), the fluid mass flux in the *y*-direction is proportional to this gradient and, hence, the solution for the condition (20) is appropriate for slip on a plane that is permeable to the diffusing species.

There is, however, evidence from both laboratory and field studies that faults are often barriers to fluid flow (Wu *et al.*, 1975; Wang and Lin, 1978; Lippincott *et al.*, 1985). This can occur because the fault zone material contains significant amounts of clay or strongly comminuted material so as to make fluid flow difficult. Alternatively, a history of extensive shearing combined with repeated dissolution and precipitation can cause development of a strongly anisotropic fabric that allows fluid flow along the fault but prevents it across the fault (Rice, 1992; Byerlee, 1990). In either case, the limiting condition of a completely impermeable fault can be imposed by requiring

$$\partial p / \partial y = 0 \tag{21}$$

on $y = 0$. Because the pore pressure field must still be antisymmetric about $y = 0$ for the shear dislocation, the pore pressure on $y = 0$ is discontinous and approaches equal and opposite values as $y = 0$ is approached from above and below. This discontinuity is the limiting case of a severe gradient that would occur across a very narrow, but finite width, relatively impermeable fault zone.

In summary, the problem is reduced to solving equations (12) and (15) for u_x, u_y and p in $y \geq 0$, subject to the boundary conditions (18), (19) and one of (20) or (21) on $y = 0$.

The formulation here is for a shear dislocation introduced instantaneously. For an opening dislocation the roles of u_x and u_y in (18) and (19) would be interchanged. For a steadily moving dislocation, time can be eliminated by viewing the problem in a coordinate system that is translating with the steadily moving dislocation. As a consequence, the derivative $\partial(..)/\partial t$ in (15) can be replaced by $-V\partial/\partial x$. Thus, the boundary condition for a shear dislocation (in the moving coordinate system) would be (18) with the step function in time eliminated.

The solution can be obtained by the application of the Fourier transform on x and the Laplace transform on t. The Fourier transform of a function $f(x,y,t)$ is defined by

$$\mathcal{F}[f(x,y,t)] = \bar{f}(\xi,y,t) = \int_{-\infty}^{\infty} f(x,y,t)\exp(-\iota\xi x)dx \tag{22}$$

and the Laplace transform by

$$\mathcal{L}[\bar{f}(\xi,y,t)] = F(\xi,y,s) = \int_{0}^{\infty} \bar{f}(\xi,y,t)\exp(-st)dt \tag{23}$$

Application of the double transform to (14) and (15) yields

$$\left[\frac{d^2}{dy^2} - \xi^2\right][(K + 4G/3)E - \zeta P] = 0 \tag{24}$$

$$\left[\frac{d^2}{dy^2} - (\xi^2 + s/c)\right][(K_u - K)E + \zeta P] = 0 \tag{25}$$

where we have used (11) to eliminate the product $B\zeta$. The solutions for the doubly transformed solid dilation E and pore pressure P are

$$E = A(\xi,s)e^{-m(\xi)y} + B(\xi,s)e^{-n(\xi,s)y} \tag{26}$$

$$\zeta P = -(K_u - K)A(\xi,s)e^{-m(\xi)y} + (K+4G/3)B(\xi,s)\exp^{-n(\xi,s)y} \tag{27}$$

where

$$m^2(\xi) = \xi^2\,,\quad n^2(\xi,s) = \xi^2 + s/c \tag{28}$$

and the branches are chosen such that $\mathrm{Re}\{m(\xi)\} > 0$ and $\mathrm{Re}\{n(\xi,s)\} > 0$. An equation for the double transform of the x direction displacement U_x can be determined from using (26) and (27) in the double transform of the first of (12) ($j=1$). The solution of the resulting equation is

$$U_x = e^{-m(\xi)y}\left[C + A\frac{(K_u+G/3)}{G}\frac{\iota\xi y}{m(\xi)}\right] + \frac{\iota\xi B}{n^2(\xi,s)-\xi^2}e^{-n(\xi,s)y} \tag{29}$$

This expression and that for E (26) can be used to obtain an expression for dU_y/dy. Integration then yields

$$U_y = -\frac{e^{-m(\xi)y}}{m(\xi)}\left\{\frac{A}{G}\left[(K_u+4G/3) + \frac{\xi^2 y(K_u+G/3)}{m(\xi)}\right] - \iota\xi C\right\} - \frac{Bn(\xi,s)}{n^2(\xi,s)-\xi^2}e^{-n(\xi,s)y} \tag{30}$$

Expressions for the doubly transformed stresses can be determined by differentiation with respect y and multiplication by $\iota\xi$. The conditions (18), (19), and either (20) or (21) can be used to determine A, B, and C. The physical stresses and displacements are obtained by inverting the transforms, which is the most difficult aspect of the solution. Details of the transform inversions for the stresses due to the instantaneous dislocations are given by Rudnicki (1987).

For the steadily moving dislocations only the Fourier transform is needed. The solution for the transformed pore pressure and displacements components is identical to that given here with s/c replaced by $-\iota\xi V/c$ in n. Details of the transform inversion for the steadily moving shear dislocations are given by Rudnicki and Roeloffs (1990).

The next two sections present some results for the pore pressure induced by the instantaneous and steadily moving dislocations. The following section derives results for the pore pressure induced by a shear dislocation on a permeable plane that moves with constant speed but starts from rest and stops.

4. Instantaneous Dislocation Solutions

The pore pressure caused by a plane strain shear dislocation introduced instantaneously at the origin in the xy plane is

$$p(x,y,t) = AbG\frac{y}{r^2}\left[1 - \exp(-\frac{r^2}{4ct})\right] \tag{31}$$

where $r = (x^2 + y^2)^{1/2}$ and A is the following combination of constants:

$$A = B(1+\nu_u)/3\pi(1-\nu_u) \tag{32}$$

This solution was first presented by Nur and Booker (1972). The solution for the stress field, assuming incompressible solid and fluid constituents, was derived by Booker (1974) using integral transforms and for arbitrarily compressible constituents by Rice and Cleary (1976) using complex variables. The pore pressure given by (1) vanishes on the plane $y = 0$ as required by the condition (20) but the gradient $\partial p/\partial y$ does not. Thus, as explained earlier, the solution is appropriate for slip on a plane that is permeable to the diffusing species.

Rudnicki (1986) presented the result for the pore pressure induced by a shear dislocation on an impermeable plane:

$$p(x,y,t) = AGb\left[\frac{y}{r^2}\mathrm{erf}(\frac{y}{\sqrt{4ct}}) + \frac{x}{r^2}\exp(-\frac{y^2}{4ct})\mathrm{D}(\frac{x}{\sqrt{4ct}})\right] \tag{33}$$

where erf(u) is the error function, D(u) is $2/\pi^{1/2}$ times Dawson's integral (Abramowitz and Stegun, 1964),

$$\mathrm{D}(u) = -\iota\exp(-u^2)\,\mathrm{erf}(\iota u) \tag{34}$$

and $\iota = \sqrt{(-1)}$. Equation (34) gives the result for $y \geq 0$; that for $y \leq 0$ is equal in magnitude and opposite in sign. Rudnicki (1987) used Fourier and Laplace transforms to derive the complete stress field.

Figure 1 plots the pore pressure predicted by (32) and (33), in the nondimensional form pr/AGb, against the nondimensional time $4ct/r^2$. In this form, the solution depends only on the angle θ and results are shown for $\theta = 0°$, $30°$ and $90°$. For the permeable fault, the pore pressure is zero on $\theta = 0°$ and the result shown for the impermeable fault is that for $y = 0^+$. Notable differences between the solutions are the slower decay for the impermeable fault, the larger pore pressure changes at low angles for the impermeable fault and the initial increase with time of the pore pressure for the impermeable fault at angles less than 90°.

Rice and Cleary (1976) also give the solution for a plane strain opening dislocation. In this case the pore pressure field is symmetric about $y = 0$. Consequently, the pore pressure has a finite nonzero value on $y = 0$ but the gradient $\partial p/\partial y$, assumed to be continuous, vanishes on $y = 0$. Rudnicki (1987) gives the corresponding solution for an opening dislocation when the pore pressure is required to be zero on the x-axis. In this case, the gradient $\partial p/\partial y$ is discontinuous on $y = 0$ and takes on values equal in magnitude and opposite in sign as the x-axis is approached from above and below. For the opening dislocation, the condition $p = 0$ on $y = 0$ is a limiting idealization of a thin layer of high permeability in the x direction so that any alterations in pore pressure are eliminated by rapid fluid mass diffusion along this plane.

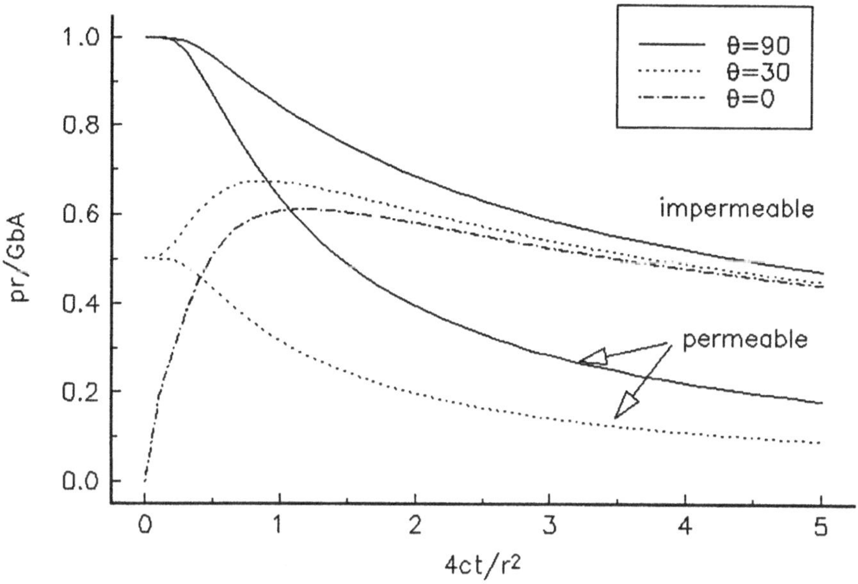

Figure 1. *Comparison of the pore pressure induced by sudden introduction of a shear dislocation on a permeable and an impermeable plane.*

The solutions for plane strain shear and opening dislocations can be written in compact form by introducing a complex displacement discontinuity (Burgers' vector), $b = b_x + \iota b_y$ and a complex position vector $z = x + \iota y$. The solution for a shear dislocation on a permeable plane and opening dislocation on an impermeable plane (continuous pore pressure and gradient) is

$$p(x,y,t) = -A\,G\,\mathrm{Im}\{b z^{-1}[1 - \exp(-r^2/4ct)]\} \tag{35}$$

where Im{...} stands for "the imaginary part of" {...}. The solution for a shear dislocation on an impermeable plane (discontinuous pore pressure) and opening dislocation on a permeable plane (discontinuous gradient) is

$$p(x,y,t) = -A\,G\,\mathrm{Im}\{b z^{-1}[1 - \mathrm{W}(x,y,t)]\} \tag{36}$$

where

$$\mathrm{W}(x,y,t) = \mathrm{erfc}[y/\sqrt{4ct}\,] + \exp(-r^2/4ct)\,\mathrm{erf}[\iota x/\sqrt{4ct}\,] \tag{37}$$

and erfc(u) is the complementary error function. Rudnicki (1987) has given the corresponding expressions for the complete stress fields.

Rudnicki *et al.* (1993) used the solutions (31) and (33) to interpret five slip-induced water level changes observed during January 1989 to July 1990 in a well 400 meters from the San Andreas fault near Parkfield, California. For the well studied and the frequency of the water level changes (see Roeloffs *et al.* (1989) for a more detailed discussion of the frequency dependence), the calculated pore pressure changes can be converted to water level changes by dividing by the weight density of water. By comparing the observed and calculated water level changes, Rudnicki *et al.* (1993) were able to infer the magnitude of the slip and the distance that the slip extended along the fault past the well. In three cases the inferred magnitude and extent of slip were consistent with that measured by creepmeters at the surface several hundred meters from the well. Discrepancies in the remaining two cases were interpreted to be the result of a depth variation of slip that was not included in the two dimensional model. Assuming the fault is impermeable required a higher diffusivity (0.15 m^2/s) than for a permeable fault (0.06 m^2/s) but otherwise had little effect.

The dislocation solutions themselves are an extreme and overly simple model of slip on actual fault systems, but, as demonstrated by Rudnicki *et al.* (1993), they are sometimes adequate to interpret the available data. More complex spatial distributions and time histories of slip can be constructed by superposition. Rudnicki and Hsu (1988) have presented results for a boxcar distribution of slip (uniform slip over a finite length) with a ramp increase of slip with time. They use the results to interpret a water level change observed by Lippincott *et al.* (1985) in a well near the Garlock fault. Lippincott *et al.* (1985) interpreted the water level changes by neglecting coupled deformation diffusion effects, using the ordinary elasticity solution for a steadily propagating distribution of slip, and assuming the pore pressure change is proportional to the negative of the mean stress change. That a satisfactory fit to the data could be achieved by either approach, with quite different slip distributions, suggests the necessity for knowing hydrologic conditions for detailed modeling.

5. Steady State Solutions

Solutions for dislocations moving steadily at a speed V can be constructed using the solutions for instantaneous dislocations and the superposition procedure described by Carslaw and Jaeger (1959). Cleary (1978) used this approach to obtain numerical results for opening and shear dislocations. Roeloffs and Rudnicki (1984/85) used the same approach to obtain the following result for the pore pressure induced by a shear dislocation moving steadily on a permeable plane:

$$p(X,Y) = AGb(Y/R^2)[1 - (VR/2c)\mathrm{K}_1(VR/2c)\exp(-VR/2c)] \tag{38}$$

where $\mathrm{K}_1(z)$ is the modified Bessel function of the second kind (Abramowitz and Stegun, 1964) and we have used upper case for the coordinates to emphasize that the origin moves with the dislocation. Roeloffs and Rudnicki (1984/85) used this solution to evaluate coupled deformation diffusion effects for propagating slip events and to re-examine a slip induced water level change reported by Johnson *et al.* (1973) and analyzed by Wesson (1981) using the ordinary elasticity solution for a steadily moving distribution of slip.

Rudnicki and Roeloffs (1990) used integral transforms to derive the complete stress and pore fluid pressure fields for shear dislocations moving on both permeable and impermeable planes. (Simons (1979) had previously formulated the problem for a shear dislocation moving steadily on a permeable plane and used Fourier transforms to obtain an expression for the shear stress induced on the plane of the dislocation.) The pore pressure induced by a shear dislocation moving steadily on an impermeable plane is (Rudnicki and Roeloffs, 1990):

$$p(X,Y) = AbG(V/2c)\exp(-VX/2c) \int_{-\cos\theta}^{1} \frac{(1+u\cos\Theta)}{(1-u^2)^{1/2}} \exp[-(VR/2c)u]\, du \tag{39}$$

where $\theta = \tan^{-1}(Y/X)$ and the result pertains for $Y \geq 0$. A surprising feature of this solution is that the pore pressure is identically zero for $Y = 0$ and $X < 0$. At $R = 0$ the pore pressure approaches a finite limit that depends on θ:

$$\lim_{R\to 0} p(X,Y) = AbG(V/2c)(\pi - \Theta + \sin\Theta\cos\Theta) \tag{40}$$

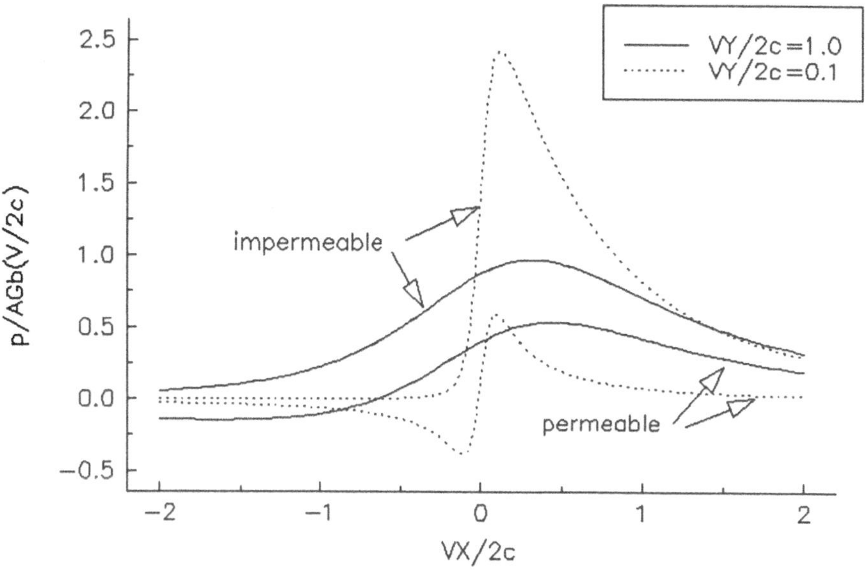

Figure 2. *Comparison of the pore pressures induced by a shear dislocation moving steadily on permeable and impermeable planes.*

Figure 2 plots the nondimensional pore pressure, $p/AGb(V/2c)$, against nondimensional distance along the slip plane, $VX/2c$, for shear dislocations moving steadily on a permeable plane and an impermeable plane. Results are shown for two values of nondimensional distance from the slip plane, $VY/2c = 0.1$ and 1.0. As for the solutions for the instantaneous dislocations, the pore pressure change is larger for the impermeable fault and decays more slowly ahead of the dislocation. Note that for the permeable fault the pore pressure change is negative behind the moving edge of the dislocation ($X < 0$). As explained by Roeloffs and Rudnicki (1984/85), this change of sign occurs because as the dislocation edge passes a fixed point the compressive mean stress decreases more quickly than diffusion can restore the fluid mass to its undisturbed value. The difference between the fluid mass change and a constant times the mean stress is proportional to the pore pressure change (3). The pore pressure for the impermeable plane does not change sign but does decrease rapidly behind the moving dislocation edge, especially for small values of $VY/2c$.

6. Non-steadily Moving Dislocations

In this section, we use the superposition procedure outlined by Carslaw and Jaeger (1959) to derive results for dislocations that move at a constant speed but start and stop at fixed times. Results are presented only for the pore pressure due to a shear dislocation moving on a permeable plane but the same procedure could be used to derive results for other stress components, for a shear dislocation moving on an impermeable plane, or for opening dislocations.

Equation (31) gives the pore pressure due to imposition of a dislocation with step function time dependence. Superposing contributions from instantaneous dislocations introduced at times τ and positions $x = X\text{-}V(t\text{-}\tau)$ yields the following expression for the pore pressure due to a dislocation that appears suddenly at $t = 0$ and begins moving at constant speed V in the positive x direction:

$$p(X,Y,t) = AbG\left[\frac{Y}{R^2} - \int_0^t \frac{Y}{4c(t-\tau)^2}\exp\left(\frac{(X-V(t-\tau))^2+Y^2}{4(t-\tau)^2}\right)d\tau\right] \tag{41}$$

where we have again used upper case letters to emphasize that the coordinate origin moves with the dislocation. A change of variable transforms the result to a more compact form:

$$p(X,Y,t) = AbG(V/2c)(Y/R^2)\left[1 - (VR/4c)^2 e^{-VX/2c}\int_0^{V^2t/4c} e^{-u-(VR/4c)^2/u}\frac{du}{u^2}\right] \tag{42}$$

In the limit $t\rightarrow\infty$, (42) approaches the steady state solution (38). Thus, for large times, it is more convenient to rewrite (42) as the steady state solution plus a correction term:

$$p(X,Y,t)=AbG\frac{V}{2c}\frac{Y}{R^2}\left[1-e^{-VX/2c}\frac{VR}{2c}K_1(VR/2c)+\left(\frac{VR}{4c}\right)^2 e^{-VX/2c}\int_{V^2t/4c}^{\infty} e^{-u-(VR/4c)^2/u}\frac{du}{u^2}\right] \tag{43}$$

Figure 3 compares the solution for a dislocation that appears at $t = 0$ and begins moving with speed V (eqn. (42) or (43)) with the solutions for steady state motion and for an instantaneous dislocation (31) rewritten in coordinates centered on the moving dislocation, i.e., setting $x = X + Vt$ in (31). Results are shown for $VY/2c = 1$ and two values of the nondimensional time $V^2t/2c$, 0.5 (a) and 1.0 (b). The solution for the dislocation that starts from rest (labelled "Moving" in Figure 3) quickly approaches the steady state solution for $X > 0$, but does so more slowly for $X<0$. For $V^2t/2c$ greater than about two, the solution is close to the steady state solution for all X.

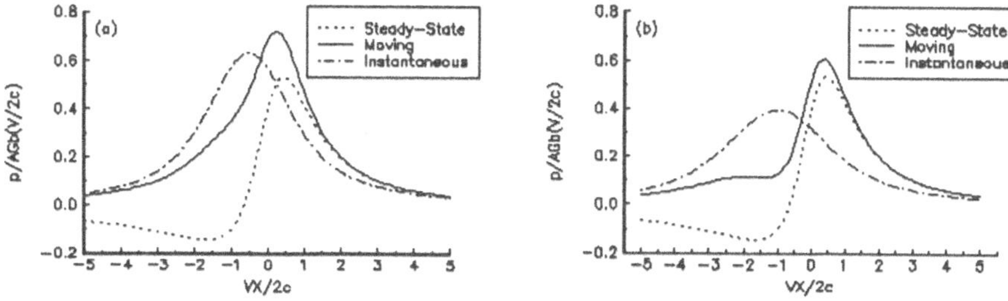

Figure 3. *Compares pore pressure induced by a shear dislocation that appears at $t=0$ and begins moving a speed V with the steady state and instantaneous solutions.*

Another view of the solution can be obtained by considering the time evolution of the pore pressure at a fixed position. Figure 4 compares the nondimensional pressure for the moving dislocation with those for the steady state and instantaneous solutions at $Vy/2c=1.0$ and two values of $Vx/2c=0.25$ (a) and 0.50 (b). The pressure is plotted against the nondimensional time $V^2t/2c$. The peak pressure change occurs when the edge of the dislocation passes the observation point and for $V^2t/2c$ greater than about one, the solution is close to the steady state value.

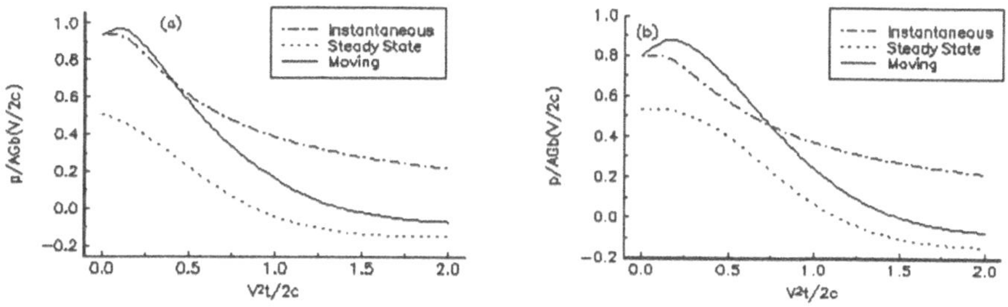

Figure 4. *Same as Figure 3 except the comparison is shown for the time history at two locations $Vy/2c = 1$ and $Vx/2c = 0.25$ (a) and 0.50 (b)*

In (42) and (43) the dislocation extends to infinity in the negative x direction. To obtain the solution for a dislocation from $x = 0$ to Vt, it is necessary to subtract the instantaneous solution. The resulting pressure, in nondimensional form, is plotted against $V^2t/2c$ for $Vy/2c=1.0$ and three values of $Vx/2c=$ 0.5, 1.0, and 1.5 in Figure 5a and for $Vx/2c = 1.0$ and three values of $Vy/2c=$ 0.5, 1.0, and 1.5 in Figure 5b. Because the length of the displacement discontinuity is Vt, there is no instantaneous pressure change at $t=0$ but the time variation is more dramatic than in Figure 4.

Thus far, the solutions above have been for dislocations that appear suddenly at the origin at $t = 0$ and then move to the right indefinitely at a constant speed V. Now,

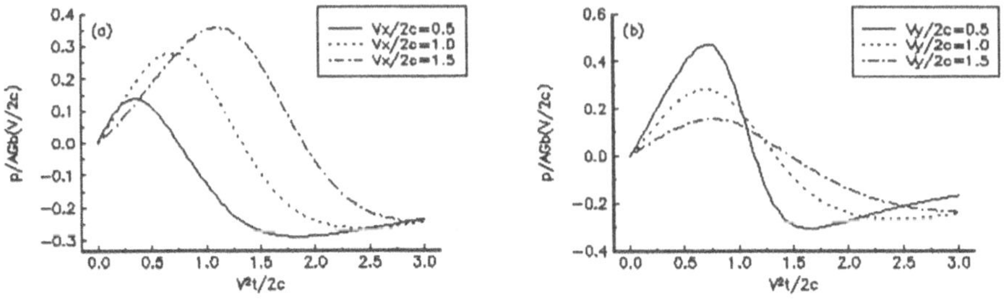

Figure 5. *Non-dimensional pore pressure for a dislocation with left end fixed at x=0 and right end at x = Vt: (a) Vy/2c = 1.0; (b) Vx/2c = 1.0.*

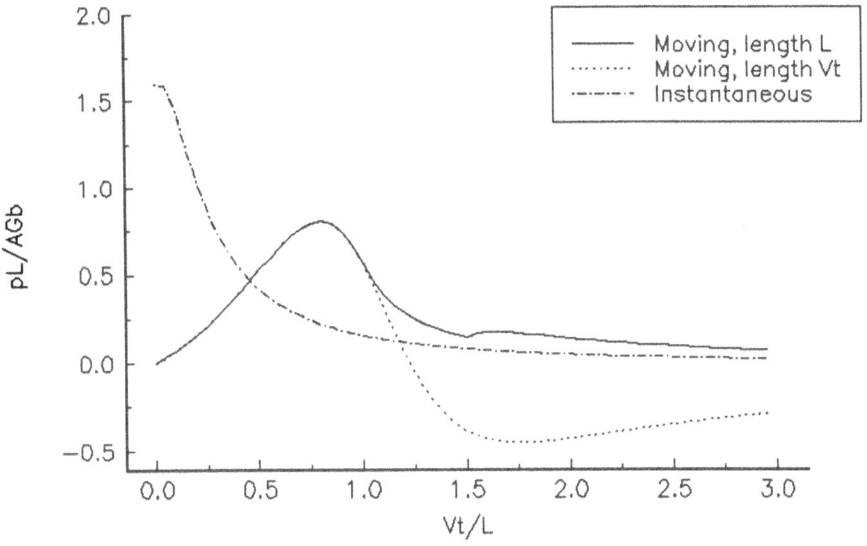

Figure 6. *Comparison of the pore pressure for the a dislocation that moves distance L at a speed V with one that appears suddenly and one that moves to the right indefinitely.*

consider the solution for a dislocation that appears suddenly at the origin at $t = 0$, moves to the right at constant speed V for a distance L, and then stops. Thus, for times greater than L/V, the distribution of displacement continuity is a boxcar extending from $x=0$ to $x=L$. This solution can be constructed from the preceding one by adding a dislocation of opposite sign and equal magnitude that appears at position $x=L$ and time $t=L/V$ and then moves to the right with constant speed cancelling the initial dislocation for $x > L$.

Figure 6 compares the solution for a dislocation that moves a distance L at a speed V with that for a stationary dislocation that appears instantaneously at $t=0$ on $y=0$, $0 < x <$ L and a dislocation that moves indefinitely to the right. The comparison is shown for $x/L = 1.0$, $y/L = 0.5$ and $VL/2c = 2.0$. The solutions for the moving dislocations are identical until $t = L/V$ and the solution for the dislocation that stops approaches that for the instantaneous dislocation. The approach to the instantaneous dislocation solution is more rapid for larger values of $VL/2c$.

The nondimensional grouping $VL/2c$ is the ratio of the slip zone length L to the diffusion length for steady propagation $2c/V$. For large values of the product VL the solution will approach the instantaneous solution (except for small times). Figure 7 shows the variation for three values of $VL/2c$, 0.5, 1.0 and 2.0 for the fixed position $x/L = 1.0$, $y/L = 0.5$.

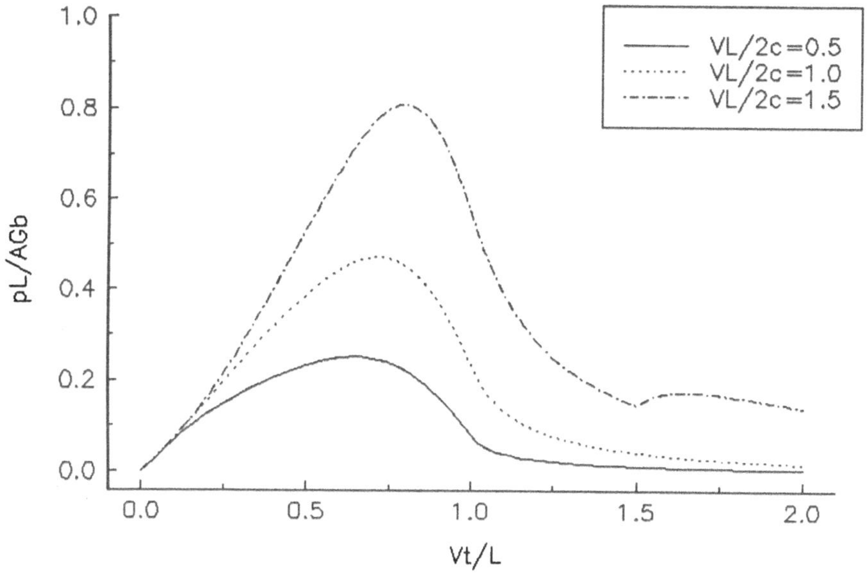

Figure 7. *Effect of VL/2c on the pore pressure induced at x/L = 1.0, y/L = 0.5 by a dislocation that moves with speed V over a distance L, then stops.*

Figure 8 shows the solution at different locations for $VL/2c = 1.0$. Figure 8a shows the solution for a fixed value of $y/L = 0.5$ and three values of $x/L = 0.5$, 1.0, and 1.5; Figure 8b shows the pore pressure response for a fixed value of $x/L = 1.0$ and three values of $y/L = 0.25$, 0.5, and 1.0. The effect of the stopping of the slip event is evident in Figure 8a as is the dramatically different pore pressure histories experienced by different positions parallel to the slip zone.

7. Numerical Values

The results have been all been plotted in nondimensional form for greater generality and to emphasize the essential dependence on nondimensional groups. But to gain some

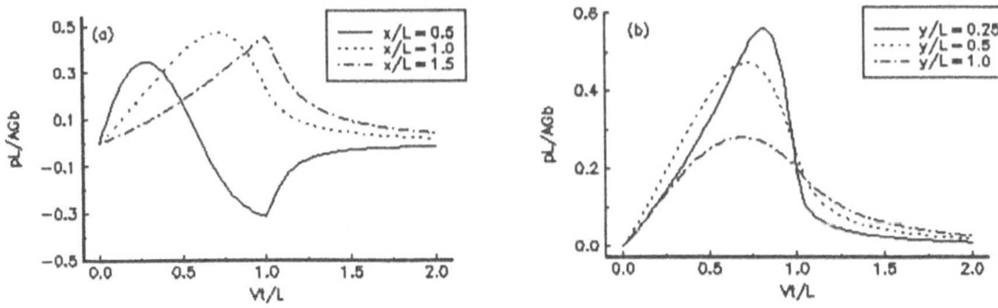

Figure 8. *Same as Figure 5 but for a fixed value of VL/2c = 1.0 and different locations: (a) y/L = 0.5; (b) x/L = 1.0.*

perspective on the magnitudes of the quantities involved, it is useful to convert nondimensional parameters to physical dimensions.

A representative velocity of creep events is one to 10 km/day, but values somewhat less and up to 50 km/day have been reported (Roeloffs and Rudnicki, 1984/85). Field values of the diffusivity c vary enormously but the range 0.1 to 1.0 m^2/s seem to be representative of earthquake regions (Rudnicki, 1984). The diffusion length $2c/V$ varies from 1.7 m to 173 m for c and V in the range cited. For this range of diffusion lengths and a slip length L ranging from 100m to 1 km, the ratio $VL/2c$ varies from 0.58 to 578. For large values of $VL/2c$ the solution for a dislocation that moves a distance L at speed V will approach that for a finite length instantaneous dislocation, except at points very near the slip plane.

The variation of the diffusion time $2c/V^2$ is even greater than that of the diffusion length because of the appearance of V^2 in the denominator. For the same range of V and c, $2c/V^2$ varies from 15 s to 4.15 hr. Water level changes observed in response to creep events typically occur over a period of several hours to tens of days. Consequently, for diffusion times near the lower end of the range cited, the difference between steady state motion and slip that started from rest could not be distinguished from the observed response of the well.

The magnitude of the pressure change due to slip depends on the product AG, in addition to the diffusion length $V/2c$. The upper bound for A of π^{-1} occurs for incompressible solid and fluid constituents ($B=1.0$ and $\nu_u = 0.5$). Roeloffs and Rudnicki (1984/85) suggest a value of $A = 0.7/\pi$, corresponding to $B = 0.9$ and $\nu_u = 0.4$, as representative of gouge and for a shear modulus $G = 3.0$ GPa, the product AG is 0.66 GPa. They suggest a smaller value of $0.5/\pi$ for rock and with $G= 30$ GPa, this yields a value of $AG = 4.73$ GPa, about seven times larger. Based on information of Roeloffs *et al.* (1989) for the Middle Mountain well near Parkfield, California, Rudnicki *et al.* (1993) infer AG in the range 0.48 to 1.01 GPa. The higher value is close to that for $A = 0.5/\pi$ and $G=6$ GPa used by Rudnicki and Hsu (1988) to reexamine the water level change reported by of Lippincott *et al.* (1985).

For the value of AG suggested by Roeloffs and Rudnicki (1984/85) for gouge (0.66 GPa) and the range of c and V cited earlier, a value of the nondimensional pressure,

$p/AbG(V/2c)$, of unity corresponds to a pressure change of 1.9 to 193 kPa per millimeter of slip. Assuming that the pressure change can be converted to water level change by dividing by weight density of water, this pressure change is equivalent to a water level change of 19 cm to 19.3 m. Water level changes observed in response to creep events range from about 4 cm to 30 cm and the inferred slip from about 1 to 15 mm. For the Middle Mountain well studied by Roeloffs *et al.* (1989), the minimum water level change that can be detected is about 1 mm.

8. Conclusions

This paper has reviewed the solutions for instantaneous and steadily moving plane strain dislocations in poroelastic materials. In these solutions, the plane containing the dislocation can be either permeable or impermeable. In addition, we have derived and presented some results for dislocations that start from rest, move with constant speed and stop. These latter results have been obtained only for shear dislocations on a permeable plane, but the same procedure could be used to derive results for an impermeable plane or for opening dislocations.

The solutions illustrate the time dependence introduced into the behavior of an elastic solid by the coupling of fluid mass diffusion with volumetric deformation. Typically, the coupling is significant within about a diffusion length of the dislocation edge. For the elementary instantaneous dislocation, there is no characteristic length scale but for the steadily moving dislocations, the diffusion length is $2c/V$. For both the instantaneous and steadily moving shear dislocations, the pore pressure changes are greater for the impermeable fault than for the permeable. The greater pore pressure change occurs because fluid mass diffusion across the fault that would diminish the pore pressure is prohibited by the impermeable fault.

The principal motivation for the solutions discussed here has been applications to water well level changes in response to fault creep, aseismic slip events. Because the data for such events are limited, the simplest dislocation solutions often suffice to model the observed water level changes. The discussion of the solutions here makes it clear, however, that the coupling of deformation with pore fluid diffusion can introduce a complexity into the pore pressure response even when the distribution of slip is simple. This complexity is particularly evident in the solutions for dislocations that start from rest, move a finite distance and then stop. These solutions demonstrate the necessity of understanding the hydrologic regime and the nature of the fault zone, whether it is permeable or impermeable, for a definitive understanding of the observations.

The solutions discussed here are two dimensional and, hence, most appropriate for slip zones that are wide compared with their length and the diffusion length. Nevertheless, the steady state and instantaneous solutions can be considered as limiting idealizations of different three dimensional slip distributions. For a vertical slip zone, like the San Andreas fault, the steady state solution is an appropriate idealization of a slip event that is relatively uniform with depth but moves down the fault (parallel to the surface) so that it reaches different points at different times. In contrast the instantaneous solution is appropriate for an event that principally propagates up from depth so that the slip reaches different points along the fault at the same time. Although Cleary (1977) has derived the three dimensional solution for a point slip in an infinite poroelastic solid, there has been no detailed study of the predicted pore pressure alterations and the solution has not been used to interpret observations.

Another effect neglected in application of these infinite space solutions to water well level measurements is flow to the free surface. Rudnicki and Wu (1993) have made some simple estimates of this effect on the basis of solutions of the one dimensional diffusion equation. For example, for an initial pore pressure distribution peaked at a depth d, they find that the half-space solution is similar to that in the full space only for depths greater than approximately 0.5 d and times less than about 0.5 $d^2/4c$, where c is the diffusivity. Rudnicki and Wu (1993) also used the ABAQUS finite element program to compare the solution for a dislocation in a half-space to the analytical solution in a full space. The comparison reveals that the pore pressure at shallow depths in the half space is initially larger than in the full space because of the larger volume strains but decays more rapidly because of flow to the free surface.

Despite the simplicity of the dislocation solutions, they provide significant insight into the nature of coupled deformation diffusion effects and the behavior of more complex models. In addition, they often suffice to interpret the available data and can be superposed to generate more elaborate models.

Acknowledgement: Financial support for work leading to some of the results reported here was provided by the U.S. Geological Survey Earthquake Hazards Reduction Program and the National Science Foundation Geophysics Program.

9. References

Abramowitz, M. and Stegun, I. A., Eds. (1964), *Handbook of Mathematical Functions*, Appl. Math. Ser. 55, National Institute of Standards and Technology, Gaithersburg, Md.

Atkinson, C. and Craster, R.V. (1991), Plane strain fracture in poroelastic media, *Proc. Royal Society London A* **434**, 605-633.

Biot, M. A. (1941), General theory of three dimensional consolidation, *J. Applied Physics* **12**, 155-164.

Booker, J. R. (1974), Time dependent strain following faulting of a porous medium, *J. Geophysical Research* **79**, 2037-2044.

Byerlee, J. D. (1990), Friction, overpressure and fault normal compression, *Geophysical Research Letters* **17**, 2109-2112.

Carslaw, H. S. and Jaeger, J. C. (1959), *Conduction of Heat in Solids*, 2nd Ed., Oxford University Press, Oxford, U. K.

Cleary, M. P. (1977), Fundamental solutions for a fluid-saturated porous solid, *Int. J. Solids and Structures* **13**, 785-806.

Cleary, M. P. (1978), Moving singularities in elasto-diffusive solids with applications to fracture propagation

Int. J. Solids and Structures **14**, 81-97.

Detournay, E. and Cheng, A. H-D. (1991a), Plane strain analysis of a stationary hydraulic fracture in a poroelastic medium, *Int. J. Solids and Structures* **27**, 1645-1662.

Detournay, E. and Cheng, A. H-D. (1991b), Fundamentals of poroelasticity, in J. A. Hudson (ed.), *Comprehensive Rock Engineering: Principles, Practice and Projects*, Vol. 2, Pergamon Press.

Detournay, E., Cheng, A. H-D., and McLennan, J. D. (1990), A poroelastic PKN hydraulic fracture model based on an explicit moving mesh algorithm. *J. of Energy Resources Technology* **112**, 224-230.

Johnson, A. G., Kovach, R. L. and Nur, A. (1973), Pore pressure changes during creep events on the San Andreas fault, *J. Geophysical Research* **78**, 851-857.

Lippincott, D. K., Bredehoeft, J. D. and Moyle, W. R. Jr. (1985), Recent movement on the Garlock Fault suggested by water level fluctuations in a well in Fremont Valley, California, *J. Geophysical Research* **90**, 1911-1924.

Nur, A. and Byerlee, J. D. (1971), An exact effective stress law for elastic deformation of rock with fluids, *J. Geophysical Research* **76**, 6414-6419.

Nur, A. and Booker, J. R. (1972), Aftershocks caused by fluid flow? *Science* **175**, 885-887.

Rice, J. R. (1992), Fault stress states, pore pressure distributions, and the weakness of the San Andreas fault, in Brian Evans and Teng-Fong Wong (eds.), *Fault Mechanics and Transport Properties of Rocks*, Academic Press Ltd., pp. 475-503.

Rice, J. R. and Cleary, M. P. (1976), Some basic stress diffusion solutions for fluid-saturated elastic porous media with compressible constituents, *Reviews of Geophysics* **14**, 227-241.

Rice, J. R. and Simons, D. A. (1976), The stabilization of spreading shear faults by coupled deformation-diffusion effects in fluid-infiltrated porous materials, *J. Geophysical Research* **81**, 5322-5344.

Roeloffs, E. A. and Rudnicki, J. W. (1984/85), Coupled-deformation diffusion effects on water level changes due to propagating creep events, *Pure and Applied Geophysics (PAGEOPH)* **122**, 560-582.

Roeloffs, E. A., Burford, S. S., Riley, F. S. and Records, A. W. (1989), Hydrologic effects on water level changes associated with episodic fault creep near Parkfield, California, *J. Geophysical Research* **94**, 12387-12402.

Rudnicki, J. W. (1984), Effects of dilatant hardening on the development of concentrated shear deformation in fissured rock masses, *J. Geophysical Research* **89**, 9259-9270.

Rudnicki, J. W. (1986), Slip on an impermeable fault in a fluid-saturated rock mass, in S. Das, J. Boatwright, and C. H. Scholz (eds.), *Earthquake Source Mechanics, Geophys. Monogr. Ser.*, vol. 37, AGU, Washington, D. C., pp. 81-89.

Rudnicki, J. W. (1987), Plane strain dislocations in linear elastic diffusive solids, *J. Applied Mechanics* **54**, 545-552.

Rudnicki, J. W. (1991), Boundary layer analysis of plane strain shear cracks propagating steadily on an impermeable plane in an elastic diffusive solid, *J. Mechanics and Physics of Solids* **39**, 201-221.

Rudnicki, J. W. and Hsu, T.-C (1988), Pore pressure changes induced by slip on permeable and impermeable faults, *J. Geophysical Research* **93**, 3275-3285.

Rudnicki, J. W. and Roeloffs, E. A. (1990), Plane strain shear dislocations moving steadily in linear elastic diffusive solids, *J. Applied Mechanics* **57**, 32-39.

Rudnicki, J. W. and Koutsibelas, D. A. (1991), Steady propagation of plane strain shear cracks on an impermeable plane in an elastic diffusive solid. *Int. J. Solids and Structures* **27**, 205-225.

Rudnicki, J. W., Yin, J. and Roeloffs, E. A. (1993), Analysis of water level changes induced by fault creep at Parkfield, California, *J. Geophysical Research* **98**, 8143-8152.

Rudnicki, J. W. and Wu, M. (1993), Pore pressure changes induced by slip in a poroelastic half-space. Draft manuscript submitted as part of the Final Report to U. S. Geological Survey for Award No. 1434-92-G-2164, *Coupled Deformation Diffusion Solutions fo the Interpretation of Slip Induced Water Well level Changes at Parkfield.*

Ruina, A. (1978), Influence of coupled deformation-diffusion effects on retardation of hydraulic fracture, in Y. S. Kim (ed.), *Proc. U. S. Symposium on Rock Mechanics*, 19th, Stateline, Nev., University of Nevada Reno, pp. 274-282.

Simons, D. A. (1979), The analysis of propagating slip zones in porous elastic media, in R. Burridge (ed.), *Fracture Mechanics, Proceedings of the symposium in applied mathematics of AMS and SIAM*, SIAM-AMS, New York, pp. 153-169.

Wang, C-Y. and Lin, W. (1978), Constitution of the San Andreas fault zone at depth, *Geophysical Research Letters* **5**, 741-744.

Wesson, R. L. (1981), Interpretation of changes in water level accompanying fault creep and implications for earthquake prediction, *J. Geophysical Research* **86**, 9259-9267.

Wu, F. T., Blatter, L. and Roberson, H. (1975), Clay gouges in the San Andreas fault system and their possible implications, *Pure and Applied Geophysics (PAGEOPH)* **113**, 87-95.

THEORETICAL ASPECTS OF FRACTURE IN POROUS ELASTIC MEDIA

R.V.CRASTER
Department of Theoretical Mechanics,
University of Nottingham,
Nottingham, NG7 2RD, U.K

AND

C.ATKINSON
Department of Mathematics,
Imperial College of Science, Technology and Medicine,
London, SW7 2BZ, U.K

Abstract. Various different poroelastic fracture problems are analysed exactly using a semi-infinite crack model. In particular cracks with fracture faces which are either permeable, or impermeable, to the diffusing pore fluid are considered. The stress intensity factors are found exactly for transient problems, and the full solution found for steadily propagating cracks. Asymptotic methods are developed for finite length cracks and the results are checked by using an invariant integral. Recent results for inhomogeneous elastic and poroelastic media are briefly indicated.

1. Introduction

The equations of linear, isotropic poroelasticity were introduced by Biot (1941), and are mathematically equivalent to the fully coupled thermoelastic equations, Biot (1955). This might lead one to suppose that the multitude of fracture problems which have been solved for thermoelastic fracture would carry straight over to poroelasticity. However, for thermoelastic materials a coupling parameter appears which is typically small, the term involving this parameter is usually ignored and this is then used to uncouple the governing equations, leading to the theory of thermal stresses. For most poroelastic materials the equivalent parameter is not small; the equations remain fully coupled and no such approximation can be made -

P. S. Selvadurai (ed.), Mechanics of Poroelastic Media, 23–45.

hence there are fundamental differences. Our aim in a series of papers has been to analyse the effect of the coupling using a variety of analytical and asymptotic methods, by solving various fracture problems. These solutions are, of course, also useful for thermoelastic problems in situations where the thermal stress reduction discussed above is not an accurate approximation.

The poroelastic theory attempts to model the mechanical response of a porous material which has a solid elastic skeleton with the pore space filled with a viscous fluid. The equations derived by Biot involve a coupling of the theories of elasticity and diffusion; there is an explicit coupling between the dilatation of the elastic skeleton and the pressure in the diffusing pore fluid. The governing equations are derived by using Darcy's equation to model the diffusion process, a mass conservation equation for the pore fluid and the equilibrium equation for the stress. The equations were derived by Biot using this phenomenological approach and have subsequently been derived using homogenisation theory, Auriault (1980), Burridge and Keller (1981) and mixture theory, Bowen (1982). Hence, the fundamental basis of the equations is well founded. The poroelastic materials differ from the corresponding quasi-static elastic solids as a time dependence is introduced into the otherwise time independent elasticity equations. This diffusing pore fluid can have a large effect; for example, for rapid loadings (compared with the diffusion time scale) the material response is stiffer than for slower loadings as the fluid has less time to diffuse away. In particular for such transient loadings the stress intensity factors which characterise the singular near crack tip stress fields for fracture problems are time dependent, and depend upon the pore pressure boundary conditions on the fracture plane. For steadily propagating fractures the stress intensity factors become velocity dependent.

The theory is relevant to the fracture of rocks and can be applied to a variety of geophysical problems: for a review see Rudnicki (1985). The work on fracture in a porous elastic material is motivated by applications to fracture and fault initiation, propagation and creep in porous rocks. Such problems of fracture where there is pore fluid interaction are believed to be important in fault creep and in enhancing the recovery of oil via the hydraulic fracturing process. In several recent papers the authors have examined a sequence of model problems with the aim of understanding and modelling the physical processes which occur. As an example, one can imagine that during the initiation of fracture in the hydraulic fracturing process, when a crack is pressurised by pumping fluid into it, the fluid in the crack may leak out into the formation. This causes a volume expansion of the material around the crack, tending to close the fracture. There will also be external stress or pore pressure fields which may affect the fracture. Therefore, there is a complicated interaction between the diffusing pore fluid and applied loading, which must be modelled correctly if fracture in fluid filled materials is to be understood. A simple model has been used by Atkinson and Cook (1993) to study strain rate effects in compressive rock tests. A more complete model would apply the poroelastic theory and analysis developed here.

The equations are quasi-static, the effect of inertia is neglected; this is because the effects which we are considering should happen on a time scale large compared with that for wave propagation. It should be noted that Biot extended his theory to include inertia, e.g. Biot (1956a,b) and that some attempts at solving the fully coupled dynamic thermoelastic crack problems are possible, Atkinson and Craster (1992b).

The governing equations used are those of Biot (1941), in the form used by Rice and Cleary (1976). The equations are characterised by five independent constants; G, ν, ν_u, κ and B (the shear modulus, drained and undrained Poisson's ratios, the permeability coefficient and Skempton's pore pressure coefficient). The stress σ_{ij} is given by

$$\sigma_{ij} = 2G\epsilon_{ij} + \frac{2G\nu}{(1-2\nu)}\delta_{ij}e - \alpha p\delta_{ij}, \tag{1.1}$$

with ϵ_{ij} as the strain tensor and $e = \epsilon_{kk}$ is the dilatation. The pore pressure, p, is related to the variation of fluid volume content, ζ, and the dilatation, e by $p = Q\zeta - \alpha Qe$. Provided there are no body forces, or fluid sources, in the material the governing equations are written as: the equilibrium equation for the stresses, $\sigma_{ij,j} = 0$, Darcy's law which relates the mass flux to the gradient of the pore pressure, $q_i = -\rho_0\kappa p_{,i}$, and a mass conservation equation for the fluid, $\partial m/\partial t = -q_{i,i}$, with m the mass of fluid per unit volume and ρ_0 as the reference density. A nomenclature which gives the interrelations between the material constants and definitions of variables is in Appendix A. The equations are also written as an elastic Navier equation with a coupling term for the pore pressure and as a diffusion equation for the pore pressure with a coupling term for the dilatation, i.e.,

$$G\nabla^2 u_i + \frac{G}{(1-2\nu)}e_{,i} - \alpha p_{,i} = 0, \quad \frac{\partial p}{\partial t} - \kappa Q\nabla^2 p = -\alpha Q\frac{\partial e}{\partial t}. \tag{1.2}$$

In previous papers by the present authors, Atkinson and Craster (1991), Craster and Atkinson (1992a,1994a), on time dependent fracture in a poroelastic material, the governing equations have been solved for an impulsively applied and spatially exponentially decaying internal stress, or pore pressure loading, on the crack faces of a semi-infinite crack. The crack faces may be either permeable or impermeable depending upon the physical situation. Exact solutions for the time dependent stress intensity factors were found using the Wiener-Hopf technique after the governing equations had been Fourier transformed in space and Laplace transformed in time. In more complicated cases, those with mixed boundary conditions for the pore pressure on the fracture plane, a matrix functional equation was solved. Fortunately, this matrix functional equation could be solved by first solving a subsidiary equation. The stress intensity factors and the coefficients of the singular pore pressure gradient as functions of the Laplace transform parameter were found and the stress intensity factors were inverted numerically. Steadily propagating cracks are also considered and exact solutions found.

In the neighbourhood of the crack tip an asymptotic analysis shows that the usual elastic eigensolutions which are deduced from $\sigma_{ij,j} = 0$ still apply here, i.e., in shear as $r \to 0$ ((r, θ) are polar coordinates centered upon the crack tip) the stresses are asymptotically

$$\sigma_{xx} \sim \frac{-K_{II}}{(2\pi r)^{\frac{1}{2}}} \sin(\frac{\theta}{2}) \left(2 + \cos(\frac{\theta}{2}) \cos(\frac{3\theta}{2})\right), \tag{1.3}$$

$$\sigma_{xy} \sim \frac{K_{II}}{(2\pi r)^{\frac{1}{2}}} \cos(\frac{\theta}{2}) \left(1 - \sin(\frac{\theta}{2}) \sin(\frac{3\theta}{2})\right), \tag{1.4}$$

$$\sigma_{yy} \sim \frac{K_{II}}{(2\pi r)^{\frac{1}{2}}} \sin(\frac{\theta}{2}) \cos(\frac{\theta}{2}) \cos(\frac{3\theta}{2}). \tag{1.5}$$

The mode II stress intensity factor is K_{II}. Corresponding elastic results (with appropriate functions of θ) hold asymptotically for the mode I situation, with the stress intensity factor denoted by K_I. These stress intensity factors are time dependent for the stationary transient situations and are velocity dependent when the crack is moving with constant velocity (the functions of θ given above also hold in the moving coordinate frame because elastic inertia is neglected). The pore pressure behaves differently as $r \to 0$ in the transient and steady problems and its behaviour is considered in later sections.

2. Steadily propagating fractures

If the crack lies on $y = 0$ and propagates at a steady velocity V in the positive x direction, then all the field variables have the dependence on x, y, t such that $g = g(x_1 - Vt, y)$ (g being any field variable) with $x = x_1 - Vt$ being a coordinate moving with the crack tip. Hence the explicit time dependence is removed by writing $\partial/\partial t = -V\partial/\partial x$.

The governing equations are unchanged except that the diffusion-like equation becomes

$$V\frac{\partial p}{\partial x} + \kappa Q \nabla^2 p = -\alpha Q V \frac{\partial e}{\partial x}, \tag{2.6}$$

The velocity is explicitly scaled out of this equation be setting

$$X = \frac{V}{c}x,\ Y = \frac{V}{c}y,\ U_i(X,Y) = \frac{V}{c}u_i(x,y), \tag{2.7}$$

and leaving the other variables untouched.

Let us consider shear fracture (the analogous tensile problems are in Atkinson and Craster (1991)), and allow the crack to have crack faces which are either permeable or impermeable to the diffusing pore fluid. Rice and Simons (1976) have previously considered the permeable crack case. A related paper is by Rudnicki and Koutsibelas (1991) who consider a shear crack embedded upon an impermeable fault plane; this problem is relevant to the reactivation of a pre-existing shear fault. The problems considered here are relevant for fracture in previously undamaged material.

By symmetry the boundary conditions ahead of the crack on $Y = 0$ are that

$$p = 0 \text{ and } U_1 = 0 \text{ for } X > 0, \tag{2.8}$$

and for no tensile stress on the X axis $\sigma_{YY} = 0$. Symmetry is used to turn the mixed boundary problem for the crack in an infinite material into a corresponding half plane problem which is then solved in $Y > 0$. As stated above we are using a moving coordinate frame centered upon the crack tip, the applied loading moves with the crack and decays as $X \rightarrow -\infty$; it is taken here to be

$$\sigma_{XY} = -\tau_0 e^{\frac{X}{a_1}}, \tag{2.9}$$

where a_1 is the non-dimensionalised length scale $\frac{aV}{c}$, and more general loadings can be generated by superposition. In particular Dugdale (1960) zones can be introduced, i.e., cohesive zones near the crack tip.

For a crack with impermeable crack faces

$$\frac{\partial p}{\partial Y} = 0 \text{ for } X < 0 \tag{2.10}$$

and for a crack with permeable crack faces

$$p = 0 \text{ for } X < 0. \tag{2.11}$$

Note that we are assuming that conditions pertain which are consistent with shear crack propagation only. For more general stress fields these solutions can be combined with the steadily propagating tensile fractures of Atkinson and Craster (1991). The impermeable crack problem leads us to a situation where the pore pressure boundary conditions are now mixed on the fracture plane and this leads to some technical difficulties. Nonetheless the full solution can be extracted; for instance the pore pressure is explicitly given by

$$p(X,Y) = \frac{2B(1+\nu_u)\overline{\eta}\tau_0}{3\left(\overline{\eta}+\frac{1}{a_1}\left(\frac{1}{a_1^{\frac{1}{2}}(1+\frac{1}{a_1})^{\frac{1}{2}}}-1\right)\right)(2\pi)^{\frac{1}{2}}}(a_1^{\frac{1}{2}}\frac{(R-X)^{\frac{1}{2}}}{R}(1-e^{-\frac{1}{2}(R+X)})$$

$$+e^{\frac{X}{a_1}}\int_X^{\infty}e^{-\frac{X'}{a_1}}\frac{(R'-X')^{\frac{1}{2}}}{R'}\left(\frac{e^{-\frac{1}{2}(R'+X')}}{a_1(1+\frac{1}{a_1})^{\frac{1}{2}}}-\frac{1}{a_1^{\frac{1}{2}}}\right)dX') \qquad (2.12)$$

for $Y > 0$, which is shown in figure 2 (non-dimensionalised by dividing through by $\frac{2B(1+\nu_u)\overline{\eta}\tau_0}{3(2\pi)^{\frac{1}{2}}}$) for $\overline{\eta} = 4$, corresponding to $\nu = 0.2, \nu_u = 0.3$. The variable R is defined as $R^2 = X^2 + Y^2$. Note the pore pressure contours intersect the crack ($X < 0$) at right angles, (which follows from the boundary condition (2.10)). The expressions $R - X$ and $R + X$ are a parabolic coordinate system related to the semi-infinite geometry and the differential equation (2.6). Note the anti-symmetry implies that $p(X,-Y) = -p(X,Y)$. The pore pressure distribution for the permeable crack is

$$p(X,Y) = \frac{2B(1+\nu_u)\overline{\eta}a_1^{\frac{1}{2}}\tau_0}{3\left(\overline{\eta}+\frac{1}{a_1}-\frac{1}{a_1^{\frac{1}{2}}}(1+\frac{1}{a_1})^{\frac{1}{2}}\right)(2\pi)^{\frac{1}{2}}}(\frac{(R-X)^{\frac{1}{2}}}{R}(1-e^{-\frac{1}{2}(R+X)})$$

$$+e^{\frac{X}{a_1}}\int_X^{\infty}\frac{e^{-\frac{X}{a_1}}(R-X)^{\frac{1}{2}}}{a_1R}(e^{-\frac{1}{2}(R+X)}-1)dX). \qquad (2.13)$$

This is shown (non-dimensionalised and for $\overline{\eta} = 4$) for $Y > 0$ in figure 1. This pore pressure profile is similar to that calculated by Rudnicki and Roeloffs (1990) for the analogous steadily propagating shear dislocation.

The stresses and displacement fields can similarly be evaluated everywhere; however, our main interest here is in the pore pressure. It is interesting to note that the first term in both pore pressure expressions coincide as $a_1 \to \infty$ with the first order term in the crack tip boundary layer analysis of Simons (1977). The pore pressure profiles for steadily propagating tensile fractures are shown for the same material parameters in figures 3,4. The pore pressure profiles obtained are quite different, with the tensile cases showing significant pore fluid suction ahead of the crack tip.

The stress intensity factors are written as $K_{II} = f(a_1)K_e$ where K_e, the corresponding elastic stress intensity factor, is given by $\sqrt{2}a_1^{\frac{1}{2}}\tau_0$. The

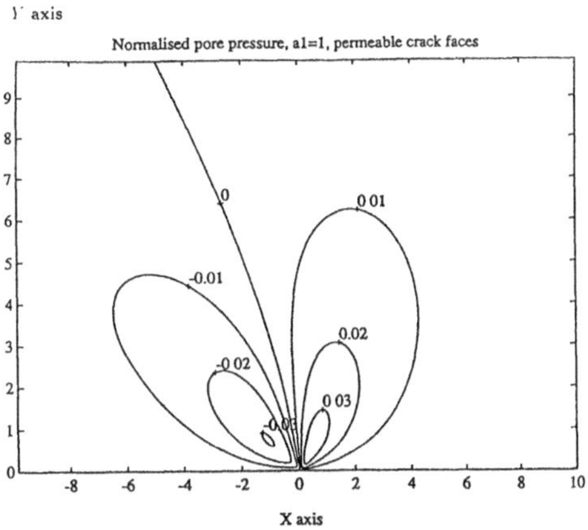

Figure 1: Normalised pore pressure contours for a steadily propagating permeable shear crack with $a_1 = 1$.

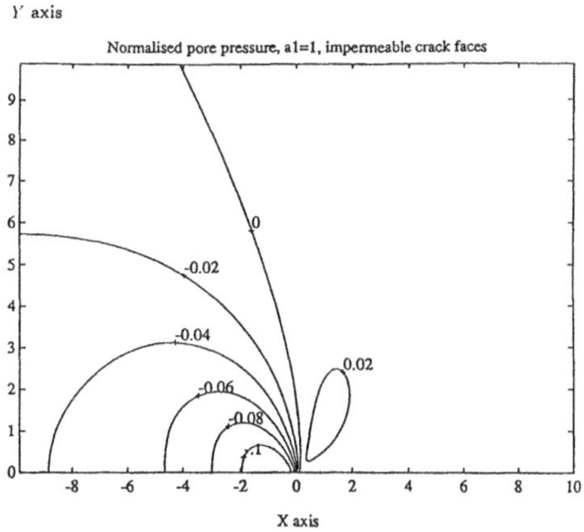

Figure 2: Normalised pore pressure contours for a steadily propagating impermeable shear crack with $a_1 = 1$.

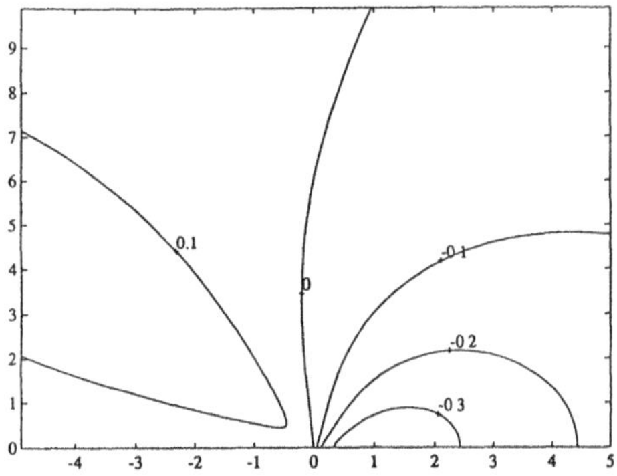

Figure 3: Normalised pore pressure contours for a steadily propagating permeable tensile crack with $a_1 = 10$.

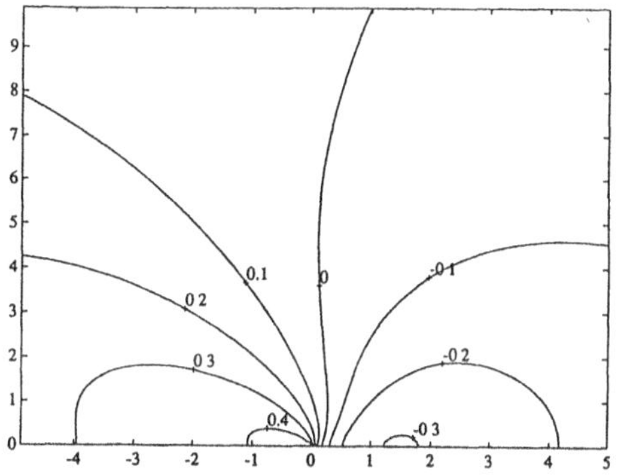

Figure 4: Normalised pore pressure contours for a steadily propagating impermeable tensile crack with $a_1 = 10$.

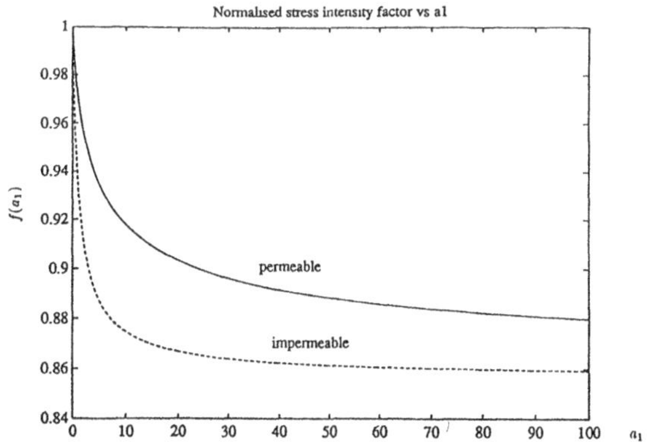

Figure 5: The functions $f(a_1)$ versus a_1 when $\nu = 0.3$, $\nu_u = 0.4$

function $f(a_1)$ is

$$f(a_1) = \frac{\overline{\eta}_u}{\overline{\eta} + \frac{1}{a_1} - \frac{1}{a_1^{\frac{1}{2}}}(1 + \frac{1}{a_1})^{\frac{1}{2}}}, \tag{2.14}$$

for the permeable case and is

$$f(a_1) = \frac{\overline{\eta}_u}{\overline{\eta} + \frac{1}{a_1}\left(\frac{1}{a_1^{\frac{1}{2}}(1+\frac{1}{a_1})^{\frac{1}{2}}} - 1\right)}, \tag{2.15}$$

for the impermeable case; these are bounded above by one and are monotonically decreasing. The functions $f(a_1)$ are shown in figure 5 for $\nu = 0.3$, $\nu_u = 0.4$. The impermeable fracture has a smaller stress intensity factor than the permeable one. This is of particular relevance to shear faults in the earth, as the faults may contain small particles which would clog the pores, making the fault effectively impermeable. The lower stress intensity factor implies increased stability for this case; this becomes particularly marked in materials for which the consolidation parameter is small e.g. clays.

In the limit as $a_1 \to \infty$ then $K_{II} \to (1 - \nu_u)/(1 - \nu)K_e$, Rice and Simons (1976), Simons (1977) give an heuristic result to support this. They argue that the near tip displacements are proportional to $\frac{(1-\nu)}{G}K_{II}$ from the drained elastic field, and $\frac{(1-\nu_u)}{G}K_e$ from the undrained elastic field. Equating these they obtain the above result. The same heuristic argument is shown to be correct for steadily propagating cracks along an interface between a porous elastic material and a rigid substrate, Craster and Atkinson (1992b).

One point to note is that the functions $f(a_1)$ defined above for shear cracks with permeable (or impermeable) crack faces are the same for the corresponding tensile problems with permeable (or impermeable) crack faces. Indeed the $f(a_1)$ for the impermeable crack in a continuous material is even the same as the $f(a_1)$ which occurs when the fracture plane is prefaulted and assumed to be impermeable for all X in shear. This suggests that for steadily propagating cracks the stress intensity factor dependence upon the diffusion process is through the pore pressure boundary condition on the crack face and is insensitive as to whether the pore pressure boundary conditions are mixed upon the fracture plane; this is not the case for stationary transient crack problems. This insensitivity does not, in general, follow through to higher order terms in the stress expansion near the crack tip, or to the other variables.

The local behaviour of the field variables in the neighbourhood of the crack tip is particularly important, and is analysed separately. The elastic eigensolutions are (1.3)-(1.5) and the pore pressure diffusion equation in the steadily propagating case is (2.6). For the impermeable crack we use the edge conditions at the crack tip i.e., that as $r \to 0$, $p \sim O(r^{\frac{1}{2}})$ and $u \sim O(r^{\frac{1}{2}})$. It is clear that the dilatation and the pore pressure are fully coupled in the neighbourhood of the crack tip, as the second and third terms of (2.6) are of the same order. A particular solution of the leading order equation is found satisfying the boundary conditions $p = 0$ on $\theta = 0$ and $\frac{\partial p}{\partial y} = 0$ on $\theta = \pi$ giving locally

$$p \sim \left(\frac{\alpha V(1-2\nu)K_{II}}{4\kappa G(2\pi)^{\frac{1}{2}}} \sin\frac{3\theta}{2} - K_2 \sin\frac{\theta}{2} \right) r^{\frac{1}{2}}. \tag{2.16}$$

Note that K_2 is an eigensolution of (2.6). The result for the permeable fault with boundary conditions $p = 0$ on $\theta = 0, \pi$ is

$$p \sim \frac{\alpha V(1-2\nu)K_{II}}{4\kappa G(2\pi)^{\frac{1}{2}}} \left(\sin\frac{3\theta}{2} + \sin\frac{\theta}{2} \right) r^{\frac{1}{2}}. \tag{2.17}$$

The pore pressure field now is dependent only on K_{II} i.e., it is completely driven by the dilatation.

The mixed boundary condition case is clearly coupled in a more complex way than the simpler permeable case. The pore pressure is driven in part by the dilatation, but also contains an independent term reflecting the mixed boundary conditions.

It is also worth noting that steadily moving and stationary problems are inherently different; conclusions drawn from one cannot necessarily be transferred directly to the other. For instance, in the steady permeable fault case the pore pressure gradients at the crack tip are singular (i.e $O(r^{-\frac{1}{2}})$) at the origin; in the impulsive permeable fault case considered in the next section, (3.20), this is not so. The fundamental difference is in the pore pressure diffusion equation (2.6), the $r^{\frac{1}{2}}$ behaviour is not complicated by the time dependence. Note that K_2 as given in (2.16) is used in the steady case to denote the coefficient of the eigensolution not the coefficient of the singular pore pressure gradient. This is because the coefficient of the singular pore pressure gradient is not independent of K_{II}.

3. Time dependent fracture

Transient fracture in poroelastic materials is important if a time dependent stress, or pore pressure, loading is applied to the material. Examples include triaxial testing of porous rock samples containing fluid, the pressurisation of fractures in the hydraulic fracture process or faults subject to time dependent loadings. Recently Atkinson and Cook (1993) have modeled strain rate effects in rock testing in terms of a model involving the initiation of tensile wing cracks from incipient shear flaws and the subsequent growth of these flaws under the combined effects of pore fluid diffusion into the flaws and the time dependent applied compressive loads. A more complete analysis would extend this to apply the poroelastic fracture analysis discussed here.

The governing equations can be solved exactly for several different semi-infinite crack problems. By assuming that the field variables are all zero for $t < 0$ the explicit time dependence is removed by applying a Laplace transform in time; defined as

$$\overline{f}(s) = \int_0^\infty f(t)e^{-st}dt, \tag{3.18}$$

where s is the transform variable. The analysis is performed in this Laplace transform domain and the inversion is performed numerically.

In order to identify the stress intensity factors in the neighbourhood of the crack tip it is necessary to know the structure of the stress and pore pressure fields there.

Considering shear loading we find from the diffusion-like term in (1.2) and for the problem with impermeable crack faces it is clear from the edge conditions that the term in $\nabla^2 p$ dominates as $r \rightarrow 0$, therefore

$$p \sim K_2(t)(\frac{r}{2\pi})^{\frac{1}{2}} \sin(\frac{\theta}{2}). \tag{3.19}$$

However, for the case when the crack faces are permeable, Laplace transforming the diffusion-like equation in (1.2), and making the assumption that the field variables are all zero for $t < 0$ (we use an overbar to distinguish the Laplace transformed variables) gives $s\overline{p} - \kappa Q \nabla^2 \overline{p} = -\alpha Q s \overline{e}$. In the limit as $r \rightarrow 0$ it is clear that the second and third terms are of the same order. The pore pressure behaviour is deduced by considering eigensolutions of $\nabla^2 \overline{p} = 0$ and a particular solution of $\kappa \nabla^2 \overline{p} = \alpha s \overline{e}$. Hence in Laplace transform space;

$$\overline{p} \sim \overline{K}_2(s) r \sin\theta + r^{\frac{3}{2}} \frac{\alpha s(1-2\nu)\overline{K}_{II}(s)}{2G\kappa\sqrt{(2\pi)}}(\sin(\frac{\theta}{2}) + \sin(\frac{3\theta}{2})). \tag{3.20}$$

The overbars denote the Laplace transformed quantities and $\overline{K}_{II}(s)$ is the Laplace transformed mode II stress intensity factor, c.f. (1.3)-(1.5). Not that the pore pressure gradient in this case is not singular. Similar considerations apply to the mode I problems with either mixed or unmixed pore pressure boundary conditions upon the fracture plane.

The behaviour of the pore pressure near the origin is somewhat difficult to analyse. Later we find that in the permeable case the crack tip can be thought of as a drained inclusion. However, in the mixed case we see there is a complex interaction between the pore fluid and the applied stress field, in which the crack tip has a complicated source or sink behaviour. The K_2 term is the leading order eigensolution of $\nabla^2 \overline{p} = 0$ which can be identified from the asymptotic behaviour of the transformed solutions.

Here semi-infinite cracks are considered which are either impermeable or permeable to the diffusing pore fluid. The applied stress loadings are all taken to be impulsively applied and spatially exponentially decreasing, i.e., in shear $\overline{\sigma}_{xy} = -\tau_0 e^{x/a}/s$ (in real time $\sigma_{xy} = -\tau_0 e^{x/a} H(t)$, $H(t)$ being the Heaviside step function) for $x < 0$. It is relatively simple to consider more complicated loadings in either space or time, or in the mixed cases to apply a pore pressure or pore pressure gradient loading.

Defining a non-dimensional time scale $t' = \frac{tc}{a^2}$, the Laplace inversion is performed with respect to this time scale. Below, the overbars denote Laplace transformed quantities and the subscripts $_{I,II}$ denote mode I and mode II loading respectively. The Laplace transformed stress intensity factors are found exactly: a) for the impermeable crack

$$\overline{K}_{II}(s) = \frac{\sqrt{2} c^{\frac{1}{2}} i \tau_0 \overline{k}_+(\frac{i}{a_1})}{(\overline{d} + \frac{1}{\overline{N}_-(0)}) s^{\frac{3}{2}} \overline{N}_+(\frac{i}{a_1}) a_1 (\frac{1}{a_1} + 1)^{\frac{1}{2}}} (\frac{s}{c})^{\frac{1}{4}}, \ \overline{K}_I(s) = \frac{\sqrt{2} \tau_0 a^{\frac{1}{2}}}{s \overline{N}_+(\frac{i}{a_1})}, \tag{3.21}$$

with the asymptotic result that as $t' \to 0$

$$K_{II}(t') \sim \frac{\tau_0 a^{\frac{1}{2}} \sqrt{2}}{(-\overline{d} + \frac{\overline{N}_+(0)}{\overline{\eta}}) \overline{\eta}} H(t'), \quad K_I(t') \sim \frac{\tau_0 a^{\frac{1}{2}} \sqrt{2}}{\overline{N}_+(0)} H(t'), \tag{3.22}$$

b) for the permeable crack

$$\overline{K}_{II}(s) = \frac{\sqrt{2} a^{\frac{1}{2}} \tau_0}{s N_+(\frac{i}{a_1})}, \quad \overline{K}_I(s) = 2^{\frac{1}{2}} e^{\frac{-i\pi}{4}} (\frac{s}{c})^{\frac{1}{4}} \tau_0 i \frac{c^{\frac{1}{2}} \Gamma_+(\frac{i}{a_1})}{s^{\frac{3}{2}} N_+(\frac{i}{a_1})} \left(\frac{(k_+(\frac{i}{a_1}) - c_0)}{(-d + \frac{N_+(0)}{\overline{\eta}})} \right). \tag{3.23}$$

The asymptotic results as $t' \to 0$ are

$$K_{II}(t') \sim \frac{\tau_0 a^{\frac{1}{2}} \sqrt{2}}{N_+(0)} H(t'), \ K_I(t') \sim \frac{\tau_0 a^{\frac{1}{2}} \sqrt{2} H(t')}{\overline{\eta}(-d + \frac{N_+(0)}{\overline{\eta}})}, \tag{3.24}$$

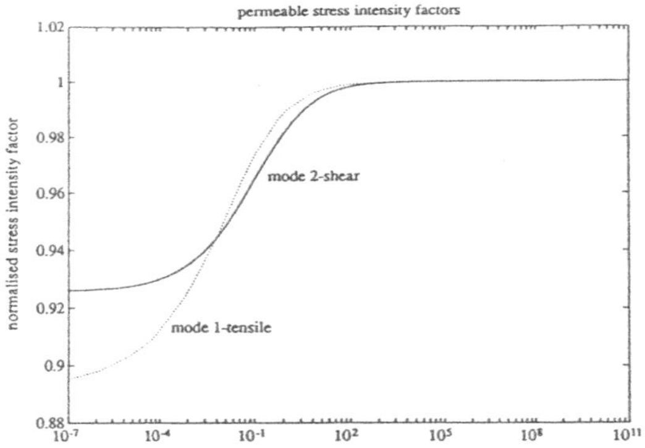

Figure 6: Normalised pore pressure stress intensity factors versus non-dimensional time scale t' for a crack with permeable faces.

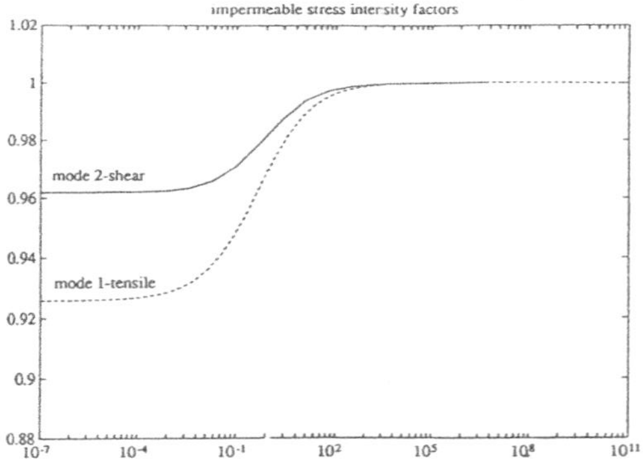

Figure 7: Normalised pore pressure stress intensity factors versus non-dimensional time scale t' for a crack with impermeable faces.

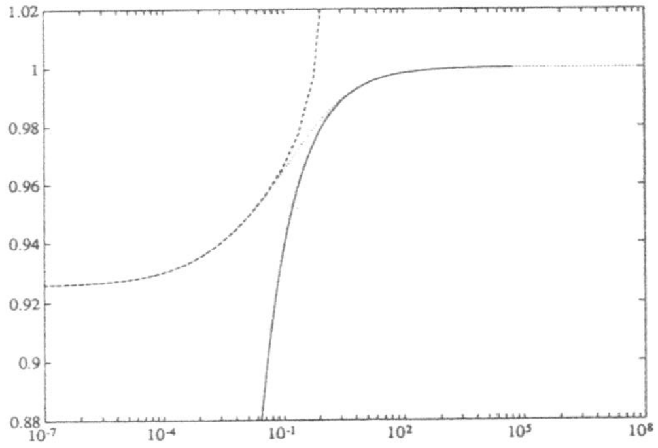

Figure 8: The non-dimensionalised mode II stress intensity factor (for $\nu = 0.3$, $\nu_u = 0.1$) versus time, for the full solution, dotted line, the small time solution, dashed line , and the large time solution, solid line.

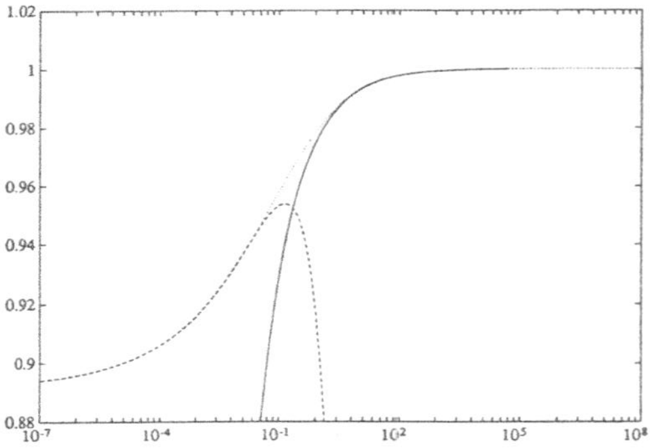

Figure 9: The non-dimensionalised mode I stress intensity factor (for $\nu = 0.3$, $\nu_u = 0.4$) versus time, for the full solution (3.23), dotted line, the small time solution, dashed line , and the large time solution, solid line

which are compared with the corresponding elastic result, $\overline{K}_{II}(s) = \sqrt{2}a^{\frac{1}{2}}\tau_0/s$. The analytic results required, i.e., the expressions for $N_+(\xi)$, $k_+(\frac{i}{a_1}) - c_0$ and d etc. are given in Appendix B. The terms $N_+(0), \overline{N}_+(0)$ are the same and are equal to $((1-\nu)/(1-\nu_u))^{\frac{1}{2}}$; the small time results involving only this term can be deduced directly using energy release rate arguments, c.f. (4.40).

It is worth emphasising that the cracks we are considering are in undamaged material: the pore pressure conditions ahead of the crack being set by symmetry. In the tensile situations we have a symmetric problem, and the pore pressure boundary condition ahead of the crack becomes $\frac{\partial p}{\partial y} = 0$. In the shear problems we consider in this paper the problem is antisymmetric and the pore pressure boundary condition is that $p = 0$ ahead of the crack tip.

If we consider, briefly, the problem of an impermeable crack subjected to an arbitrary (in plane) stress field. This arbitrary field can be split into tensile and shear components, therefore, in general we would need both stress intensity factors from (3.21) to characterise the near crack tip stress fields. The material ahead is continuous hence the pore pressure conditions ahead are combinations of those we have taken above.

The permeable and impermeable normalised stress intensity factors (we have normalised by dividing the poroelastic results by $\sqrt{2}\tau_0 a^{\frac{1}{2}}$, so 1 is the corresponding elastic result) are plotted against time in figures 6,7 for $\overline{\eta} = 3.5$ (corresponding to $\nu = 0.3, \nu_u = 0.4$), where the Laplace transform inversion is performed numerically.

It is worthwhile to emphasise the four distinct problems we see here; for the tensile case, symmetry requires $\frac{\partial p}{\partial y} = 0$ on the x axis ahead of the crack, and for the shear case anti-symmetry requires $p = 0$ ahead of the crack. The crack faces are either permeable or impermeable depending on the physical situation. The following two cases have not been considered in Atkinson and Craster (1991): the shear crack embedded on an impermeable plane, and the tensile crack embedded on a permeable plane. These problems require the pore pressure and pore pressure gradient respectively to be discontinuous ahead of the crack and are therefore unphysical in a continuous material, such assumptions are not consistent with the anti-symmetry or symmetry of the problems. However these problems may have relevance to the initiation of fracture in a prefaulted medium, (c.f. Rudnicki and Koutsibelas (1991) for a steadily propagating crack). For the impulsively applied and spatially exponentially decaying stress loading the stress intensity factors for a shear crack on an impermeable plane and a tensile crack upon a permeable plane are

$$\overline{K}_{II}(s) = \frac{\tau_0 a^{\frac{1}{2}}\sqrt{2}}{s\overline{N}_+(\frac{i}{a_1})}, \quad \overline{K}_I(s) = \frac{\tau_0 a^{\frac{1}{2}}\sqrt{2}}{sN_+(\frac{i}{a_1})}, \tag{3.25}$$

respectively. Other loadings can be considered by superposition.

The results above were deduced directly from the governing equations and the appropriate boundary conditions. An alternative, and equivalent, way of considering crack problems is via singular integral equations. For a semi-infinite, permeable, shear crack under an applied shear loading

$$\int_{-\infty}^{0} \frac{b(x',s)}{s}\Big(\frac{1}{(x-x')} + \frac{1}{\overline{\eta}}\Big(\frac{-2\epsilon^2}{(x-x')^3} + \frac{2\epsilon K_1\left(\frac{|x-x'|}{\epsilon}\right)}{(x-x')|x-x'|}$$

$$+ \frac{K_0\left(\frac{|x-x'|}{\epsilon}\right)}{(x-x')}\Big)\Big)dx' = \begin{cases} \overline{\sigma}_{12}(x,0,s) & \text{for } x \geq 0 \\ -\frac{\tau_0 e^{\frac{x}{a}}}{s} & \text{for } x < 0 \end{cases} . \tag{3.26}$$

The integral above must be interpreted in the Cauchy principal value sense, the expressions $K_0(z), K_1(z)$ denote modified Bessel's functions, and $b(x,s), \overline{\sigma}_{12}$ represent the unknown dislocation density in the crack and stress distribution ahead of the crack. We have defined ϵ as $\epsilon = (\frac{c}{s})^{\frac{1}{2}}$. In the cases where the pore pressure boundary conditions are mixed on the fracture plane, i.e., a permeable crack opening in tension, we study the stress and pore pressure fields in $y > 0$ the boundary conditions on $y = 0$ are:

$$p(x,y,t) = 0\ ,\ \sigma_{22}(x,y,t) = -\tau_0 e^{\frac{x}{a}} H(t) \text{ for } x < 0, \tag{3.27}$$

$$\sigma_{12}(x,y,t) = 0\ \forall\ x, \quad \frac{\partial p(x,y,t)}{\partial y} = 0\ ,\ u_2(x,y,t) = 0 \text{ for } x > 0. \tag{3.28}$$

Then in the Laplace transformed domain coupled integral equations are obtained

$$\frac{3\overline{p}(x,0,s)}{2B(1+\nu_u)}H(x) = \frac{1}{s}\int_{-\infty}^{0} -q(x',s)K_0(\frac{|x-x'|}{\epsilon})+$$

$$b(x',s)\left(\frac{\text{sgn}(x-x')}{\epsilon}K_1(\frac{|x-x'|}{\epsilon}) - \frac{1}{(x-x')}\right)dx', \tag{3.29}$$

$$\frac{1}{s}\int_{-\infty}^{0} \frac{q(x',s)}{\overline{\eta}}\left(\frac{-\epsilon^2}{(x-x')^2} + \frac{\epsilon}{|x-x'|}K_1(\frac{|x-x'|}{\epsilon}) + K_0(\frac{|x-x'|}{\epsilon})\right)$$

$$+\frac{b(x',s)}{(x-x')} + \frac{b(x',s)}{\overline{\eta}}\Big(\frac{2\epsilon^2}{(x-x')^3} - \frac{2\epsilon K_1(\frac{|x-x'|}{\epsilon})}{(x-x')|x-x'|} - \frac{K_0(\frac{|x-x'|}{\epsilon})}{(x-x')}$$

$$- \operatorname{sgn}(x - x')\frac{K_1(\frac{|x-x'|}{\epsilon})}{\epsilon}\Big) dx' = \begin{cases} \overline{\sigma}_{22}(x,0,s) & x \geq 0 \\ -\frac{\tau_0 e^{\frac{x}{a}}}{s} & x < 0 \end{cases} . \tag{3.30}$$

In the above $q(x',s)$ and $b(x',s)$ are unknown distributions of pore pressure gradient discontinuities and dislocation densities respectively, and $\overline{p}(x,0,s)$, $\overline{\sigma}_{22}(x,0,s)$ are the unknown pore pressure and stress ahead of the crack. Setting $\overline{\sigma}_{22}(x,0,s) + 3\overline{p}(x,0,s)/2B(1+\nu_u) = \overline{m}(x,0,s)$ gives the simpler integro-differential equation

$$\frac{-1}{s}\int_{-\infty}^{0}(q(x',s) + b(x',s)\frac{\partial}{\partial x})(K_0(\frac{|x-x'|}{\epsilon})(1-\frac{1}{\eta}) - \frac{1}{\eta}(\frac{\epsilon K_1(\frac{|x-x'|}{\epsilon})}{|x-x'|}$$

$$- \frac{\epsilon^2}{(x-x')^2}))dx' = \begin{cases} \overline{m}(x,0,s) & \text{for } x \geq 0 \\ \frac{-\tau_0 e^{\frac{x}{a}}}{s} & \text{for } x < 0 \end{cases} . \tag{3.31}$$

Solutions for the mixed problems were found mainly due to this uncoupling. Considering the problems in terms of the integral equations leads to some analytical advantages, Craster and Atkinson (1994a), where an asymptotic method for small times is developed. The accuracy of the method is demonstrated in figures 8,9. The method identifies explicitly, in real time, the sequence of events which occur when the crack is loaded and how each event drives a particular piece of the stress intensity factor. For a permeable crack subjected to an applied shear loading the asymptotic method allows us to interpret the results in the following manner. The outer field is initially that of an undrained elastic material with the same applied loading. To the next order this is corrected for (in order to maintain the permeable boundary condition on the crack faces) by a distribution of pore fluid dipoles related to this initial loading. These dipoles in turn redistribute the load on the crack, allowing a further approximation to be performed. The final order outer solution is a combination of the loading induced by the earlier dipoles inducing a correction to this dipole distribution, together with a dilatation term which balances the pore pressure induced on the fracture plane by the elastic loading. The solutions are identified in real time thereby removing the need for numerical inversion of the Laplace transforms.

A modification of this technique has recently been used in Atkinson and Craster (1994a) to solve, asymptotically in the limit of small times, the corresponding integral equations for a finite length crack (in those cases which are unmixed in the pore pressure boundary condition along the fracture plane).

4. Invariant integrals

If the field variables are assumed zero for $t < 0$ and the governing equations (1.2) are Laplace transformed so that they become

$$\overline{\sigma}_{ij,j} = 0, \;\; \kappa Q \nabla^2 \overline{p} = s(\alpha Q \overline{e} + \overline{p}), \tag{4.32}$$

then invariant integrals are deduced as follows. Consider the Lagrangian L

$$L = \frac{-1}{2}\overline{t}_{ij}\overline{\epsilon}_{ij} + \alpha\overline{p}\;\overline{u}_{i,i} + \frac{\kappa}{2s}\overline{p}_{,i}\;\overline{p}_{,i} + \frac{\overline{p}^2}{2Q}, \tag{4.33}$$

where the elastic part of the stress tensor $\overline{t}_{ij} = \overline{\sigma}_{ij} + \alpha\delta_{ij}\overline{p}$. The Euler-Lagrange equations are then equivalent to (4.32). i.e.,

$$\frac{\partial}{\partial x_j}\left(\frac{\partial L}{\partial \overline{u}_{i,j}}\right) - \frac{\partial L}{\partial \overline{u}_i} = 0 \;\rightarrow \overline{\sigma}_{ij,j} = 0, \tag{4.34}$$

$$\frac{\partial}{\partial x_j}\left(\frac{\partial L}{\partial \overline{p}_{,j}}\right) - \frac{\partial L}{\partial \overline{p}} = 0 \;\rightarrow \kappa Q\overline{p}_{,jj} = s(\alpha Q\overline{u}_{i,i} + \overline{p}). \tag{4.35}$$

In Atkinson (1991) a 'pseudo' energy momentum tensor, $\overline{P}_{lj}$, is defined as

$$\overline{P}_{lj} = \frac{\partial L}{\partial \overline{u}_{i,j}}\overline{u}_{i,l} + \frac{\partial L}{\partial \overline{p}_{,j}}\overline{p}_{,l} - L\delta_{lj}. \tag{4.36}$$

giving from (4.33) that $\overline{P}_{lj} = -\overline{\sigma}_{ij}\overline{u}_{i,l} + \frac{\kappa}{s}\overline{p}_{,j}\overline{p}_{,l} - L\delta_{lj}$. As L does not depend explicitly upon x_l, $\overline{P}_{lj,j} = 0$, and hence the integrals

$$F_l = \int_S \overline{P}_{lj} n_j dS \text{ for } l = 1,2,3 \tag{4.37}$$

are zero when S is a closed surface enclosing no singularities.

These equations are analogous to the usual formulation for elasticity, see Eshelby (1970); however, the equations are all in the transform domain so have no direct physical significance. The invariant integral F_1 is of great value as it can be used to check limiting cases of the results found in previous sections. In some very special circumstances it can be used directly to deduce the stress intensity factors. In the Laplace transform domain the tensor $\overline{P}_{lj}$ is used with the poroelastic equations in much the same way as in thermoviscoelasticity (Atkinson and Smelser 1982). However, as described there, the simplest invariant F_1 leads in the most difficult cases to a result for the near field which involves a combination (as the sum of the squares) of the stress intensity factor and the coefficient of the singular pore pressure (temperature) gradient term at the crack tip.

As a first example let us consider a finite width strip of poroelastic material with the boundary conditions

$$\overline{u}_1(x,\pm h,s)=0,\ \overline{u}_2(x,\pm h,s)=\pm u_2(s),\ \ \overline{p}(x,\pm h,s)=p(s), \tag{4.38}$$

which contains a stress free semi-infinite crack with impermeable crack faces centered in the strip, it lies on $y=0$ for $x<0$. The boundary conditions are that $\overline{\sigma}_{12}(x,0,s)=0,\ \ \overline{\sigma}_{22}(x,0,s)=0,\ \ \partial\overline{p}(x,0,s)/\partial y=0$ for $x<0$. This problem is symmetric about $y=0$ and is considered as a half strip problem in $0<y<h$, the conditions along the fracture plane $y=0,\ x>0$ are $\overline{u}_2(x,0,s)=0, \overline{\sigma}_{12}(x,0,s), \partial\overline{p}(x,0,s)/\partial y=0$. As the problem is symmetric and the pore pressure boundary condition on $y=0$ is unmixed; we use the invariant F_1 to extract the Laplace transformed mode 1 stress intensity factor directly. By using the invariant integral around a contour consisting of a vanishingly small arc around the crack tip and straight ligaments (as shown in figure 10) one obtains directly that

$$\frac{\overline{K}_1^2(s)(1-\nu)}{2G}=$$

$$\frac{\left(\frac{2G(1-\nu)}{(1-2\nu)}u_2^2(s)+\frac{(\nu_u-\nu)\kappa p^2(s)}{(1-\nu_u)(1-2\nu)}(\frac{c}{s})^{\frac{1}{2}}\tanh^2(\frac{s}{c})^{\frac{1}{2}}h-2\alpha p(s)u_2(s)(\frac{c}{s})^{\frac{1}{2}}\tanh(\frac{s}{c})^{\frac{1}{2}}h\right)}{\left(h+\frac{(\nu_u-\nu)}{(1-\nu_u)(1-2\nu)}((\frac{c}{s})^{\frac{1}{2}}\tanh(\frac{s}{c})^{\frac{1}{2}}h-h)\right)}. \tag{4.39}$$

The only contribution to the integral, apart from the small arc around the crack tip, is from the ligaments along AB and DE, in the limit as $x\to\pm\infty$ the governing equations become one dimensional and hence are solved there exactly. To obtain the stress intensity factor by solving the full equations would be a quite complicated task, however it should be noted that the problem is very special: the loadings applied on $y=\pm h$ do not vary spatially, and there is no mode mix at the crack tip (in this case we would obtain an expression for the sum of the squares of the intensity factors). Also the result is still in the Laplace transform domain so further work is required to see the variation of the stress intensity factor in real time. Nonetheless, these results are useful, in Atkinson and Craster (1992a), Craster and Atkinson (1994a) various finite width strip problems are considered when the pore pressure boundary condition is mixed and an expression for the sum of the squares of the stress intensity factor and the pore pressure gradient intensity factor. These results are then related to the semi-infinite crack problems in an infinite domain.

Note that where the pore pressure condition is not mixed, the stress intensity factors $K_I(t')$ from (3.22) and $K_{II}(t')$ from (3.24) are identical for small times; this might be expected as the interaction between the pore pressure and stress field is no longer so marked, and the fixed boundary condition partially uncouples the equations. We visualise the crack tip as an elastic inclusion, with Poisson's ratio ν i.e. drained, within an elastic medium with Poisson's ratio ν_u i.e undrained. The problem of a semi-infinite crack penetrating an inclusion with different material properties is considered as follows.

Let the inclusion have shear modulus μ_i and Poisson's ratio ν_i and the exterior material have shear modulus μ_o and Poisson's ratio ν_o. Under conditions of plane strain, define K_i, K_o to be the stress intensity factors associated with the inner and outer material respectively then consider the energy release rate $\overline{G}$. If the inclusion is sufficiently small we write this as

$$\overline{G} = \frac{K_i^2(1-\nu_i)}{2\mu_i} = \frac{K_o^2(1-\nu_o)}{2\mu_o}, \tag{4.40}$$

since the local energy release rate into the crack tip embedded in the small inclusion (moduli ν_i, μ_i) must be the same approximately for small times as that due to the energy release rate (calculated in the far field) due to the crack in the outer medium (moduli ν_o, μ_o). From this we find that;

$$K_i \sim \left(\frac{(1-\nu_o)\mu_i}{(1-\nu_i)\mu_o}\right)^{\frac{1}{2}} K_o \tag{4.41}$$

which when we take $\mu_i = \mu_o = G$ and $\nu_i = \nu$, $\nu_o = \nu_u$, $K_i = K_I$ or K_{II} and $K_o = K_e$ corresponds to the results for $K_I(t')$ in (3.22) and $K_{II}(t')$ from (3.24).

The mixed cases differ in the small time regime only in the terms $d, \overline{d}$ which come from the sum split term. The results for the stress intensity factors, and the coefficient of the singular pore pressure gradient above, can be checked in particular situations (i.e., a purely pore pressure gradient loaded crack) using the properties of an invariant integral.

The first order outer solution is given by a purely elastic solution with undrained coefficients. The inner limit of this outer solution is a stress and displacement field of the usual inverse square root form (for the stresses) characterised by a stress intensity factor given by $K_e = \tau_0 a^{\frac{1}{2}}\sqrt{2}/s$. The first order inner solution is given by the poroelastic results in the limit as $s \to \infty$ (equivalent in real time to $t \to 0$).

For the inner solution we require the stress intensity factor results from the semi-infinite poroelastic crack and the coefficient of the singular pore pressure gradient (c.f. (3.19)) which can be deduced from Craster and Atkinson (1992a) in the limit as $s \to \infty$. These are $\overline{K}_{II}(s)$ deduced from

(3.22) and

$$\overline{K}_2(s) \sim \frac{2B(1+\nu_u)\tau_0 a^{\frac{1}{2}}}{3} \frac{2\sqrt{2}}{\sqrt{sc}} \frac{(\overline{\chi} + \frac{1}{2\overline{N}_-(0)})}{(\overline{d} + \frac{1}{\overline{N}_-(0)})}, \tag{4.42}$$

respectively. The functions $\overline{d}, \overline{\chi}$ etc are given in Appendix B. If we take F_1 around the contour shown in figure 11 then we can deduce that in this limit

$$\frac{\overline{K}_{II}^2(1-\nu)}{2G} - \frac{\overline{K}_2^2\kappa}{8s} = \frac{K_e^2(1-\nu_u)}{2G}. \tag{4.43}$$

Substituting in the above results for $\overline{K}_2, \overline{K}_{II}$ gives the following relation

$$\frac{(1-\nu)}{(1-\nu_u)} - \overline{\eta}(\overline{\chi} + \frac{1}{2\overline{N}_-(0)})^2 = \overline{\eta}^2(-\overline{d} + \frac{\overline{N}_+(0)}{\overline{\eta}})^2. \tag{4.44}$$

This is then verified numerically for different values of ν, ν_u and provides a valuable independent check on the results in the mixed case.

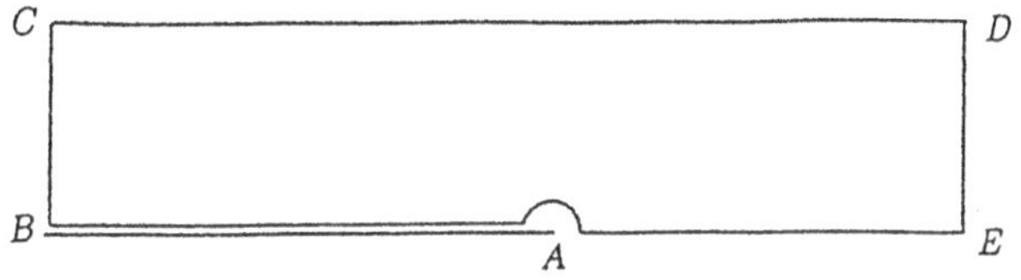

Figure 10: The contour required for problems involving a finite width strip.

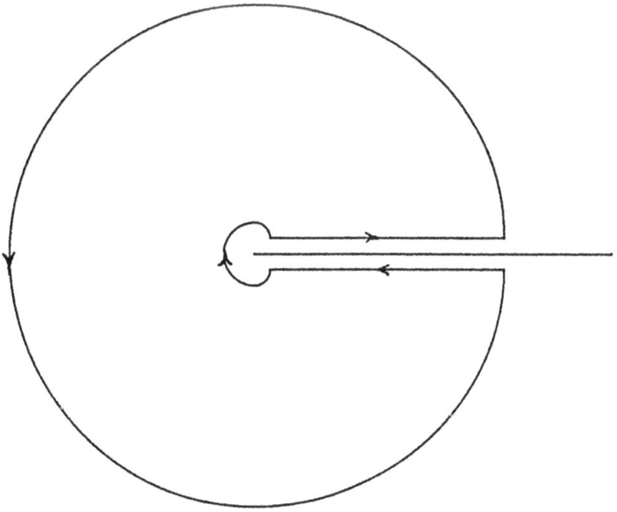

Figure 11: The contour used for the invariant in the limit of small times.

5. Conclusion

Poroelastic fracture problems can be solved exactly to extract the variables of interest. The main results in the steady cases are that the stress intensity factors are insensitive to whether the pore pressure boundary conditions on the fracture plane are mixed, or not. The full solutions can be found explicitly.

The time dependent stationary crack problems are different and conclusions drawn from the steady cases do not necessarily carry straight over. In particular the stress intensity factors do now depend upon whether the crack faces have different pore pressure boundary conditions upon them than the fracture plane ahead of the crack. The resulting coupled integral equations can be solved exactly by reducing the problems to a single subsidiary equation which is solved separately. Similar ideas carry over to finite length cracks which are approached using asymptotic methods.

Invariant integrals can be found and these are particularly valuable in both verifying and deducing results in complicated, limiting, situations.

There is one further tool available which we have not discussed here; the poroelastic reciprocal theorem. This reciprocal theorem can be used to either deduce near field stress intensity factors from far field numerical information, or deduce poroelastic weight functions allowing any loading to be considered. The details are in Craster and Atkinson (1994a), Atkinson and Craster (1992a).

The situations we have discussed in the text refer to major defects such as cracks in an otherwise homogeneous poroelastic medium (the methods developed here can also be applied in principle to contact or inclusion problems). If the medium is spatially inhomogeneous somewhat different techniques are necessary. For special types of spatial variations in elastic moduli methods have been developed, Craster and Atkinson (1994b), Atkinson and Craster (1994b), for the analysis of fracture in inhomogeneous media. For the poroelastic case dual variational methods in the Laplace transformed domain have been used by Atkinson and Appleby (1994) to derive 'effective homogeneous medium' results for inhomogeneous poroelastic media and to find bounds on the stress intensity factors for some special fracture problems.

6. References

Atkinson C. 1991, "Boundary elements and fracture stress analysis", *Engineering Analysis with Boundary Elements*, **8**, 16-23

Atkinson C. and Appleby S. 1994, "Variational principles developed for, and applied to, porous elastic solids", *Int. J. Engng. Sci*, **32**, 955-977

Atkinson C. and Cook J.M., 1993, "The influence of loading rate on crack propagation under compressive stress in a saturated porous material", *J. Geophy. Res.*,

98, 6383-6393
Atkinson C. and Craster R.V. 1991, "Plane strain fracture in poroelastic media", *Proc. Roy. Soc.*, **434**, 605-633
Atkinson C. and Craster R.V. 1992a, "The application of invariant integrals in elastic diffusive solids", *Phil. Trans. Roy. Soc.*, **A339**, 231-262
Atkinson C. and Craster R.V. 1992b, "Some fracture problems in fully coupled dynamic thermoelasticity", *J. Mech. Phys. Solids*, **40**, 1415-1432
Atkinson C. and Craster R.V. 1994a, "A singular perturbation approach to integral equations occuring in poroelasticity", *IMA J Appl. Maths*, **52**, 221-252
Atkinson C. and Craster R.V. 1994b, "Interfacial fracture of a radially inhomogeneous elastic bimaterial", *Int. J. Solids and Structures*, **31**, 9, 1207-1223
Atkinson C. and Smelser R.E. 1982, "Invariant integrals of thermoviscoelasticity", *Int. J. Solids Structures* , **18**, 533-549
Auriault J.L. 1980, "Dynamic behaviour of a porous medium saturated by a Newtonian fluid", *Int.J.Engng.Sci*, **18**, 775-785
Biot M.A. 1941, "General theory of three-dimensional consolidation", *J. Appl. Phys.*, **12**, 155-164
Biot M.A. 1955, "Thermoelasticity and irreversible thermodynamics", *J. Appl. Mech*, **27**, 240-253
Biot M.A. 1956a,b, "The theory of propagation of elastic waves in a fluid saturated porous solid, I. The low frequency range, II. The high frequency range", *J.Acoust.Soc.Am.*, **28**, 168-178, 179-191
Bowen R.M. 1982, "Compressible porous media models by use of the method of mixtures", *Int.J.Engng.Sci*, **20**, 697-735
Burridge R. and Keller J.B. 1981, "Poroelasticity equations derived from microstructure", *J.Acoust.Soc.Am*, **70**, 1140-1146
Craster R.V. and Atkinson C. 1992a, "Shear cracks in thermoelastic and poroelastic media", *J.Mech. Phys. Solids*, **40**, 4, 887-924
Craster R.V. and Atkinson C. 1992b,"Interfacial fracture in diffusive elastic media", *Int J Solids Struct.*, **29**, 1463-1498
Craster R.V and Atkinson C. 1994a,"Crack problems in a poroelastic material: an asymptotic approach", *Phil. Trans. Roy. Soc. Lond.*, **A346**, 387-428
Craster R.V and Atkinson C. 1994b, "Mixed boundary value problems in nonhomogeneous materials", *QJMAM*, **47**, 2, 183-206
Dugdale D.S. 1960, "Yielding of steel sheets containing slits", *J.Mech. Phys. Solids*, **8**, 100-104
Eshelby J.D. 1970, "Energy relations and the energy momentum tensor in continuum mechanics". In Inelastic behaviour of solids (ed. M.F. Kanninen *et al*) 77-115, New York, McGraw-Hill
Rice J.R. and Cleary M.P. 1976, "Some basic stress-diffusion solutions for fluid saturated elastic porous media with compressible constituents," *Rev. Geophys.Space Phys.*, **14**, 227-241
Rice J.R. and Simons D.A. 1976, "The stabilization of spreading shear faults by coupled deformation-diffusion effects in fluid-infiltrated porous materials", *J. Geophys. Res*, **81**, 5322-5344
Rudnicki J.W. 1985, "Effcct of pore fluid diffusion on deformation and failure of rock", *Mechanics of Geomaterials*, (Ed. Z.Bazant), John Wiley and Sons Ltd
Rudnicki J.W. and Koutsibelas D.A. 1991, "Steady propagation of plane strain shear cracks on an impermeable plane in an elastic diffusive solid", *Int.J.Solids Structures*, **27**, 205-225

Rudnicki J.W. and Roeloffs E.A. 1990, "Plane-strain shear dislocations moving steadily in linear elastic diffusive solids", *J. appl. Mech.*, **57**, 32-39
Simons D.A. 1977, "Boundary-layer analysis of propagating mode II cracks in porous elastic media", *J.Mech. Phys. Solids*, **25**, 99-115
Skempton A. 1954, "The pore pressure coefficients A and B," *Geotechnique*,**4**, 143-147

A. Nomenclature

α Biot's coefficient of effective stress, i.e. the ratio of fluid volume to the volume change of solid allowing the fluid to drain, where $0 < \alpha \leq 1$,
B Skempton's pore pressure coefficient, (Skempton 1954) i.e. the ratio of induced pore pressure to the variation of mean normal compression under undrained conditions,
c generalised consolidation coefficient,
δ_{ij} Kronecker delta,
e dilatation,
ϵ_{ij} components of the strain tensor, $e = \epsilon_{kk}$,
κ permeability coefficient,
G is the shear modulus,
$\frac{d\zeta}{dp}|_\epsilon = \frac{1}{Q}$ is a measure of the change in fluid content generated in a unit reference volume during a change of pressure with the strains kept constant,
m mass of fluid per unit volume,
ν, ν_u drained and undrained Poisson ratios, where $\nu \leq \nu_u \leq 0.5$
p perturbation pore pressure, i.e. the increase in fluid pressure from a reference pressure p_0,
q_i mass flux vector,
ρ_0 reference density,
σ_{ij} stress tensor,
u_i displacement vector,
ζ variation of fluid content per unit reference volume i.e. mass of fluid per unit volume/initial density ρ_0, i.e., $m = \rho_0\zeta$,
and the following relations are used in the text.

$$\alpha = \frac{3(\nu_u - \nu)}{B(1-2\nu)(1+\nu_u)}, \quad Q = \frac{2GB^2(1-2\nu)(1+\nu_u)^2}{9(\nu_u - \nu)(1-2\nu_u)}, \tag{A.1}$$

$$\eta = \frac{1-\nu}{1-2\nu}, \quad ,\overline{\eta}_u = \frac{(1-\nu_u)}{2(\nu_u-\nu)}, \quad \overline{\eta} = \frac{(1-\nu)}{2(\nu_u-\nu)}, \tag{A.2}$$

$$2G_u = \frac{2GB(1+\nu_u)}{3(1-\nu_u)}, \quad c = \frac{2\kappa B^2 G(1-\nu)(1+\nu_u)^2}{9(1-\nu_u)(\nu_u-\nu)}, \tag{A.3}$$

$$N_0 = \frac{1}{2}(1-2\overline{\eta}). \tag{A.4}$$

B. Some results required in the text

The following expressions are required

$$\Gamma_+^{\frac{1}{2}} = (\xi + i)^{\frac{1}{2}} \tag{B.5}$$

the subscripts +(−) are used to denote functions analytic in the upper (lower) halves of the complex ξ plane. The function Γ_+ has a branch cut from $-i$ to $-i\infty$.

$$\log\left(\frac{N_-(\xi)}{-N_0}\right) = -\frac{1}{\pi}\int_0^1 \arctan\left(\frac{p(1-p^2)^{\frac{1}{2}}}{p^2-\overline{\eta}}\right)\frac{dp}{p+i\xi}. \tag{B.6}$$

Although initially defined in the lower half plane, by analytic continuation this defines a function valid in the whole complex plane except for the branch cut $i0_+$ to i. (The 'plus' function is given from this by $N_+(\xi) = -N_-(-\xi)/N_0$).

$$k_+(\xi) - c_0 = \frac{N_+(0)}{\overline{\eta}\Gamma_+(\xi)\xi_+^{\frac{1}{2}}} + \frac{1}{i\pi}\int_0^1 \frac{dy}{(y-i\xi)y^{\frac{1}{2}}(1-y)^{\frac{1}{2}}}\left(\frac{1}{N_-(-iy)} - \frac{N_+(0)}{\overline{\eta}}\right) \tag{B.7}$$

$$c_0 = \frac{-1}{i\pi}\int_0^1 \frac{dy}{y^{\frac{3}{2}}(1-y)^{\frac{1}{2}}}\left(\frac{1}{N_-(-iy)} - \frac{N_+(0)}{\overline{\eta}}\right), \tag{B.8}$$

$$d = \frac{-1}{\pi}\int_0^1 \frac{dy}{y^{\frac{1}{2}}(1-y)^{\frac{1}{2}}}\left(\frac{1}{N_-(-iy)} - \frac{N_+(0)}{\overline{\eta}}\right), \tag{B.9}$$

$$\log\left(\frac{\overline{N}_-(\xi)}{N_0}\right) = \frac{1}{\pi}\int_0^1 \arctan\left(\frac{p^3}{(p^2+\overline{\eta})(1-p^2)^{\frac{1}{2}}}\right)\frac{dp}{p+i\xi} + \log\left(\frac{\alpha_1+i\xi}{1+i\xi}\right). \tag{B.10}$$

The result for $\overline{N}_+(\xi)$ is $\overline{N}_-(-\xi)/N_0$.

$$\alpha_1 = \left(\overline{\eta}\frac{(2-\overline{\eta}) + (\overline{\eta}^2+4\overline{\eta})^{\frac{1}{2}}}{2(2\overline{\eta}-1)}\right)^{\frac{1}{2}}. \tag{B.11}$$

$$\overline{k}_+(\xi) = \frac{\Gamma_+(\xi)}{\xi_+^{\frac{3}{2}}\overline{N}_-(0)} + \frac{i}{\pi}\int_0^1 \frac{(1-y)^{\frac{1}{2}}dy}{(y-i\xi)y^{\frac{3}{2}}}\left(\frac{1}{\overline{N}_-(-iy)} - \frac{1}{\overline{N}_-(0)}\right) \tag{B.12}$$

$$\overline{d} = \frac{-1}{\pi}\int_0^1 \frac{(1-y)^{\frac{1}{2}}}{y^{\frac{3}{2}}}\left(\frac{1}{\overline{N}_-(-iy)} - \frac{1}{\overline{N}_-(0)}\right)dy, \tag{B.13}$$

$$\overline{\chi} = \frac{1}{\pi}\int_0^1 \frac{(1-y)^{\frac{1}{2}}}{y^{\frac{1}{2}}}\left(\frac{1}{\overline{N}_-(-iy)} - \frac{1}{\overline{N}_-(0)}\right)dy. \tag{B.14}$$

On Theories of Mixtures and Their Applications to Dynamics of Fluid Saturated Porous Media

Mahantesh S. Hiremath* and Ranbir S. Sandhu**
The Ohio State University
Department of Civil Engineering
2070 Neil Avenue, Columbus, Ohio 43210

Abstract A finite strain theory for motion and deformation of fluid-saturated solids is proposed. A convected coordinate frame of reference is preferred for the solid particles that are enclosed in a material region that moves with them. The fluid is taken to be in relative motion with respect to these solid particles. This may be regarded as an extension of Gibson's quasistatic one-dimensional theory to dynamics of two-phase mixture in three dimensions. Reduction to small strains yields Biot's equations of motion but without Biot's simplifying assumptions of existence of energy functions. Numerical examples include wave propagation analysis of an elastic-perfectly-plastic solid and a fluid-saturated elastic-plastic strain-hardening soil column subjected to dynamic excitation.

1 Introduction

The response of fluid-saturated porous solids subjected to time-dependent loads has been investigated since early 1950s. However, reliable procedures to describe the resulting changes in pore pressure and stress field in the solid matrix have yet to be developed. Representation of a fluid-saturated medium as a single-phase material has been virtually discarded. Similarly, the study of fluid mechanics, although well explored, has provided few clues to the modeling problems that arise when solid and fluid phases exist together and are in relative motion. In this case, the equations of motion for individual constituents may involve interaction terms and the stress could depend on kinematics of both.

Biot (1941) was the first to propose a general theory of three-dimensional deformation of porous fluid-saturated solids. His quasistatic theory was later (1956, 1961) extended to wave propagation in saturated geological media. This work has been the basis for analytical (Garg et al. 1974, Simon et al. 1984) as well as numerical (Ghaboussi and Wilson, 1973, Hiremath et al. 1988, Morland et al. 1987, Sandhu et al. 1987, Simon et al. 1986) solution procedures. For an isotropic saturated material, Biot (1956, 1961) postulated the kinetic energy of the mixture to be quadratic in velocities of the solid and fluid and included a coupling term. For the dissipative case, a dissipative function quadratic in relative velocity was introduced. Further, in stating the constitutive equations, Biot (1941, 1956, 1961) maintained that an energy function quadratic in solid strain and change in water content (or fluid strain) exists.

---

* *Associate Engineer, ARES Corporation, San Francisco, California*
** *Professor of Civil Engineering, The Ohio State University, Columbus, Ohio*

A. P. S. Selvadurai (ed.), Mechanics of Poroelastic Media, 47–71.

Theories of the mechanical behavior of mixtures based on the principles of mechanics were developed by Truesdell and Toupin (1960) and Green and Naghdi (1965, 1967, 1969). These were specialized to the case of immiscible mixtures of solid and fluid to model the dynamics of fluid-saturated solids (Garg 1971, Garg et al. 1974, Morland 1972). Truesdell and Toupin (1960) introduced the notion of a mixture as a continuum and postulated that the form of equations of balance of a mixture is the same as for its constituents. Green and Naghdi (1965, 1967, 1969) derived the balance equations based on the material frame invariance of a rate of energy equality. Both Truesdell and Toupin (1960) and Green and Naghdi (1965) considered the kinetic energies of the constituents to be additive. In view of this, the drawbacks in Biot's theory were even more apparent, particularly his assumption of existence of a quadratic function and the form of kinetic energy of the mixture.

Sandhu (1985) reviewed various approaches to the description of the constituents and the mixture as well as the balance of mass, linear and angular momentum, and rate of energy equality. He observed that for a mixture in which the solid and fluid constituents are in relative motion, the mixture has only momentary existence as a continuum; this pointed to the need for a theory based on balance equations written for a reference set of particles of the solid. Gibson et al. (1967) presented such a procedure for one-dimensional analysis applicable to the quasistatic case.

Proposed here is its extension to three dimensions while allowing for inertia effects. The procedure is different from that of Green and Naghdi (1965) and Bowen (1976) who wrote the balance laws for a fixed volume in space. Instead, a material region in space is followed through its motion from the reference state so that its encloses the same set of solid particles throughout the process of deformation. The fluid is referred to with respect to the current position of the solid. For mathematical simplicity, the only mixture particle considered is the current instantaneous particle that is in form of a superposed particle of solid with a known reference state and a fluid particle currently associated with it. Biot's (1956, 1961) theory is shown to be a special case of the general theory proposed. Further, reduction of the balance laws to one dimension and elimination of inertial terms leads to the equations of motion proposed by Gibson et al. (1967) . A framework for three-dimensional analysis of the dynamics of a fluid-saturated solid admitting material as well as geometric nonlinearity is thus developed. The mathematical preliminaries are given elsewhere (Green and Zerna 1968). Numerical implementation for small strain analysis of the dynamics saturated soil column is also presented.

2 Choice of Reference Frame

For the phenomenon described above, an appropriate choice of the frame of reference is extremely important. The deformed state is unknown to begin with and, since the displacements are the quantities sought and are the functions of this deformed state, the Eulerian description becomes particularly unsuitable. The Lagrangian form, on the other hand, relates the quantities to the known initial configuration and may be able to identify a mixture particle in the reference state. However, such a particle exists only momentarily, since the constituents become separated in the course of time due to relative motion. Thus identification of the same particle in the current state is impossible. The stresses are related to the current deformed state and need to be referred to the initial state in a way that is physically artificial but mathematically consistent. All these difficulties are circumvented by the use of a convected coordinate system, which combines the advantages of both the Eulerian and Lagrangian forms. The strain is related to the known initial state and the stress to the current deformed state. Further, for the development of constitutive relations, the rates of Piola-Kirchoff tensor and Green's strain tensor are related in Lagrangian analysis. In the study of a binary mixture, with constituents in relative motion, this description has particularly important role. The coordinate system moves along with the material volume that contains the same solid particles, the initial unstrained state of which is known.

The present work is based on this approach and the balance laws are developed in convected coordinates. This requires rigorous use of tensor calculus, and the algebra becomes quite complicated. The averaging concept in the theories of mixtures based on the assumption of superposed continua are used to associate quantities with the individual constituents of the mixture and for the continuity equations of the solid and fluid phases. In deriving the dynamical equilibrium equations for the fluid phase and the mixture, physical concepts were preferred.

3 Notation and Some Preliminaries

Standard indicial notation is used throughout. Latin indices describe the range 1 to 3. Repeated indices imply summation over the range. A superposed dot indicates the time derivative. A subscripted comma implies differentiation with respect to the coordinates defined by the subscripts following the commas.

Development of a theory of the dynamics of two-phase media requires scrutiny of such physical quantities as density, porosity, stress, etc. for individual constituents in light of the assumption of superposed continua. The partial and intrinsic quantities are defined following Garg et al. (1974) as: the total mass, M, and volume, V, of the mixture are considered to be the sum of corresponding quantities associated with the solid and fluid, which are denoted by the superscripts 1 and 2, respectively. The volume fractions $n^{(1)}$ and $n^{(2)}$ are defined by:

$$n^{(1)} + n^{(2)} = 1, \quad \frac{V^{(1)}}{V} = n^{(1)}, \quad \frac{V^{(2)}}{V} = n^{(2)} \tag{1}$$

The intrinsic densities are defined as:

$$\rho^{(k)*} = \frac{M^{(k)}}{V^{(k)}} \qquad k = 1,2 \tag{2}$$

which in turn are related to the partial densities by the relations:

$$\rho^{(k)*} = \frac{M^{(k)}}{V^{(k)}} \qquad k = 1,2 \tag{3}$$

This leads to the mixture density:

$$\rho = \rho^{(1)} + \rho^{(2)} \tag{4}$$

Some investigators (Ghaboussi and Wilson 1973, Simon et al. 1986) have taken the total stress t_{ij} to be the sum of the partial stresses $t_{ij}^{(1)}$ and $t_{ij}^{(1)}$, viz.:

$$t_{ij} = t_{ij}^{(1)} + t_{ij}^{(2)} \tag{5}$$

If the fluid is isotropic, then:

$$t_{ij}^{(2)} = \pi \delta_{ij} \tag{6}$$

where π is the fluid pressure. The intrinsic stresses $t_{ij}^{(1)*}$ and $t_{ij}^{(2)*}$ are defined by their relationships with the partial stresses, as follows:

$$t_{ij}^{(1)} = n^{(1)}\, t_{ij}^{(1)*} \tag{7}$$

$$t_{ij}^{(2)} = n^{(2)}\, t_{ij}^{(2)*} = \pi\, \delta_{ij} = n^{(2)}\, \pi^{*}\, \delta_{ij} \tag{8}$$

where π^{*} is the intrinsic fluid pressure.

4 Balance Laws For A Fluid-Saturated Porous Solid

The equations of balance of mass and linear and angular momentum are summarized. The detailed derivations are given elsewhere (Hiremath, 1987).

The strain in the solid $\gamma_{ij}^{(1)}$ is defined as:

$$\gamma_{ij}^{(1)} = E_{ij}^{(1)} = \frac{1}{2}\,[u_{i,j}^{(1)} + u_{j,i}^{(1)} + u_{m,i}^{(1)}\,u_{m,j}^{(1)}] \tag{9}$$

where $E_{ij}^{(1)}$ is the Green strain tensor and $u_i^{(1)}$ are components of solid displacement referred to the Cartesian system in the initial state C_0. For isotropic fluids, considering only the infinitesimal deformations, the components of strain are given by:

$$\gamma_{ij}^{(2)} = e_{ij}^{(2)} = u_{k,k}^{(2)} \tag{10}$$

4.1 Balance of Mass for the Solid Phase

If $\rho_o^{(1)}$ and $\rho^{(1)}$ denote the mass densities in the configurations C_0 and C, respectively, for the solid phase, then from (3):

$$\rho_0^{(1)} = n_0^{(1)}\,\rho_0^{(1)*} \tag{11}$$

and

$$\rho^{(1)} = n^{(1)}\,\rho^{(1)*} \tag{12}$$

in which $n_0^{(1)}$ and $n^{(1)}$ refer to the solid volume fractions in the initial and current configurations, respectively. Since the volume embraces the same solid particles, balance of mass implies:

$$\int_{V_0} n_0^{(1)}\,\rho_0^{(1)*}dV_0 = \int_V n^{(1)}\,\rho^{(1)*}dV \tag{13}$$

Here V_0 and V denote the volume elements in C_0 and C, respectively. Use of $dV = \sqrt{G}\,dV_0$ and noting that V_0 is an arbitrary volume leads to:

$$n_0^{(1)}\,\rho_0^{(1)*} = \sqrt{G}\,n^{(1)}\,\rho^{(1)*} \tag{14}$$

In one dimension (Green and Zerna 1968), (14) takes the form:

$$\sqrt{G} = \frac{\partial z}{\partial x} \tag{15}$$

$$n_0^{(1)} \rho_0^{(1)*} = \frac{\partial z}{\partial x} n^{(1)} \rho^{(1)*} \tag{16}$$

This is same as the equation of continuity by Gibson et al. (1967) for the solid phase.

4.2 Balance of mass for the fluid Phase

Motion of the fluid is relative to that of the solid, and the fluid itself does not have a reference state C_0 as does the solid. For deriving the equation of mass continuity, the physical concept illustrated in Fig. 1 is preferred. A rectangular parallelepiped at point P_0 in the reference state C_0 becomes a skewed parallelepiped in the deformed state C. For example, the surface formed by sides dx^2 and dx^3 becomes an area formed by $\mathbf{G}_2 dx^2$ and $\mathbf{G}_3 dx^3$ in the state C. Denoting the velocity vectors for the constituents by:

$$\boldsymbol{v}^{(1)} = v^{(1)m} \boldsymbol{G}_m = v_m^{(1)} \boldsymbol{G}^m = \dot{u}_m^{(1)} \boldsymbol{e}_m \tag{17}$$

and

$$\boldsymbol{v}^{(2)} = v^{(2)m} \boldsymbol{G}_m = v_m^{(2)} \boldsymbol{G}^m = \dot{u}_m^{(2)} \boldsymbol{e}_m \tag{18}$$

where $\mathbf{e}_m$ and $\mathbf{G}_m$ (or $\mathbf{G}^m$) are the base vectors in the undeformed and deformed state, respectively, the net mass fluid inflow is expressed as (Hiremath, 1987):

$$\frac{\partial}{\partial x^i}\{n^{(2)} \rho^{(2)*} [\dot{u}_i^{(2)} - \dot{u}_i^{(1)}]\}\ dx^1 dx^2 dx^3 \tag{19}$$

Equating this to the rate of increase of the fluid within the deformed parallelepiped, the mass continuity yields:

$$\frac{d}{dt}[n^{(2)} \rho^{(2)*} \sqrt{G} + \frac{\partial}{\partial x^i}\{n^{(2)} \rho^{(2)*} [\dot{u}_i^{(2)} - \dot{u}_i^{(1)}]\} = 0 \tag{20}$$

In one dimension, this reduces to:

$$\frac{d}{dt}\,[n^{(2)}\,\rho^{(2)*}\,\frac{\partial z}{\partial x}\,+\,\frac{\partial}{\partial x}\{n^{(2)}\,\rho^{(2)*}\,[\dot{u}^{(2)}\,-\,\dot{u}^{(1)}]\}\,=\,0 \tag{21}$$

which is the same as that of Gibson et al. (1967).

4.3 Equilibrium equation for the fluid phase

Consider an infinitesimal rectangular parallelepiped (Fig. 2) that is transformed into a skewed parallelepiped after motion and deformation. The internal forces on the fluid phase in this deformed body may be expressed by considering the forces on the surfaces of the body. For examples, the net force on the fluid phase across the surface formed by the vectors $\mathbf{G}_2dx^2$ and $\mathbf{G}_3dx^3$ is given by (Green and Zerna, 1968) - $\mathbf{T}_1^{(2)}\, dx^2dx^3$. Similar expressions for the net forces across other faces can be written (Hiremath, 1987).

Denote the components of the body force, the inertia force, and the viscous coupling force vectors by $\mathbf{F}_i^{(2)}$, $\mathbf{f}_i^{(2)}$, and $(\mathbf{v}_i^{(2)} - \mathbf{v}_i^{(1)})$, respectively; the corresponding forces acting over the deformed volume $\sqrt{G}\, dx^1dx^2dx^3$ are given by:

$$[\rho^{(2)}F_1^{(2)},\ \rho^{(2)}F_2^{(2)},\ \rho^{(2)}F_3^{(2)}]\ \sqrt{G}\ dx^1dx^2dx^3 \tag{22}$$

$$[-\ \rho^{(2)}f_1^{(2)},\ -\ \rho^{(2)}f_2^{(2)},\ -\ \rho^{(2)}f_3^{(2)}]\ \sqrt{G}\ dx^1dx^2dx^3 \tag{23}$$

$$[-\ D\ (v_1^{(2)}\ -\ v_1^{(1)}),\ -\ D\ (v_2^{(2)}\ -\ v_2^{(1)}),\ -\ D\ (v_3^{(2)}\ -\ v_3^{(1)})]\ \sqrt{G}\ dx^1dx^2dx^3 \tag{24}$$

Here, $D = \mu / \kappa\, n^{(2)}n^{(2)}$ is the viscous coupling term used in the literature (Garg et al. 1974). κ denotes the specific permeability (dimensions L^2) and μ is the viscosity of the fluid (dimensions $ML^{-2}T$). Summing up the forces and setting the total equal to zero for equilibrium yields:

$$\frac{\partial}{\partial x^i}\,T_i^{(2)}\,+\,\sqrt{G}\ \rho^{(2)}\ F^{(2)}\,+\,\sqrt{G}\ \rho^{(2)}\ f^{(2)}\,-\,\sqrt{G}\ D\ [v^{(2)}\,-\,v^{(1)}]\,=\,0 \tag{25}$$

For the binary mixture, the following identities hold (Hiremath 1987):

$$v^{(k)}\,=\,v^{(k)m}\ G_m\,=\,v_m^{(k)}\ G^m\,=\,\dot{u}_m^{(k)}\ e_m\quad (k\,=\,1,2) \tag{26}$$

$$\mathbf{f}^{(k)} = f^{(k)m}\, \mathbf{G}_m = f^{(k)}_m\, \mathbf{G}^m = \ddot{u}^{(k)}_m\, \mathbf{e}_m \quad (k = 1,2) \tag{27}$$

$$\mathbf{F}^{(k)} = F^{(k)m}\, \mathbf{G}_m = F^{(k)}_m\, \mathbf{G}^m = \hat{F}^{(k)}_m\, \mathbf{e}_m \quad (k = 1,2) \tag{28}$$

$$\hat{F}^{(k)}_m = z_{j,m}\, F^{(k)j} \qquad (k = 1,2) \tag{29}$$

The equilibrium in the initial state C_0 is given by:

$$[S^{(2)i}_m]_{,i} + \sqrt{G}\, \rho^{(2)}\, \hat{F}^{(2)}_m = \sqrt{G}\, \rho^{(2)}\, \ddot{u}^{(2)}_m + \sqrt{G}\, D\, (\dot{u}^{(2)}_m - \dot{u}^{(1)}_m) \tag{30}$$

Further, for isotropic fluids, $S_m^{(2)}$ may be replaced by $\pi\delta_m^{\,i}$ and recalling (8), (30) may be expressed as:

$$\pi^*_{,m} + \sqrt{G}\, \rho^{(2)}\, \hat{F}^{(2)}_m = \sqrt{G}\, \rho^{(2)*}\, \ddot{u}^{(2)}_m + \sqrt{G}\, \frac{\mu}{\kappa}\, n^{(2)}\, (\dot{u}^{(2)}_m - \dot{u}^{(1)}_m) \tag{31}$$

Neglecting the inertia term, (31) is similar to Darcy's law. In the one-dimensional case, (31) takes the form:

$$\frac{\partial \pi^*}{\partial x} + \frac{\partial z}{\partial x}\, \rho^{(2)}\, \hat{F}^{(2)} = \frac{\partial z}{\partial x}\, \rho^{(2)*}\, \ddot{u}^{(2)} + \frac{\partial z}{\partial x}\, \frac{\mu}{\kappa}\, n^{(2)}\, (\dot{u}^{(2)} - \dot{u}^{(1)}) \tag{32}$$

The equation by Gibson et al. (1967) for quasistatic analysis is recovered if the inertial term in (32) is discarded.

4.4 Dynamic equilibrium of the mixture in the deformed body

The solid phase associated with any configuration is the same, but the instantaneous mixture particle in the current state does not remain the same an instant later. The boundary conditions in an initial-boundary value problem of fluid-saturated media are usually given in terms of the total stress rather than the partial stresses. Hence, it is essential to consider the overall equilibrium of the deformed parallelepiped instead of the equilibrium of the solid phase.

As in the case of fluid equilibrium, the internal forces on the six bounding surfaces of the skewed parallelepiped (Fig. 3) are given by:

$$\boldsymbol{T}_i\, dx^j dx^k,\ \boldsymbol{T}_i\, dx^j dx^k + \frac{\partial}{\partial x^i}\boldsymbol{T}_i\, dx^j dx^k \tag{33}$$

where $\mathbf{T}_i$ is associated with the mixture in the deformed state.

Denoting the body force vectors by $\mathbf{F}_i$ and acceleration vectors of the two constituents by $\mathbf{f}_i^{(k)}$ (k = 1,2), the expressions for these forces acting on the deformed volume $\sqrt{G}\, dx^1dx^2dx^3$ are given by:

$$[\rho\, F_1,\ \rho\, F_2,\ \rho\, F_3]\ \sqrt{G}\ dx^1dx^2dx^3 \tag{34}$$

$$\begin{aligned}[-\ \{\rho^{(1)}f_1^{(1)} + \rho^{(2)}f_1^{(2)}\},\ -\ \{\rho^{(1)}f_2^{(1)} + \rho^{(2)}f_2^{(2)}\},\\ \{\rho^{(1)}f_3^{(1)} + \rho^{(2)}f_3^{(2)}\}]\ \sqrt{G}\ dx^1dx^2dx^3\end{aligned} \tag{35}$$

Note that the coupling terms do not appear here as they cancel out, being of opposite signs for the two constituents. Summing up the forces:

$$\frac{\partial}{\partial x^i}\ \boldsymbol{T}_i + \sqrt{G}\ \rho\ \boldsymbol{F} = \sqrt{G}\ \rho^{(1)}\ \boldsymbol{f}^{(1)} + \sqrt{G}\ \rho^{(2)}\ \boldsymbol{f}^{(2)} \tag{36}$$

Referring all quantities to the reference state C_0, two alternative forms of (36) are (Hiremath 1987):

$$[S^{ij}\ z_{m,j}]_{,i} + \sqrt{G}\ \rho\ \hat{F}_m = \sqrt{G}\ \rho^{(1)}\ \ddot{u}_m^{(1)} + \sqrt{G}\ \ddot{u}_m^{(2)} \tag{37}$$

$$S^i_{m,i} + \sqrt{G}\ \rho\ \hat{F}_m = \sqrt{G}\ \rho^{(1)}\ \ddot{u}_m^{(1)} + \sqrt{G}\ \ddot{u}_m^{(2)} \tag{38}$$

In one dimension, the above is written as:

Ignoring the inertia terms, the equation of motion by Gibson et al. (1967) for the bulk, viz.:

$$[S_x^x]_{,x} + \frac{\partial z}{\partial x} \rho \hat{F}_x = \frac{\partial z}{\partial x} \rho^{(1)} \ddot{u}^{(1)} + \frac{\partial z}{\partial x} \rho^{(2)} \ddot{u}^{(2)} \qquad (39)$$

$$\frac{\partial S^x}{\partial x} + \rho \hat{F}_x \frac{\partial Z}{\partial x} = 0 \qquad (40)$$

is recovered.

4.5 Balance of angular momentum for the mixture

From Fig. 3, the angular moments of the forces in (33) with respect to the deformed axes along $\mathbf{G}_i$ are given by:

$$- (\boldsymbol{T}_i \ x \ \boldsymbol{R}) \ dx^j dx^k, \ (\boldsymbol{T}_i \ x \ \boldsymbol{R}) \ dx^j dx^k + \frac{\partial}{\partial x^i} (\boldsymbol{T}_i \ x \ \boldsymbol{R}) \ dx^j dx^k \quad (no\ sum\ on\ i,\ i \neq j \neq k) \qquad (41)$$

where $\mathbf{R}$ is the position vector of the point P in the deformed state. Similar expressions for angular moments of body forces in (34) and inertial forces in (35) are easily derived (HIremath, 1987). Summation of these angular moments along with the use of (36) and noting $\mathbf{R}_{,i} = \mathbf{G}_i$ yields:

$$\boldsymbol{T}_i \ x \ \boldsymbol{G}_i = 0 \qquad (42)$$

which is same as:

$$\epsilon_{ijk} \ \delta^{ij} = 0 \qquad (43)$$

Alternative forms include:

$$\tau^{ij} = \tau^{ji} \qquad (44)$$

and

$$s^{ij} = s^{ji} \qquad (45)$$

The bulk stresses τ^{ij} and s^{ij} are symmetric. This agrees with Green and Naghdi's (1965) assertion that the partial stresses need not be symmetric but the total stress is.

5 Specialization to Biot's Theory

Biot's (1956, 1961) equations for small deformation theory are embedded in the general theory presented here. The continuity equations are not required specifically as the changes in densities are small. All quantities are referred to the initial state with rectangular cartesian system as a frame of reference. All the stress tensors are symmetric and the distinction between the covariant and contravariant components disappears.

The dynamic equilibrium equations in terms of the bulk stress t_{ij} and the partial pore pressure π^* are obtained from (38) and (31), respectively, as:

$$t_{ij,j} + \rho \hat{F}_i = \rho^{(1)} \ddot{u}_i^{(1)} + \ddot{u}_i^{(2)} \tag{46}$$

$$\pi^*_{,i} + \rho^{(2)} F_i = \rho^{(2)*} \ddot{u}_i^{(2)} + \frac{\mu}{\kappa} n^{(2)} (\dot{u}_i^{(2)} - \dot{u}_i^{(1)}) \tag{47}$$

These equations correspond to those of Biot's (1956, 1961); however, they are derived without the assumption of existence of energy functions. Alternatively, the partial solid stress may also be used in the equilibrium equations derived from (46) and (47) as follows:

$$t_{ij,j}^{(1)} + \rho^{(1)} F_i = \rho^{(1)*} \ddot{u}_i^{(1)} - \frac{\mu}{\kappa} n^{(2)} n^{(2)} (\dot{u}_i^{(2)} - \dot{u}_i^{(1)}) \tag{48}$$

Comparison with Biot's (1956, 1961) corresponding equations shows that (48) does not include any mass coupling terms.

6 Constitutive Equations

The dynamical theory summarized above is based on the movement of nonpenetrating particles. In this theory, the constitutive relations are required only for the partial stresses and the diffusive resistance. Porosity is a quantity directly related to the deformation of the set of particles and need not be treated as an additional variable as some investigators (Garg et al. 1974, Morland 1972) do. The relative movement between the pore water and the reference set of soil particles would be the principal kinematical variable related to the interaction force. For linear theory, this relationship would reduce to Darcy's law. The stresses in the reference set of particles must be described by means of convected coordinates just as the deformation and its rate are. For validation of the constitutive relations, the need for more data on the behavior of saturated media under large deformation is apparent.

7 Numerical Examples

Equations (46) and (47) along with small strain kinematics were used to set up a Galerkin formulation, which in semi-discretized form becomes (Hiremath, 1987):

$$\boldsymbol{M}\,\ddot{\boldsymbol{U}} + \boldsymbol{C}\,\dot{\boldsymbol{U}} + \boldsymbol{K}\,\boldsymbol{U} = \boldsymbol{R} \tag{49}$$

An incremental form, allowing for elastic-plastic stress-strain relations for the solid phase, is:

$$\begin{aligned} &\boldsymbol{M}\,\Delta\ddot{\boldsymbol{U}}(t) + \boldsymbol{C}\,\Delta\dot{\boldsymbol{U}}(t) + \boldsymbol{K}^{P}(t)\,\Delta\boldsymbol{U}(t) + \boldsymbol{K}^{**}\,\Delta\boldsymbol{U}(t) \\ &= \boldsymbol{R}(t+\Delta t) - \boldsymbol{M}\,\ddot{\boldsymbol{U}}(t) - \boldsymbol{C}\,\dot{\boldsymbol{U}}(t) - \boldsymbol{K}^{**}\,\boldsymbol{U}(t) - \boldsymbol{F}_{s}^{*}(t) \end{aligned} \tag{50}$$

Here $F_s^*(t)$ represents the internal resisting force at time t. $\boldsymbol{K}^P(t)$ is the stiffness matrix dependent upon the nonlinear solid stress-strain relations. The above equation is solved iteratively for $\Delta U(t)$ using Wilson's step-forward procedure in the time domain.

Two computer programs were developed. One implemented bilinear hardening of elastic-perfectly plastic type (Figs. 4a and 4b) with one-dimensional elements. The other incorporated Singh's (1972) cap model (Fig. 5) applicable to strain-hardening soils and used two-dimensional elements.

In Example 1, the computer programs were checked for the wave propagation problem under uniaxial strain conditions. The solid column shown in Fig. 6 was assumed to be elastic-perfectly plastic. The excitation consisted of a stress pulse that decayed over a short time. The analytical solution for this single material problem is due to Wood (1952). The magnitude of the peak stress is ten times that of the elastic yield limit under uniaxial strain. The elastic wave velocity was 8.0 and the plastic wave velocity 6.56. The material properties are given in Table 1. The total time of analysis was such that reflection of waves was possible. The two-dimensional finite element analysis was restricted to t = 4 because the computational effort increased dramatically with spatial and temporal refinement. The CPU time for a mesh with 640 linear one-dimensional elements and 1600 time-steps of 0.0125 was 48 minutes. For 320 linear quadrilateral elements along with 160 time-steps of 0.0125, it was 68 minutes. The stress wave profiles are shown in Figs. 7 and 8 at t = 4 and 16. Spatial and temporal refinements yielded numerical solutions that matched very closely the profiles and sharp wave fronts of the incident and reflected waves.

In Example 2, these two codes, based on different algorithms and constitutive models, were employed to perform the wave propagation analysis for a fluid-saturated elastic-perfectly plastic solid. No analytical solutions are available. Fig. 9 shows the one-dimensional fluid-saturated solid subjected to dynamic excitation at the free end. The stress profiles for the solid and pore fluid were plotted from the code at t = 4, 8 (Figs. 10 through 13). Table 1 gives the material properties. The numerical solution from the two codes appear to agree very well. Two types of stress waves were seen to propagate in the solid. The faster wave has an amplitude equal to the yield stress under uniaxial strain. The amplitude of the slower wave was slightly greater. A similar representation is seen for the pore fluid stress.

Finite element procedures can produce satisfactory results provided the element size and time-step size are carefully chosen. In general, sharp gradients in wave profiles can be reproduced more effectively if the element size is such that the faster wave travels no more than one element length in a single time-step.

8 Conclusions

A three-dimensional theory admitting material as well as geometric nonlinearity for dynamics of fluid-saturated deformable porous solids is proposed. Current theories based on simplifying assumptions are shown to be special cases of this general theory. To describe the motion and deformation of a binary mixture, a convected coordinate system is preferred for the solid particles that are enclosed in a material region that moves with them, as opposed to the fixed volume in space. The solid has a reference state, but the fluid is assumed to be in relative motion with respect to the known solid configuration. Numerical examples based on small strain theory showed the versatility of finite element analysis to effectively reproduce the wave fronts of reflecting waves.

Acknowledgement: The work presented herein was supported by the U.S. Air Force Office of Scientific Research, Air Force Systems Command, Grant No. AFOSR-83-0055.

References

Biot, M. A. (1941) General theory of three-dimensional consolidation. J. Appl. Phys., 12: 155-164.

Biot, M. A. (1956) Theory of propagation of elastic waves in fluid-saturated porous solids I and II. J. Acous. Soc. Amer., 28: 168-178 and 179-191.

Biot, M. A. (1961) Mechanics of deformation and acoustic propagation in porous media. J. Appl. Phys., 13: 1482-1498.

Bowen, R. M. (1976) Theories of mixtures. Continuum Physics III, Mixtures and Engineering Mechanics field theories, Ed. A. C. Eringen, Academic Press, New York: 1-127.

Garg, S. K. (1971) Wave propagation effects in fluid-saturated porous solids. J. Geophys. Res., 76: 7947-7962.

Garg, S. K., Nayfeh, A. H., and Good, A. J. (1974) Compressional waves in fluid-saturated elastic porous media. J. Appl. Phys., 45: 1968-1974.

Ghaboussi, J., and Wilson, E. L. (1973) Variational formulation of dynamics of fluid-saturated porous elastic solids. ASCE, Eng. Mech. Div., 98: 947-956.

Gibson, R. E., England, R. L., and Hussey, M. J. L. (1967) The theory of one-dimensional consolidation of saturated clays. Geotechnique, 17: 261-263.

Green, A. E., and Naghdi, P. M. (1965) A dynamical theory of interacting continua. Int. J. Eng. Sci., 3: 231-241.

Green, A. E., and Naghdi, P. M. (1967) A theory of mixtures, Arch. Rat. Mech. Anal. 24: 243-263.

Green, A. E., and Zerna, W. (1968) Theoretical Elasticity. Clarendon Press, Oxford.

Green, A. E., and Naghdi, P. M. (1969) On basic equations of mixtures, Quar. J. Mech. Appl. Math. XXII 4: 427-438.

Hiremath, M. S. (1987) A dynamical theory of deformable fluid-saturated porous solids. Ph. D. Dissertation, The Ohio State University, Columbus, Ohio.

Hiremath, M. S., Sandhu, R. S., and Wolfe, W. E. (1988) Analysis of One-Dimensional Wave Propagation in Fluid-Saturated Finite Soil Column, Int. J. Num. Anal. Meth. Geomech. 12: 121-139.

Morland, L. W. (1972) A theory of slow fluid flow through a porous media. J. Geophys. Res., 77: 890-900.

Morland, L. W., Sandhu, R. S., Wolfe, W. E., and Hiremath, M. S. (1987) Wave propagation in fluid-saturated elastic layer. Report to Air Force Office of Scientific Research Washington D. C.

Simon, B. R., Zienkiewicz, O. C., and Paul, D. K. (1984) An analytical solution for transient response of saturated porous elastic solid. Int. J. Num. Anal. Meth. Geomech., 8: 381-394.

Simon, B. R., Wu, J. S. S., and Zienkiewicz, O. C. (1986) Evaluation of higher order, mixed, and hermitean finite element procedures for dynamic analysis of saturated porous media using one-dimensional models, Int. J. Num. Anal. Meth. Geomech. 10: 483-499.

Sandhu, R. S., (1985) Mechanical behavior of saturated soils - a review. Report to Air Force Office of Scientific Research Washington D. C.

Sandhu, R. S., Wolfe, W. E., and Hiremath, M. S. (1987) Finite element analysis of wave propagation in linear elastic fluid-saturated solids. Report to Air Force Office of Scientific Research Washington D. C.

Singh, R. D. (1972) Mechanical characterization and finite element analysis of elastic-plastic work-hardening soils. Ph. D. Dissertation, The Ohio State University, Columbus, Ohio.

Truesdell, C., and Toupin, R. A. (1960), The classical field theories. Handbuch der Physik, III/1, Ed. S. Flugge, Springer-Verlag.

Wood, D. S. (1952) On longitudinal plane waves of elastic-plastic strain in solids. J. Appl. Mech., 25: 521-525.

Table 1

Material Properties for Numerical Examples

Description	Ex. 1	Ex. 2
Modulus of elasticity (E)	42.0352	100.0
Poisson's ratio (ν)	0.3372	0.25
Shear modulus (μ)	15.718	40.0
Mixture mass density (ρ)	1.0	1.0
Partial mass density of fluid ($\rho^{(2)}$)	0.0	0.2
Porosity ($n^{(2)}$)	0.0	0.2
Solid compressibility (α)	1.0	0.68
Fluid compressibility (M)	0.0	54.65
Constant (μ / κ)	0.001	100.0
Drucker's α	0.0	0.0
Drucker's κ	0.5773	0.5773
Ellipse ratio (R)	3.5	3.5
Initial position of ellipse (P_0)	-10^{10}	-10^{10}
Constant A	0.0025	0.0025
Elastic strain limit (uniaxial stress) Y_0 / E	0.0237	0.01
Elastic stress limit (uniaxial stress) Y_0	1.0	1.0
Elastic strain limit (uniaxial strain) Y_0 / 2G	0.0318	0.125
Elastic stress limit (uniaxial strain) [K/2G + 2/3] Y_0	2.0356	1.5

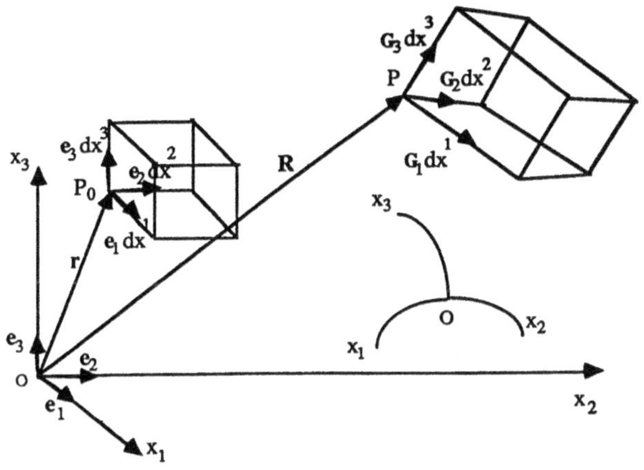

Figure 1. Geometry of an Infinitesimal Parallelepiped

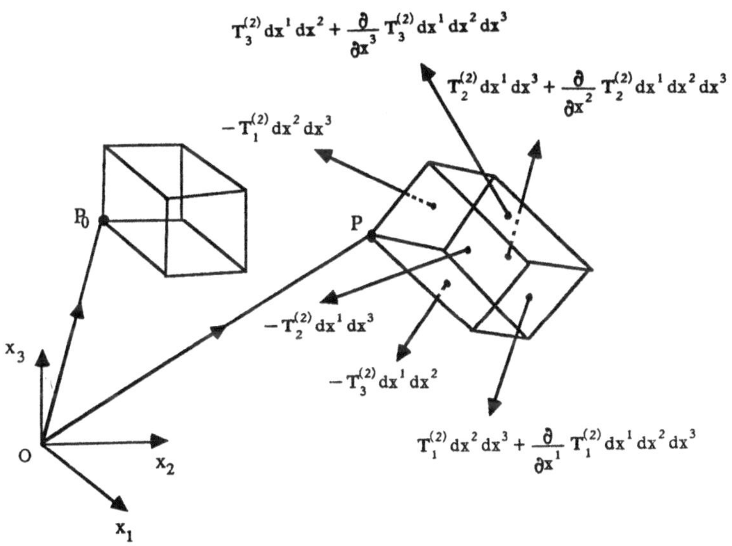

Figure 2. Fluid Equilibrium in the Deformed State

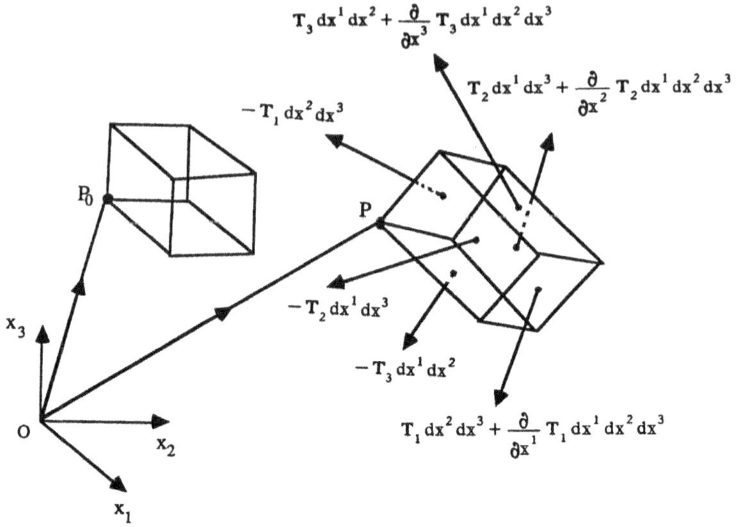

Figure 3. Equilibrium of the mixture in the Deformed State

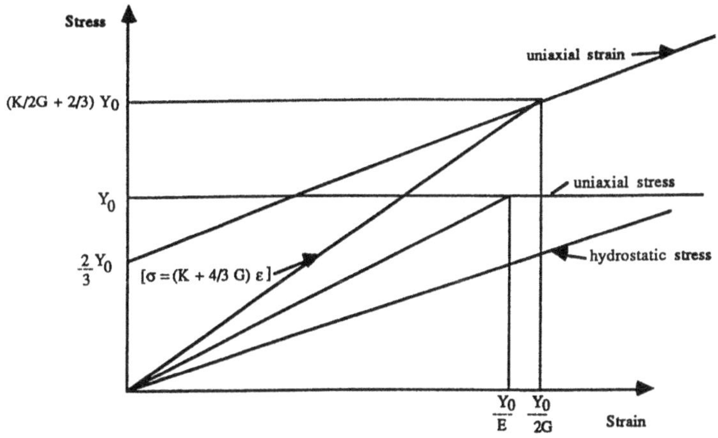

Figure 4a. Stress-strain relations for elastic-perfectly plastic behavior

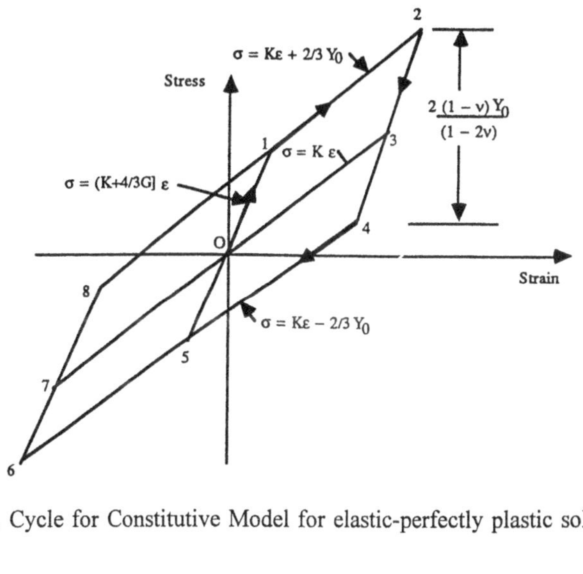

Figure 4b. Cycle for Constitutive Model for elastic-perfectly plastic solid

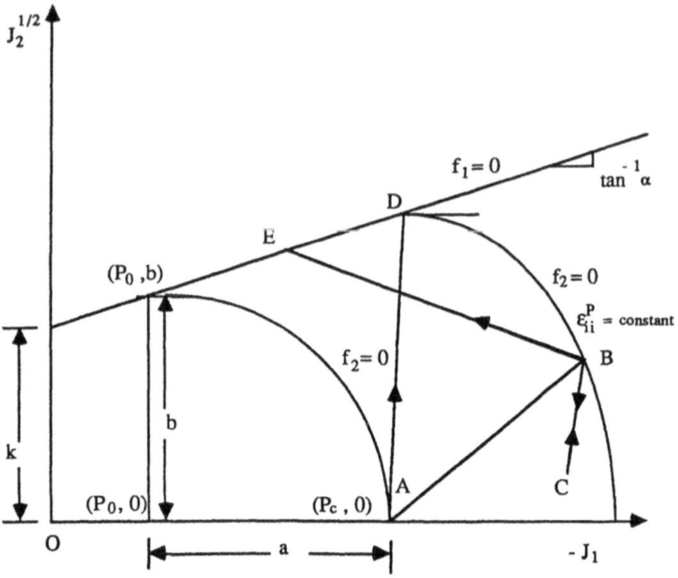

Figure 5. Singh's mechanical model

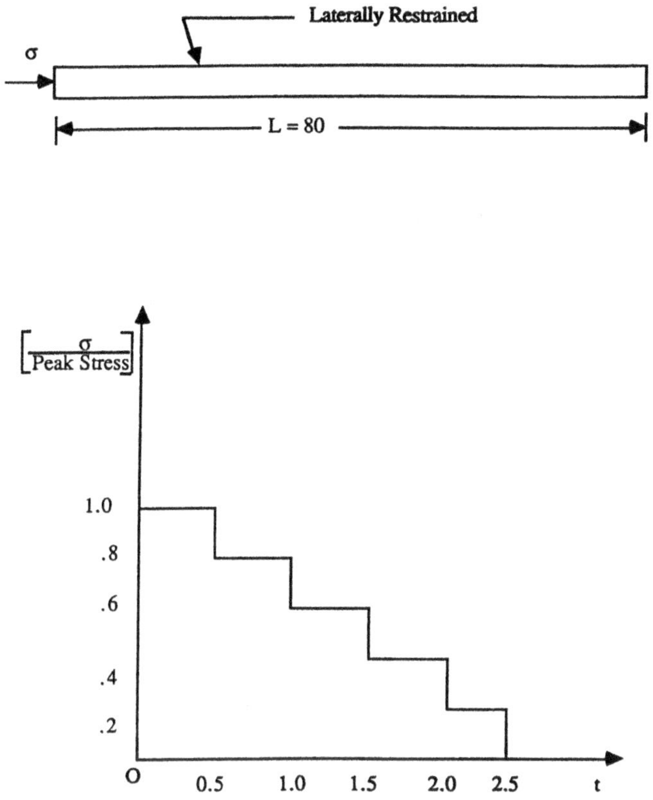

Figure 6. Solid column under pulse of short duration

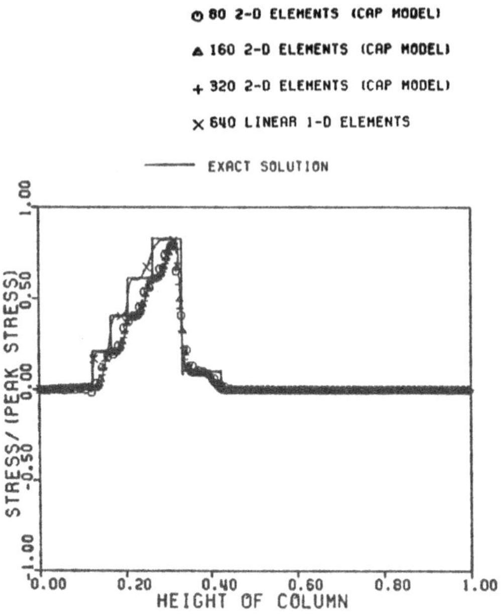

Figure 7. Example 1, single material problem, stress profile at t = 4

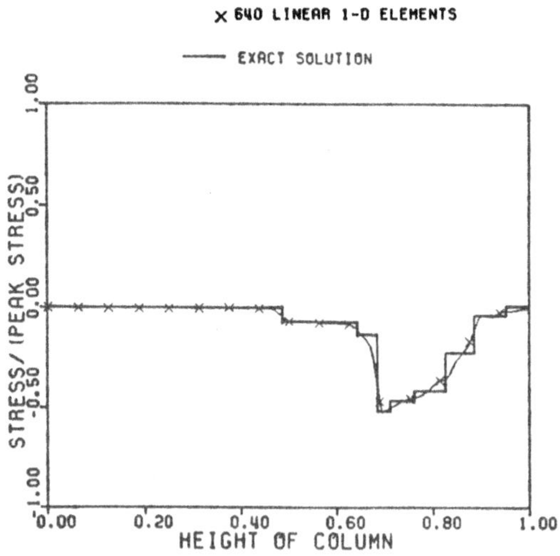

Figure 8. Example 1, single material problem, stress profile at t = 16

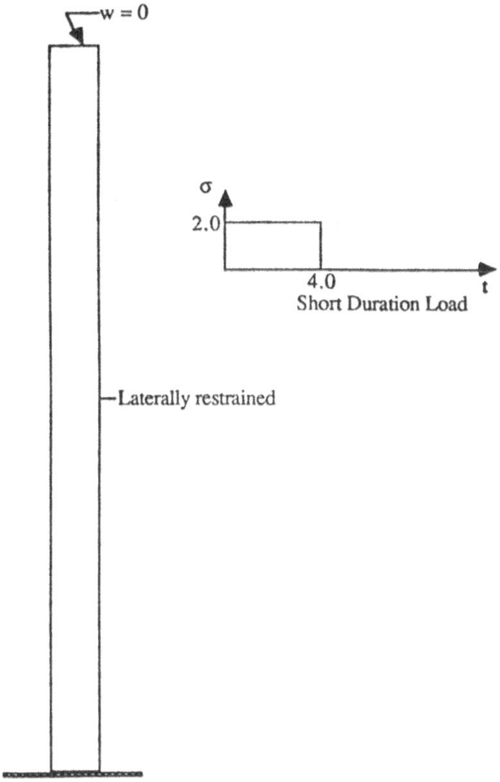

Figure 9. Example 2, saturated soil column under dynamic loading

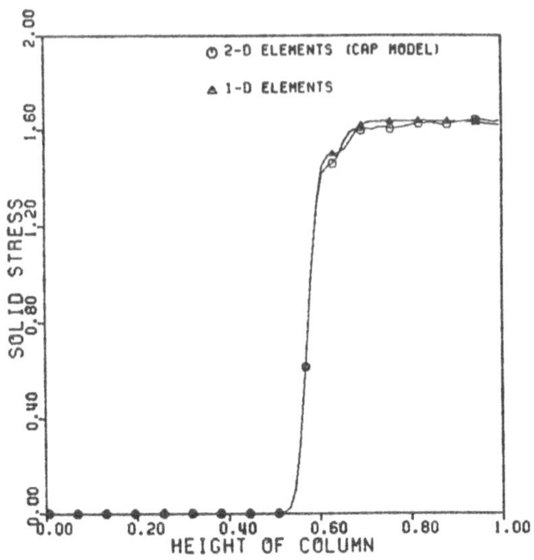

Figure 10. Example 2, solid stress profile at t = 4

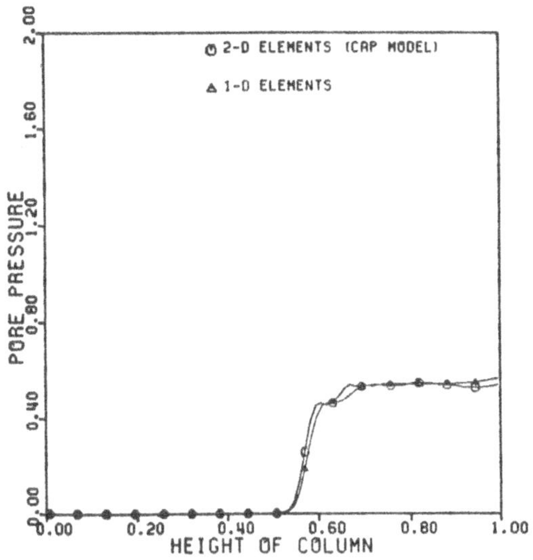

Figure 11. Example 2, pore pressure profile at t = 4

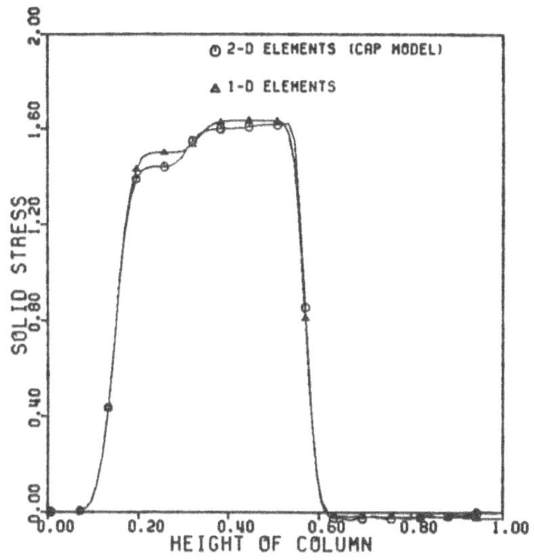

Figure 12. Example 2, solid stress profile at t = 8

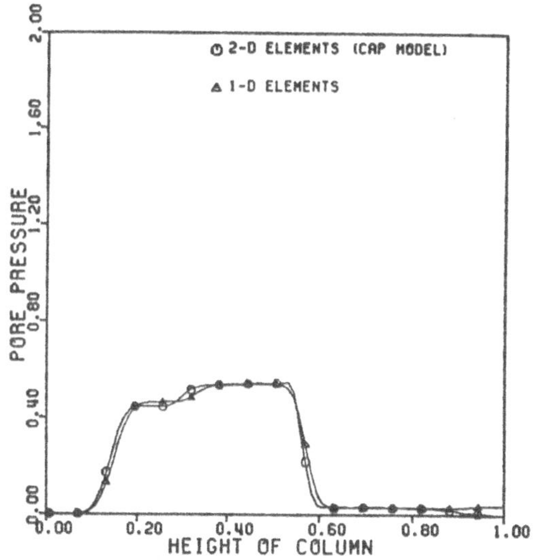

Figure 13. Example 2, pore pressure profile at t = 8

Filtration of bubbly fluids

J.-L. Auriault
Laboratoire Sols, Solides, Structures, UJF, INPG, CNRS URA 1511
Domaine Universitaire, BP 53X, 38041 Grenoble Cedex, France.
and
C. Boutin
Laboratoire Géomatériaux, DGCB, CNRS URA 1652, ENTPE, rue Maurice Audin, 69518 Vaulx En Velin, France.

Abstract: We investigate the macroscopic seepage law of a bubbly fluid through a rigid porous medium when the capillary forces are taken into account. For this purpose we use a method of multiple scale asymptotic developments. Starting from the pore scale it gives the macroscopic equivalent description without any prerequisite at the macroscopic scale. It makes use of the presence of well-separated scales and it permits the investigation of the interactions between them. We study two cases. The first one concerns bubble sizes of the order of the pore size. This case presents only two characteristic separated scales, the pore scale and the macroscopic scale. The second one considers bubble sizes much smaller than the pore size. It presents three characteristic scales. The two cases give Darcy laws at the macroscopic scale. But the large bubble case, the more frequent, does not show an intrinsic permeability and the conductivity depends on the capillary pressure. The macroscopic description is not classical and is strongly non linear.

1. Introduction

This paper deals with the filtration of liquids containing bubbles at finite concentration through rigid porous media. The seepage law for such a mixture of fluids is generally introduced directly at the macroscopic level. It assumes the existence of a macroscopic equivalent description, i.e., a description valid at a scale very large compared to the pore scale. The mixture of fluids is considered as a single compressible fluid, that permits the use of the classical Darcy law:

$$\mathrm{v} = -\,\mathbf{K}\ \mathbf{grad}\ \mathrm{p} \qquad (1)$$

v is the mean fluid velocity, p is the pressure and **K** is the conductivity tensor. The latter can be put in the form:

$$\mathbf{K} = \frac{\mathbf{K}^*}{\mu} \qquad (2)$$

μ is the fluid viscosity and $\mathbf{K}^*$ is the intrinsic permeability and depends solely on the pore geometry. Therefore $\mathbf{K}^*$ is independent on the fluid mixture and is a constant of the porous medium under consideration. It is identical to the intrinsic permeability for an incompressible single liquid. This represents an important advantage to macroscopic

P. S. Selvadurai (ed.), Mechanics of Poroelastic Media, 73–89.
1996 *Kluwer Academic Publishers.*

modelling. The description is completed by the mass balance which is written for permanent flows in the form

$$\mathrm{div}(\rho\, \mathbf{v}) = 0 \tag{3}$$

We show hereafter that (1-3) is generally not valid in the case of bubbly flow.

We address the filtration law by using a homogenization process. We start from the description at the pore scale where the two fluids are described by the Navier-Stoke equations and their interface by the Laplace relation for the capillary effects. The upscalin is performed by a double scale asymptotic development [1], [2], [3]. The process assume the existence of very different characteristic lengths appearing in the problem unde consideration. They are at least two: l is the representative length of the pore scale and L i the representative length of the macroscopic scale, with l / L = $\varepsilon << 1$. Typically l is the siz of the representative elementary volume and L the size of the macroscopic sample or i characteristic of the macroscopic flow. If $\mathbf{X}$ denotes the space variable, these tw characteristic lengths introduce two dimensionless space variables, $\mathbf{y} = \mathbf{X} / l$ and $\mathbf{x} = \mathbf{X} / L$ The consequence is that each physical quantity $\boldsymbol{\Phi}$ does depend on these two dimensionles space variables, among eventually other ones:

$$\boldsymbol{\Phi} = \boldsymbol{\Phi}(\mathbf{x}, \mathbf{y}, \ldots),\ \mathbf{y} = \varepsilon\, \mathbf{x}$$

The presence of the small parameter ε incites us to look for $\boldsymbol{\Phi}$ in the form of the followin asymptotic development:

$$\boldsymbol{\Phi} = \overset{0}{\boldsymbol{\Phi}}(\mathbf{x}, \mathbf{y},\ldots) + \varepsilon\, \overset{1}{\boldsymbol{\Phi}}(\mathbf{x}, \mathbf{y},\ldots) + \varepsilon^2\, \overset{2}{\boldsymbol{\Phi}}(\mathbf{x}, \mathbf{y},\ldots) + \ldots \tag{4}$$

The analysis of the successive terms in the above development leads to the macroscop description, if it exists. For more details concerning the methodology to be used, see [3]. I particular the $\overset{i}{\boldsymbol{\Phi}}$ possess a local invariance by translation. They are stationary or period with respect to the variable $\mathbf{y}$, depending on the random or the periodic structure of th porous medium, respectively. Since the macroscopic description and its existence do n depend on the nature of the porous structure, we will assume it as periodic, without loss generality.

In the section 2 we analyze the physics at the micro-scale. The section 3 is devoted the investigation of the case where the bubble sizes are of the same order as the pore siz This case concerns most of the applications. In the section 4, we address the case when th bubble sizes are small compared to the pore size. This case is interesting since corresponds to the above presented macroscopic description. It introduces three separat scales and the asymptotic development will therefore be a three scale development in pla of two.

2. The physics at the microscopic scale

We examine the physics of a bubbly fluid, at the bubble scale. We assume an isotherm flow and a large concentration of the bubbles. We consider the mixture flow as disturbance from an equilibrium state where the pressures in the two fluids and the capill pressure are of the same order of magnitude. The equilibrium density and the equilibrium pressure of the fluid α are denoted $\rho_{0\alpha}$ and $p_{0\alpha}$. $\alpha = f$ stands for the continuous fluid and $\alpha = g$ stands for the bubbly fluid. For the simplicity of the analysis we assume $\rho_{0\alpha}$ and $p_{0\alpha}$ as **x** and **y** independent. The non linear terms are negligible when the velocity is small. The velocity $\mathbf{v}_\alpha$, and the increments of the pressure p_α and of the density ρ_α in the fluid α are given by the Navier-Stokes equation, the mass balance and the equation of state:

$$- \mathbf{grad}\ p_\alpha + \mu_\alpha \left(\mathbf{grad} \left(\operatorname{div} \mathbf{v}_\alpha \right) + \Delta \mathbf{v}_\alpha \right) = \left(\rho_{0\alpha} + \rho_\alpha \right) \frac{\partial \mathbf{v}_\alpha}{\partial t} \tag{5}$$

$$\frac{\partial \rho_\alpha}{\partial t} + \operatorname{div} \left(\left(\rho_{0\alpha} + \rho_\alpha \right) \mathbf{v}_\alpha \right) = 0 \tag{6}$$

$$p_\alpha = A_\alpha \rho_\alpha \tag{7}$$

The state equation is assumed as linear. We are interested in analyzing situations corresponding to flow generated by permanent conditions at the boundary of the macroscopic sample. Nevertheless the presence of bubbles can render the different fields at the pore scale non permanent. This explains the presence of time derivatives in (5) and (6). Remark that a quantity that is **y** independent, will be also time independent.

On the surface S of the pores, the velocity verifies the adherence condition:

$$\mathbf{v}_f = 0 \tag{8}$$

On the surface Γ of a bubble, we have the continuity of the velocity (9), the capillary discontinuity of the stress (10) and the Laplace relation (11):

$$\mathbf{v}_f = \mathbf{v}_g \tag{9}$$

$$\left(\sigma_f - \sigma_g \right) \mathbf{n} = p_c\ \mathbf{n} \tag{10}$$

$$p_c = - 2\, \sigma \frac{r}{R_0 \left(R_0 + r \right)} \tag{11}$$

where we have:

$$\sigma_\alpha = - p_\alpha \mathbf{I} + 2\, \mu_\alpha \mathbf{E}(\mathbf{v}_\alpha), \quad \mathbf{E}(\mathbf{v})_{ij} = \frac{1}{2} \left(v_{i,j} + v_{j,i} \right)$$

σ_α, p_c and $r / R_0 (R_0 + r)$ are the increments of the stress, the capillary pressure and the curvature, respectively, from the equilibrium state where the considered bubble has a constant curvature radius R_0. σ is the surface tension.

As an example, we consider an air-water mixture. Typical values of the viscosities μ_α, the densities ρ_α and the incompressibilities K_α, are given in the following table.

μ_f	μ_g	ρ_{0f}	ρ_{0g}	K_f	K_g
(Pa.s)	(Pa.s)	(kg/m^3)	(kg/m^3)	(Pa)	(Pa)
10^{-3}	$19.\ 10^{-6}$	10^{3}	1.2	$2.\ 10^9$	$1.\ 10^5$

3. Large bubble size

The bubble sizes are of the order of the pore size. Such a situation is shown in the figure 1.

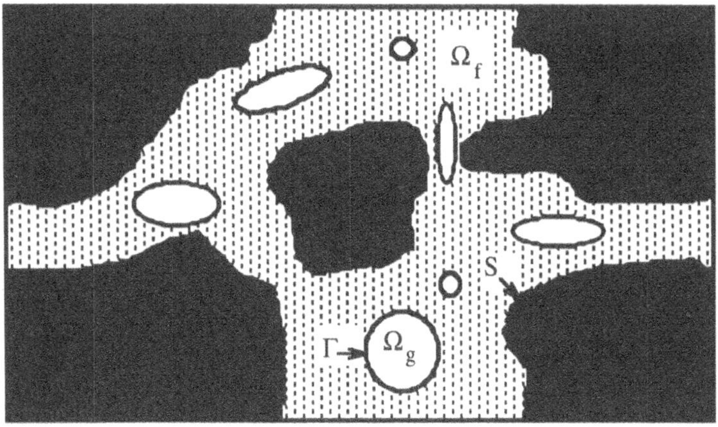

Figure1. Large bubbles, the period Ω.

We assume that the porous medium and the mixture are periodic without lossof generality The characteristic length of the period Ω is l and L stands for the characteristic macroscopic length. The value of the small parameter ε depends on these two lengths. For permanent macroscopic boundary value problems, it is generally very small. We assume in the following $\varepsilon \approx 10^{-4}$. This will simplify our analysis. It is worthwhile noticing that changing the value of ε can change the values of the dimensionless numbers with respect to the powers of ε. Therefore it can change the macroscopic description. It is nevertheless possible to show that in our problem the main results are preserved even if the value of ε is modified Due to our selection of ε, $\mu_f = O(\mu_g)$ and

$$A_f = O(\frac{K_f}{\rho_{0f}}) = O(\frac{K_g}{\rho_{0g}}) = O(A_g)$$

Obviously each bubble changes its geometry during its travel in the porous medium. W therefore have $r = O(R_0)$. An other important fact is that the flow is not permanent at th pore scale because the bubble positions are changing in the period during their travel. Th associated characteristic time is the transit time $\tau = O(l / v)$. Nevertheless, a quantity that is shown to be **y** independent, will be also time independent. On an other hand, the driving force for the flow is a macroscopic gradient of pressure O(p / L). It balances the viscous term in (5):

$$\frac{p}{L} = O(\mu \frac{v}{l^2})$$

We are now able to investigate the dimensionless numbers in the description (5-11). We use l to make the equations dimensionless. The Navier-Stokes equation reveals two dimensionless numbers:

$$P_l = \frac{|\mu_\alpha (\mathbf{grad} (\operatorname{div} \mathbf{v}_\alpha) + \Delta \mathbf{v}_\alpha)|}{|\mathbf{grad}\, p_\alpha|} = O(\frac{\mu v}{p l}) = O(\frac{l}{L}) = O(\varepsilon)$$

and the transient Reynolds number

$$R_{tl} = \frac{\left|(\rho_{0\alpha} + \rho_\alpha) \frac{\partial \mathbf{v}_\alpha}{\partial t}\right|}{|\mu_\alpha (\mathbf{grad} (\operatorname{div} \mathbf{v}_\alpha) + \Delta \mathbf{v}_\alpha)|} = O(\frac{\rho_{0\alpha} l^2}{\mu \tau}) = O(\frac{\rho_{0\alpha} v l}{\mu})$$

For most problems, v and l are very small. We will consider $R_{tl} = O(\varepsilon)$.

The mass balance (6) introduces the Strouhal number

$$S_l = \frac{\left|\frac{\partial \rho_\alpha}{\partial t}\right|}{|\operatorname{div} ((\rho_{0\alpha} + \rho_\alpha) \mathbf{v}_\alpha)|}$$

The value of this number is directly related to the magnitude of the time dependence of the density. From [4] we know that homogenizable situations correspond to $S_l \leq O(\varepsilon)$, only. We therefore start the analysis with $S_l = O(\varepsilon)$.

Finally the equation (10) shows the Weber number. We assume at the equilibrium state a Weber number O(1)

$$\frac{p_{0\alpha}}{p_{0c}} = O(\frac{p_{0\alpha} R}{2\sigma}) = O(1)$$

Assuming $p_\alpha = O(p_{0\alpha})$ gives, with $r = O(R_0)$:

$$W = O(\frac{p_\alpha}{p_c}) = O(\frac{p_\alpha R_0 (R_0 + r)}{2 \sigma r}) = O(1)$$

Using for simplicity the same notations for dimensionless as for dimensional quantities, with the exception of the space variables, we obtain the following dimensionless description at the pore level:

$$- \mathbf{grad}_y\, p_\alpha + \varepsilon\, \mu_\alpha\, (\, \mathbf{grad}_y\, (\, \mathrm{div}_y\, \mathbf{v}_\alpha) + \Delta_y \mathbf{v}_\alpha) = \varepsilon^2\, (\rho_{0\alpha} + \rho_\alpha)\, \frac{\partial \mathbf{v}_\alpha}{\partial t} \quad (12)$$

$$\varepsilon \frac{\partial \rho_\alpha}{\partial t} + \mathrm{div}_y((\rho_{0\alpha} + \rho_\alpha)\, \mathbf{v}_\alpha) = 0 \quad (13)$$

$$p_\alpha = A_\alpha\, \rho_\alpha \quad (14)$$

on Γ

$$\mathbf{v}_f = \mathbf{v}_g \quad (15)$$

$$(\, \sigma_f - \sigma_g\,)\, \mathbf{n} = p_c\, \mathbf{n} \quad (16)$$

$$\sigma_\alpha = -\, p_\alpha\, \mathbf{I} + 2\, \varepsilon\, \mu_\alpha\, \mathbf{E}_y(\mathbf{v}_\alpha) \quad (17)$$

$$p_c = -\, 2\, \sigma \frac{r}{R_0\, (R_0 + r\,)} \quad (18)$$

and on S

$$\mathbf{v}_f = 0 \quad (19)$$

The subscript y shows that the operators are now defined with respect to the variable **y**. The details of the homogenization process are given in the Appendix A. The macroscopic instantaneous behavior is (A.14), (A.15)

$$\widetilde{\mathbf{v}^0} = -\, \mathbf{K}(\mathbf{x}, t, p_c^0)\, \mathbf{grad}_x\, p_f^0 \qquad \mathbf{K}(\mathbf{x}, t, p_c^0) = \widetilde{\mathbf{J}_f} + \widetilde{\mathbf{J}_g} \quad (A.14)$$

$$\mathrm{div}_x(\rho^{eff}\, \widetilde{\mathbf{v}^0}) + Q = 0 \qquad \rho^{eff} = ((\, \rho_{0f} + \rho_f^0\,)\widetilde{\mathbf{J}_f} + (\, \rho_{0g} + \rho_g^0\,)\widetilde{\mathbf{J}_g}\,)\mathbf{K}^{-1} \quad (A.15)$$

where the effective density is of tensorial character. The mass balance (A.15) also shows non-classical source term Q.

The conductivity tensor **K** is fluctuating with time, with a periodicity $O(\tau)$. An averag behavior of the Darcy law can be defined by averaging over a period $T >> \tau$

$$\overline{\widetilde{\mathbf{v}^0}} = -\, \overline{\mathbf{K}}\, \mathbf{grad}_x\, p_f^0 \qquad \overline{\Phi} = \frac{1}{T} \int_t^{t+T} \Phi\, d\Omega$$

Clearly, $\overline{\mathbf{K}}$ cannot be put in the form $\frac{\mathbf{K}^*}{\mu}$: firstly it is not possible to define an effectiv viscosity of the mixture and secondly $\overline{\mathbf{K}}$ depends on p_{0c}. Therefore, an eventual $\mathbf{K}^*$ woul not be intrinsic. Note that the dependence of $\overline{\mathbf{K}}$ on p_c^0 renders the Darcy law in (A.1 strongly non linear. Finally the time average is applied to eq. (A.15)

$$\mathrm{div}_x({}^{eff}\rho\ \overline{\tilde{\mathbf{v}}^0}) + \overline{Q} = 0 \qquad {}^{eff}\rho = \overline{\rho^{eff}\mathbf{K}}\ \overline{\mathbf{K}}^{-1}$$

4. Small bubble size

The bubble sizes are now very small compared to the pore size, figure 2. We assume

$$\frac{R}{l} = O(\frac{l}{L}) = O(\varepsilon)$$

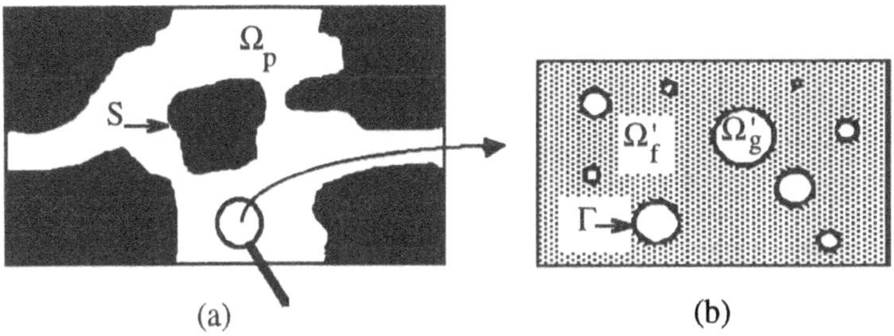

Figure 2. Small bubbles. (a) Period Ω. (b) PeriodΩ'.

The problem shows three dimensionless space variables, $\mathbf{x} = \mathbf{X} / L$, $\mathbf{y} = \mathbf{X} / l$ and $\mathbf{z} = \mathbf{X} / R$. The expansions to be applied are then three scale expansions. The medium is doubly periodic, i.e., y periodic with a period Ω and z periodic with a period Ω'. l and l' are the characteristic lengths of Ω and Ω', respectively. Using $l' = O(R)$ to make the equations dimensionless, these expansion are in the form

$$\Phi = \Phi^0(\mathbf{z}, \mathbf{y}, \mathbf{x},...) + \varepsilon\ \Phi^1(\mathbf{z}, \mathbf{y}, \mathbf{x},...) + \varepsilon^2\ \Phi^2(\mathbf{z}, \mathbf{y}, \mathbf{x},...) + ... \tag{20}$$

with $\qquad \mathbf{y} = \varepsilon\ \mathbf{z} \qquad \mathbf{x} = \varepsilon^2\ \mathbf{z}$

The description at the bubble scale is again given by the set (3-9). The flow is governed as in section 3 by

$$\frac{P}{L} = O(\mu \frac{V}{l^2})$$

The dimensionless numbers are now evaluated using l' as the characteristic length. It is easy to obtain following the reasoning given in section 3

$$P_{l'} = O(\frac{\mu\, v}{p\, l'}) = O(1)$$

$$R_{tl'} = O(\frac{\rho_\alpha\, l'^2}{\mu \tau}) = O(\varepsilon^3)$$

$$S_{l'} = O(\frac{l'}{v\, \tau}) = O(\varepsilon^2)$$

Let us now investigate the Weber number for the increments of pressure. While comparing with the case in section 3, the "rigidity" σ / R of the capillary membrane is increased by a factor ε^{-1}, whereas the rigidity of the mixture is unchanged. The Weber number at equilibrium is therefore increased by the same factor ε^{-1}

$$W_0 = O(\varepsilon^{-1}) \qquad\qquad p_{0f} = O(\varepsilon\, p_{0g})$$

A disturbance $p_f = O(p_{0f})$ then gives

$$r = O(\varepsilon\, R_0) \qquad\qquad \rho_\alpha = \varepsilon\, \rho_{0\alpha}$$

With the same convention for notations as in section 3, the dimensionless set of equations is written

$$\mathrm{div}_z\, \sigma_\alpha = \varepsilon^3\, (\rho_{0\alpha} + \varepsilon\, \rho_\alpha) \frac{\partial \mathbf{v}_\alpha}{\partial t} \tag{21}$$

$$\sigma_\alpha = -\, p_\alpha\, \mathbf{I} + 2\, \mu_\alpha\, \mathbf{E}_z(\mathbf{v}_\alpha) \tag{22}$$

$$\varepsilon^2 \frac{\partial \rho_\alpha}{\partial t} + \mathrm{div}_z((\rho_{0\alpha} + \varepsilon\, \rho_\alpha)\, \mathbf{v}_\alpha) = 0 \tag{23}$$

$$p_\alpha = A_\alpha\, \rho_\alpha \tag{24}$$

on Γ

$$\mathbf{v}_f = \mathbf{v}_g \tag{25}$$

$$(\sigma_f - \sigma_g)\, \mathbf{n} = \varepsilon\, p_c\, \mathbf{n} \tag{26}$$

$$p_c = -\, 2\, \sigma \frac{r}{R_0\, (R_0 + \varepsilon\, r)} \tag{27}$$

and on S

$$\mathbf{v}_f = 0 \tag{28}$$

The differential operators are now defined with respect to **z**.

The details of the homogenization process are in the Appendix B. The macroscopic behavior is (B.13) and (B.14)

$$\mathrm{div}_x\,(\widetilde{\mathbf{v}^0}) = 0 \qquad \widetilde{\mathbf{v}^0} = -\,\mathbf{K}\,\mathbf{grad}_x p^0$$

There is no more time dependence. The mass balance is classical. The bubbly fluid flows out like a single phase fluid with generally an anisotropic viscosity. The compressibility of the mixture is small. It does not intervene at the first order of approximation. In the case where the mixture viscosity μ_{eff} is isotropic and can be considered as a constant, we recover an intrinsic permeability

$$\mathbf{K} = \frac{\mathbf{K}^*}{\mu_{eff}}$$

APPENDIX A

As described in the introduction, we introduce into the set (12-19) asymptotic expansions for each variable. Due to the double scale dependence,

$$\Phi = \Phi(\mathbf{y}, \mathbf{x}) = \Phi(\mathbf{y}, \varepsilon\, \mathbf{y})$$

and $\dfrac{\partial}{\partial y_i}$ is replaced by $\dfrac{\partial}{\partial y_i} + \varepsilon\,\dfrac{\partial}{\partial x_i}$. By collecting together terms corresponding to the same powers of ε, we obtain for the first order terms:

In the liquid and in the gas

$$\mathbf{grad}_y\, p_\alpha^0 = 0$$

$$\mathbf{grad}_x\, p_\alpha^0 + \mathbf{grad}_y\, p_\alpha^1 - \mu_\alpha\,(\mathbf{grad}_y\,(\mathrm{div}_y\, \mathbf{v}_\alpha^0) + \Delta_y \mathbf{v}_\alpha^0) = 0 \tag{A.1}$$

- - - - - - - - - - - - - - - - - - - -

$$\mathrm{div}_y((\rho_{0\alpha} + \rho_\alpha^0)\, \mathbf{v}_\alpha^0) = 0$$

$$\frac{\partial \rho_\alpha^0}{\partial t} + \mathrm{div}_y((\rho_{0\alpha} + \rho_\alpha^0)\, \mathbf{v}_\alpha^1 + \rho_\alpha^1\, \mathbf{v}_\alpha^0) + \mathrm{div}_x((\rho_{0\alpha} + \rho_\alpha^0)\, \mathbf{v}_\alpha^0) = 0 \tag{A.2}$$

- - - - - - - - - - - - - - - - - - - -

$$p_\alpha^0 = A_\alpha\, \rho_\alpha^0$$

$$p_\alpha^1 = A_\alpha\, \rho_\alpha^1 \tag{A.3}$$

- - - - - - - - - - - - - - - - - - - -

$$\sigma_\alpha^0 = - p_\alpha^0\, \mathbf{I}, \quad \sigma_\alpha^1 = - p_\alpha^1\, \mathbf{I} + 2\, \mu_\alpha\, \mathbf{E}_y(\, \mathbf{v}_\alpha^0\,) \tag{A.4}$$

At the interface Γ

$$\mathbf{v}_f^0 = \mathbf{v}_g^0, \quad \mathbf{v}_f^1 = \mathbf{v}_g^1 \tag{A.5}$$

$$p_g^0 - p_f^0 = p_c^0, \quad (\, \sigma_f^1 - \sigma_g^1\,)\, \mathbf{n} = p_c^1\, \mathbf{n} \tag{A.6}$$

$$p_c^0 = - 2\, \sigma \frac{r^0}{R_0\, (R_0 + r^0\,)} \qquad p_c^1 = - 2\, \sigma \frac{r^1}{(R_0 + r^0\,)^2} \tag{A.7}$$

At the pore surface S

$$\mathbf{v}_f^0 = 0 \qquad \mathbf{v}_f^1 = 0 \tag{A.8}$$

Clearly, the first orders of the pressures and the densities are y independent and therefore time independant:

$$p_\alpha^0 = p_\alpha^0(\mathbf{x}) \quad \rho_\alpha^0 = \rho_\alpha^0(\mathbf{x}) \quad p_c^0 = p_c^0(\mathbf{x}) = p_g^0(\mathbf{x}) - p_f^0(\mathbf{x}) \tag{A.9}$$

As a consequence the bubbles are spherical at the first order:

$$r^0 = r^0(\mathbf{x}) \text{ and } R = R_0 + r^0 = R(\mathbf{x})$$

Since ρ_α^0 is y independent, the time derivative in (A.2.2) cancels out. For simplicity we assume bubbles of uniform size at the first order. The result will not change if different bubble sizes are present.
The basic boundary value problem on the periodic cell appears at the next order:

$$\mathbf{grad}_x\, p_\alpha^0 + \mathbf{grad}_y\, p_\alpha^1 - \mu_\alpha\, \Delta_y \mathbf{v}_\alpha^0 = 0 \tag{A.10}$$

$$\text{div}_y\, \mathbf{v}_\alpha^0 = 0 \text{ in the liquid and the gas,}$$

$$\mathbf{v}_f^0 = 0 \text{ on S}$$

$$(\, \sigma_f^1 - \sigma_g^1\,)\, \mathbf{n} = p_c^1\, \mathbf{n} \qquad p_c^1 = - 2\, \sigma \frac{r^1}{R^2}$$

$$\mathbf{v}_f^0 = \mathbf{v}_g^0 \qquad \text{on } \Gamma$$

where all the quantities are Ω periodic with respect to the variable **y**.

Obviously, the position and the shape of the bubbles in the period are changing with time. We consider now a fixed time and a fixed position $\mathbf{x}$, with given locations of the interfaces Γ, and we look for the small displacement $\mathbf{u}^0$ of the mixture during the small time increment δt

$$\mathbf{u}^0 = \mathbf{v}^0 \, \delta t$$

We study the above differential system using its variational formulation. We multiply (A.10) by any test field of the vectorial space $\mathcal{W}$ defined in the pore domain Ω_p by

$$\mathcal{W} = \{ \ \mathbf{w} \text{ defined on } \Omega_p, \ \Omega\text{-periodic}, \ \mathrm{div}_y(\mathbf{w}) = 0, \ \mathbf{w} = 0 \text{ on } S \}$$

Then we integrate over Ω_p. Using the solenoidal, the adherence and the interface conditions gives

$$\int_{\Omega_p} \delta t \ \mathbf{w} \ \mathbf{grad}_x \, p^0 \, d\Omega - \int_{\Gamma} \delta t \ p_c^1 \ \mathbf{w} \ \mathbf{n}_f \, d\Gamma + \int_{\Omega_p} \mu \ \mathbf{grad}_y \, \mathbf{u}^0 \ \mathbf{grad}_y \, \mathbf{w} \, d\Omega = 0$$

$\mathbf{n}_f$ is the unit outer normal to the fluid domain. For a given bubble, following [5, relation 30C], we have as a first approximation:

$$p_c^1 = \frac{2\,\sigma}{R^2} \, (\, \mathbf{u}_f^0 - \mathbf{u}_c) \, \mathbf{n}_f$$

$\mathbf{u}_c$ is the average displacement in the bubble.
Introducing this last relation and putting

$$< \mathbf{u}, \mathbf{w} > = \int_{\Omega_p} \mu \ \mathbf{grad}_y \, \mathbf{u} \ \mathbf{grad}_y \, \mathbf{w} \, d\Omega - \frac{2\,\sigma\,\delta t}{R^2} \int_{\Gamma} ((\, \mathbf{u}_f^0 - \mathbf{u}_c) \ \mathbf{n}_f) \ (\, \mathbf{w} \ \ \mathbf{n}_f) \, d\Gamma$$

the variational formulation becomes

$$\forall \mathbf{w} \in \mathcal{W}, \ < \mathbf{u}^0, \mathbf{w} > = - \int_{\Omega_f} \delta t \ \mathbf{w} \ \mathbf{grad}_x \, p_f^0 \, d\Omega - \int_{\Omega_g} \delta t \ \mathbf{w} \ \mathbf{grad}_x \, p_g^0 \, d\Omega$$

A sufficiently small value of δt insures the coercivity of the form in the left hand member of the above formulation. The existence and uniqueness of $\mathbf{u}^0$ follow from the Lax-Milgram lemma. The linearity of the problem gives $\mathbf{u}^0_\alpha$ in the form

$$\mathbf{u}^0_\alpha = - \mathbf{J}_{\alpha f}(\mathbf{x}, \mathbf{y}, t, \delta t)\, \delta t\, \mathbf{grad}_x\, p^0_f - \mathbf{J}_{\alpha g}(\mathbf{x}, \mathbf{y}, t, \delta t)\, \delta t\, \mathbf{grad}_x\, p^0_g$$

The second order tensors $\mathbf{J}_{\alpha f}$ and $\mathbf{J}_{\alpha g}$ depend on $\mathbf{y}$, δt and on the mixture geometry, i.e., the time. Note that they are not inversely proportional to some viscosity. In the limit $\delta t \to 0$, we have

$$\mathbf{v}^0_\alpha = - \mathbf{J}_{\alpha f}(\mathbf{x}, \mathbf{y}, t, 0)\, \mathbf{grad}_x\, p^0_f - \mathbf{J}_{\alpha g}(\mathbf{x}, \mathbf{y}, t, 0)\, \mathbf{grad}_x\, p^0_g \tag{A.11}$$

Therefore (A.10) gives

$$p^1_\alpha = - \tau_{\alpha f}(\mathbf{x}, \mathbf{y}, t)\, \mathbf{grad}_x\, p^0_f - \tau_{\alpha g}(\mathbf{x}, \mathbf{y}, t)\, \mathbf{grad}_x\, p^0_g \tag{A.12}$$

The second order tensors $\mathbf{J}_{\alpha f}$, $\mathbf{J}_{\alpha g}$, $\tau_{\alpha f}$ and $\tau_{\alpha g}$ depend on $\mathbf{x}$, $\mathbf{y}$, on the mixture geometry, i.e., the time. Note that $\mathbf{J}_{\alpha f}$ and $\mathbf{J}_{\alpha g}$, are not inversely proportional to some viscosity. Averaging (A.11) would give coupled Darcy laws, as for the seepage of two immiscible connected fluids [6]. Unfortunately the volume averages of the velocities $\mathbf{v}^0_\alpha$ are not fluxes. On an other hand the first order approximation of the mass conservation of a bubble can be written

$$\left(1 + \frac{r^0}{R_0}\right)^3 - 1 = \frac{p^0_g}{A_g\, \rho_{0g}}$$

Eliminating r^0 between (A.7.1) and the above equality shows a relation between p^0_c and p^0_g Finally, differentiation of (A.9.3) gives $\mathbf{grad}_x\, p^0_g$ as a linear function of $\mathbf{grad}_x\, p^0_f$ and as a non linear function of p^0_g

$$\mathbf{grad}_x p^0_g = \left(1 + \frac{3\, R_0\, K_g}{2\, \sigma}\left(\frac{p^0_g}{K_g} + 1\right)^{4/3}\right) \mathbf{grad}_x p^0_f \tag{A.13}$$

(A.11) can now be rewritten in the form

$$\mathbf{v}^0_\alpha = - \mathbf{J}_\alpha(\mathbf{x}, \mathbf{y}, t, p^0_c)\, \mathbf{grad}_x\, p^0_f$$

The $\mathbf{J}_\alpha$ are depending on $\mathbf{x}$, $\mathbf{y}$ and t, but also on p^0_c.
The macroscopic mass balance is obtained by integration of (A.2.2) over Ω_p . It needs the knowledge of ρ^1_α and $\mathbf{v}^1_\alpha$. p^1_α is given by (A.12). $\mathbf{v}^1_\alpha$ needs the investigation of the following orders to (A.1.2). Without entering into details, the instantaneous macroscopic balance is in the form

$$\mathrm{div}_x((\rho_{0f} + \rho_f^0)\,\widetilde{\mathbf{v}_f^0} + (\rho_{0g} + \rho_g^0)\,\widetilde{\mathbf{v}_g^0}) + Q = 0$$

$$\widetilde{\mathbf{v}_\alpha^0} = -\,\widetilde{\mathbf{J}_\alpha}(\mathbf{x}, t, p_c^0)\,\mathbf{grad}_x\, p_f^0$$

where a tilde means a volume average

$$\widetilde{\Phi} = \frac{1}{|\Omega_p|}\int_{\Omega_p} \Phi \, d\Omega$$

and Q is a non linear functional of p_f^0, p_c^0 and the $\rho_{0\alpha}$.
It is also possible to define a Darcy velocity by

$$\widetilde{\mathbf{v}^0} = -\,\mathbf{K}(\mathbf{x}, t, p_c^0)\,\mathbf{grad}_x\, p_f^0 \qquad \mathbf{K}(\mathbf{x}, t, p_c^0) = \widetilde{\mathbf{J}_f} + \widetilde{\mathbf{J}_g} \tag{A.14}$$

Introducing this Darcy velocity into the mass balance gives:

$$\mathrm{div}_x(\rho^{eff}\,\widetilde{\mathbf{v}^0}) + Q = 0 \qquad \rho^{eff} = ((\rho_{0f} + \rho_f^0)\widetilde{\mathbf{J}_f} + (\rho_{0g} + \rho_g^0)\widetilde{\mathbf{J}_g})\mathbf{K}^{-1} \tag{A.15}$$

Note that the effective density ρ^{eff} is of tensorial character.

APPENDIX B

We introduce in the set (21-28) the expansion (20). Due to the triple scale dependence, $\Phi = \Phi(\mathbf{z}, \mathbf{y}, \mathbf{x}) = \Phi(\mathbf{z}, \varepsilon\,\mathbf{z}, \varepsilon^2\,\mathbf{z})$ and $\frac{\partial}{\partial z_i}$ is replaced by $\frac{\partial}{\partial z_i} + \varepsilon\frac{\partial}{\partial y_i} + \varepsilon^2\frac{\partial}{\partial x_i}$. Collecting like powers of ε gives

$$\mathrm{div}_z\, \sigma_\alpha^0 = 0 \qquad \mathrm{div}_z\, \sigma_\alpha^1 + \mathrm{div}_y\, \sigma_\alpha^0 = 0$$
$$\mathrm{div}_z\, \sigma_\alpha^2 + \mathrm{div}_y\, \sigma_\alpha^1 + \mathrm{div}_x\, \sigma_\alpha^0 = 0 \tag{B.1}$$

- -

$$\sigma_\alpha^0 = -\,p_\alpha^0\,\mathbf{I} + 2\,\mu_\alpha\,\mathbf{E}_z(\mathbf{v}_\alpha^0)$$
$$\sigma_\alpha^1 = -\,p_\alpha^1\,\mathbf{I} + 2\,\mu_\alpha\,(\mathbf{E}_z(\mathbf{v}_\alpha^1) + \mathbf{E}_y(\mathbf{v}_\alpha^0)) \tag{B.2}$$
$$\sigma_\alpha^2 = -\,p_\alpha^2\,\mathbf{I} + 2\,\mu_\alpha\,(\mathbf{E}_z(\mathbf{v}_\alpha^2) + \mathbf{E}_y(\mathbf{v}_\alpha^1) + \mathbf{E}_x(\mathbf{v}_\alpha^0))$$

- -

$$\mathrm{div}_z\,(\,\rho_{0\alpha}\,\mathbf{v}^0_\alpha) = 0$$

$$\mathrm{div}_z\,(\,\rho^0_\alpha\,\mathbf{v}^0_\alpha + \rho_{0\alpha}\,\mathbf{v}^1_\alpha\,) + \mathrm{div}_y\,(\rho_{0\alpha}\,\mathbf{v}^0_\alpha) = 0 \tag{B.3}$$

$$\frac{\partial \rho^0_\alpha}{\partial t} + \mathrm{div}_z\,(\,\rho^1_\alpha\,\mathbf{v}^0_\alpha + \rho^0_\alpha\,\mathbf{v}^1_\alpha + \rho_{0\alpha}\,\mathbf{v}^2_\alpha)$$
$$+ \mathrm{div}_y\,(\,\rho^0_\alpha\,\mathbf{v}^0_\alpha + \rho_{0\alpha}\,\mathbf{v}^1_\alpha\,) + \mathrm{div}_x\,(\,\rho_{0\alpha}\,\mathbf{v}^0_\alpha) = 0$$

- -

$$p^0_\alpha = A_\alpha\,\rho^0_\alpha \qquad p^1_\alpha = A_\alpha\,\rho^1_\alpha \tag{B.4}$$

- -

On Γ

$$\mathbf{v}^0_f = \mathbf{v}^0_g, \qquad \mathbf{v}^1_f = \mathbf{v}^1_g, \qquad \mathbf{v}^2_f = \mathbf{v}^2_g \tag{B.5}$$

- -

$$(\,\sigma^0_f - \sigma^0_g\,)\,\mathbf{n} = 0 \qquad (\,\sigma^1_f - \sigma^1_g\,)\,\mathbf{n} = p^0_c\,\mathbf{n} \tag{B.6}$$

- -

$$p^0_c = -\,2\,\sigma\,\frac{r^0}{R_0^2} \tag{B.7}$$

- -

and on S

$$\mathbf{v}^0_f = 0 \qquad \mathbf{v}^1_f = 0 \tag{B.8}$$

The set (B.1.1), (B.2.1), (B.3.1), (B.4.1), (B.5.1) and (B.6.1) defines on Ω' a bounda[ry] value problem for the Ω' periodic unknowns $\mathbf{v}^0_\alpha$, p^0_α and ρ^0_α. It is a homogeneous set, th[e] solution of which is obvious. Consequently, all the unknowns are z independent

$$\mathbf{v}^0_\alpha = \mathbf{v}^0(\mathbf{y}, \mathbf{x}, t), \;\; p^0_\alpha = p^0(\mathbf{y}, \mathbf{x}, t), \;\; \text{and } \rho^0_\alpha = \rho^0_\alpha(\mathbf{y}, \mathbf{x}, t)$$

Integrating (B.1.2) over Ω', using the Ω' periodicity and (B.6.2) gives

$$\mathrm{div}_y\,\sigma^0_\alpha = -\,\mathbf{grad}_y p^0_\alpha = \int_\Gamma p^0_c\,\mathbf{n}_f\,d\Gamma$$

The surface integral cancels out because of the equilibrium of the capillary membrane at [the] first order. p^0_α and therefore ρ^0_α (through (B.4.1)) are **y** independent, and therefore ti[me] independent

$$p_f^0 = p_g^0 = p^0(\mathbf{x}), \quad \rho_\alpha^0 = \rho_\alpha^0(\mathbf{x})$$

With these results, (B.3.2) simplifies into

$$\mathrm{div}_z \mathbf{v}_\alpha^1 + \mathrm{div}_y \mathbf{v}^0 = 0$$

Integrating now this last relation over Ω', using the Ω' periodicity and (B.5.2) gives

$$\mathrm{div}_y \mathbf{v}^0 = 0 \tag{B.9}$$

and therefore

$$\mathrm{div}_z \mathbf{v}_\alpha^1 = 0 \tag{B.10}$$

This last result implies that ρ_α^1 is $\mathbf{z}$ independent

$$\rho_\alpha^1 = \rho_\alpha^1(\mathbf{y}, \mathbf{x}, t)$$

The basic cell problem on Ω'

The basic cell problem is given by (B.1.2), (B.2.2), (B.10), (B.5.2) and (B.6.2)

$$\mathrm{div}_z\, \sigma_\alpha^1 = 0$$

$$\sigma_\alpha^1 = -\, p_\alpha^1\, \mathbf{I} + 2\, \mu_\alpha\, (\, \mathbf{E}_z(\mathbf{v}_\alpha^1) + \mathbf{E}_y(\mathbf{v}^0)\,)$$

$$\mathrm{div}_z \mathbf{v}_\alpha^1 = 0$$

On Γ

$$\mathbf{v}_f^1 = \mathbf{v}_g^1$$

$$(\, \sigma_f^1 - \sigma_g^1\,)\, \mathbf{n} = p_c^0\, \mathbf{n}$$

This is a linear problem for the Ω' periodic $\mathbf{v}_\alpha^1$ and p_α^1. The solution can be put in the form

$$\mathbf{v}_\alpha^1 = \chi(\mathbf{z}, \ldots)\, \mathbf{E}_y(\mathbf{v}^0) + \overline{\mathbf{v}^1}(\mathbf{y}, \mathbf{x}, t)$$

$$p_f^1 = \zeta(\mathbf{z}, \ldots)\, \mathbf{E}_y(\mathbf{v}^0) + \overline{p^1}(\mathbf{y}, \mathbf{x}, t) \qquad p_g^1 = p_f^1 + p_c^0$$

where χ and ζ are a third and a second order tensor with zero valued volume averages $\widehat{\chi}$ and $\widehat{\zeta}$ over Ω', respectively. $\overline{\mathbf{v}^1}$ and $\overline{p^1}$ are arbitrary and z independent.

Taking the volume average $\widehat{\sigma_\alpha^1}$ of σ_α^1 over Ω' gives

$$\widehat{\sigma_\alpha^1} = -\overline{p^{*1}}\,\mathbf{I} + 2\,\mu_{eff}(\mathbf{x},\mathbf{y})\,\mathbf{E}_y(\mathbf{v}^0)\,) \qquad \overline{p^{*1}} = -\overline{p^1} + \frac{\Omega'_g}{\Omega'}\,p_c^0 \tag{B.11}$$

The relations (B.9) and (B.11) represent the behavior of the bubbly mixture at the pore scale, when it flows through the porous medium. Despite its compressibility it classically appears to be incompressible at the pore scale at the first order of approximation [4]. Remark that its effective viscosity μ_{eff} is a second order tensor. It *a priori* depends on $\mathbf{y}$. The circumflex accent over a quantity shows the volume average over Ω'. A tilde will mean in the following a volume average on Ω, as in section 3.

The basic cell problem on Ω

Averaging (B.1.3) over Ω' and using the equilibrium of the capillary membrane at the second order gives

$$\mathrm{div}_y\,\widehat{\sigma_\alpha^1} - \mathbf{grad}_x\,p^0 = 0 \tag{B.12}$$

We add to (B.12) the relations (B.11), (B.9) and (B.8.1)

$$\widehat{\sigma_\alpha^1} = -\overline{p^{*1}}\,\mathbf{I} + 2\,\mu_{eff}\,\mathbf{E}_y(\mathbf{v}^0)\,) \tag{B.11}$$

$$\mathrm{div}_y\mathbf{v}^0 = 0 \tag{B.9}$$

$$\mathbf{v}^0 = 0 \quad \text{on } S \tag{B.8.1}$$

Despite the tensorial character of μ_{eff} and its y dependence, this is a classical problem for the filtration of an incompressible fluid through a porous medium. It comes

$$\mathbf{v}^0 = -\,\mathbf{J}(\mathbf{x},\mathbf{y})\,\mathbf{grad}_x p^0$$

Obviously, the second order tensor $\mathbf{J}$ is time independent, as μ_{eff}.
Taking into account the above results, (B.3.3) reduces to

$$\mathrm{div}_z\,(\,\mathbf{v}_\alpha^2) + \mathrm{div}_y\,(\,\mathbf{v}_\alpha^1\,) + \mathrm{div}_x\,(\,\mathbf{v}^0) = 0$$

Finally, we successively take an average of this relation over Ω' and Ω. By using the Ω' periodicity the first averaging gives

$$\mathrm{div}_y\,(\,\widehat{\mathbf{v}^1}\,) + \mathrm{div}_x\,(\,\mathbf{v}^0) = 0$$

And the second volume averaging leads to

$$\mathrm{div}_x\,(\,\widetilde{\mathbf{v}^0}) = 0 \qquad \text{(B.13)}$$

$$\widetilde{\mathbf{v}^0} = -\,\mathbf{K}(\mathbf{x})\,\mathbf{grad}_x p^0 \qquad\qquad \mathbf{K} = \widetilde{\mathbf{J}} \qquad \text{(B.14)}$$

References

1 A.Bensoussan, J.L. Lions, G. Papanicolaou, *Asymptotic analysis for periodic structures*, North-Holland, Amsterdam (1978).
2 E. Sanchez-Palencia, *Non-homogeneous media and vibration theory*, Springer-Verlag, Lecture Notes in Physics 127, Berlin (1980).
3 J.-L. Auriault, Effective macroscopic description for heat conduction in periodic composites, *Int. J. Heat Mass Transfer* , **26**, 6 861-869 (1983).
4 J.L. Auriault, T. Strzelecki, J. Bauer, S. He, Porous deformable media saturated by a very compressible fluid : quasi-statics, *Eur.J.Mech.,A/Solids* **9**, 4, 373-392 (1990).
5 C. Boutin, J.-L. Auriault, Acoustics of a bubbly fluid at large bubble concentration, *Eur. J. Mech., B/Fluids*, **12**, 3, 367-399 (1993).
6 J.-L. Auriault, Nonsaturated deformable porous media: quasi-statics, *TiPM*, **12**, 3, 45-64 (1987).

SECTION 2

POROELASTICITY:

NUMERICAL MODELLING ASPECTS

On Singular Integral Equations of Poroelasticity

A. H-D. Cheng
Department of Civil Engineering, University of Delaware
Newark, Delaware 19716

E. Detournay
Department of Civil and Mineral Engineering, University of Minnesota
Minneapolis, Minnesota 55455

Abstract

The singular integral equations for the direct and indirect boundary element methods for linear quasi-static poroelasticity are examined. The direct integral equations are derived based on the principle of reciprocity of work. Analogous to the simple and double layer methods in potential theories, two indirect methods, a stress discontinuity and a displacement discontinuity method, are obtained from the direct integral equations.

Introduction

In boundary element literature the 'direct methods' are referred to methods that are derived from integral equations based on generalized Green's theorem. The parameters modeled in the equations are 'physical' quantities such as potential, flux, displacement, stress, etc. On the other hand, the 'indirect methods' are based on the distribution of singular solutions such as source, dipole, point force, etc., at 'fictitious' densities.

The theoretical link between the direct and the indirect boundary integral equations for the potential,[1,2] the diffusion,[3] and the elasticity operators[1,4] are well known. It has been shown that the so called fictitious densities are actually jumps between the exterior and the interior solution under the same boundary condition.

The present paper is intended to provide a link which ties the direct boundary element method with the indirect method for the linear, quasi-static poroelasticity.[5–8] In the following, the direct integral equations are first derived from the principle of reciprocity of work. The indirect integral equations then follow by adding together direct integral equations respectively representing an interior and an exterior domain problem, subject to the same boundary conditions. Two methods, a stress discontinuity and a displacement discontinuity, which are analogous to the simple and double layer method in

A. P. S. Selvadurai (ed.), *Mechanics of Poroelastic Media*, 93–108.

potential theory, are derived. The displacement discontinuity method is closely related to the dislocation method used in elasticity to model fracture problems.[9,10] Through the derivation of integral equations, we also demonstrate the connection among the various singular solutions.

Governing Differential Equations

The governing equations of linear, quasi-static poroelasticity can be expressed as follows[6,11]

- constitutive relations

$$\sigma_{ij} = M_{ijkl}e_{kl} - \alpha_{ij}p \tag{1}$$

$$p = M(\zeta - \alpha_{ij}e_{ij}) \tag{2}$$

- equilibrium equation

$$\sigma_{ij,j} = -F_i \tag{3}$$

- Darcy's law

$$q_i = -\kappa_{ij}(p_{,j} - f_j) \tag{4}$$

- continuity equation

$$\frac{\partial \zeta}{\partial t} + q_{i,i} = \gamma \tag{5}$$

where σ_{ij} is the total stress tensor, e_{ij} the total strain tensor, p the pore pressure, ζ the variation of fluid content per unit volume of porous media, q_i the specific discharge vector, M_{ijkl} the drained elastic modulus tensor, α_{ij} the Biot effective stress tensor, M the Biot modulus, κ_{ij} the permeability tensor, F_i the bulk body force, f_i the fluid body force, and γ the fluid source. We also note the following kinematic relations:

$$e_{ij} = \frac{1}{2}(u_{i,j} + u_{j,i}) \tag{6a}$$

$$v_i = \int_0^t q_i \, dt \tag{6b}$$

$$Q = \int_0^t \gamma \, dt \tag{6c}$$

where u_i is the solid displacement vector, v_i the fluid relative displacement vector, and Q the volume of injected fluid from a source.

Direct Method

The constitutive equations, (1) and (2), are constructed under the assumption of a reversible thermodynamic process.[5,6] It can therefore be proven that the theory of reciprocity of work exists (see Nowacki[12] for the equivalent relation in coupled thermoelasticity)

$$\sigma_{ij}^{(1)} e_{ij}^{(2)} + p^{(1)}\zeta^{(2)} = \sigma_{ij}^{(2)} e_{ij}^{(1)} + p^{(2)}\zeta^{(1)} \tag{7}$$

where superscripts (1) and (2) denote quantities under two independent stress and strain states at different spatial and time coordinates. For the present purpose, the following form is taken:

$$\begin{aligned} &\sigma_{ij}^{(1)}(\chi,\tau)\, e_{ij}^{(2)}(\chi-\boldsymbol{x},t-\tau) + p^{(1)}(\chi,\tau)\,\zeta^{(2)}(\chi-\boldsymbol{x},t-\tau) \\ &= \sigma_{ij}^{(2)}(\chi-\boldsymbol{x},t-\tau)\, e_{ij}^{(1)}(\chi,\tau) + p^{(2)}(\chi-\boldsymbol{x},t-\tau)\,\zeta^{(1)}(\chi,\tau) \end{aligned} \tag{8}$$

Integrating (8) over the solution domain and time, and performing integration by parts, we arrive at the reciprocal integral equation[13–15]

$$\begin{aligned} &\iint_\Gamma (\sigma_{ij}^{(1)} n_j u_i^{(2)} - \sigma_{ij}^{(2)} n_j u_i^{(1)})\, d\chi\, d\tau - \iint_\Gamma (p^{(1)} v_i^{(2)} n_i - p^{(2)} v_i^{(1)} n_i)\, d\chi\, d\tau \\ &+ \iint_\Omega (F_i^{(1)} u_i^{(2)} - F_i^{(2)} u_i^{(1)})\, d\chi\, d\tau + \iint_\Omega (f_i^{(1)} v_i^{(2)} - f_i^{(2)} v_i^{(1)})\, d\chi\, d\tau \\ &+ \iint_\Omega (p^{(1)} Q^{(2)} - p^{(2)} Q^{(1)})\, d\chi\, d\tau + \iint_\Omega \left(p^{(1)}\, \zeta^{(2)}\Big|_{\tau=0} - p^{(2)}\, \zeta^{(1)}\Big|_{\tau=0}\right) d\chi\, d\tau \\ &= 0 \end{aligned} \tag{9}$$

in which Γ is the bounding surface of a domain Ω, and n_i the component of the unit outward normal of Γ.

To obtain singular integral equations of Somigliana type, the following substitutions, respectively corresponding to an instantaneous point force in the x_k-direction and an instantaneous fluid volume dilatation, are made

$$\begin{aligned} F_{ik}^{(2)} &= \delta_{ik}\, \delta(\chi - \boldsymbol{x})\, \delta(t-\tau) \\ Q^{(2)} &= \delta(\chi - \boldsymbol{x})\, \delta(t-\tau) \end{aligned} \tag{10}$$

where δ_{ik} is the Kronecker delta, and $\delta()$ the Dirac delta function with singularity located at point $\boldsymbol{x}$ and time t. The above substitutions yield the following

$$\begin{aligned} \beta u_k(\boldsymbol{x},t) = \int_0^t \int_\Gamma &\Big[u_{ik}^{fi}(\chi-\boldsymbol{x},t-\tau)\sigma_{ij}(\chi,\tau)n_j(\chi) \\ &-\sigma_{ijk}^{fi}(\chi-\boldsymbol{x},t-\tau)n_j(\chi)u_i(\chi,\tau)\Big]\, d\chi\, d\tau \\ - \int_0^t \int_\Gamma &\Big[v_{ik}^{fi}(\chi-\boldsymbol{x},t-\tau)p(\chi,\tau)n_i(\chi) - p_k^{fi}(\chi-\boldsymbol{x},t-\tau)v_i(\chi,\tau)n_i(\chi)\Big]\, d\chi\, d\tau \end{aligned}$$

$$
\begin{aligned}
&+\int_0^t\int_\Omega \Big[u_{ik}^{fi}(\chi-x,t-\tau)F_i(\chi,\tau)+v_{ik}^{fi}(\chi-x,t-\tau)f_i(\chi,\tau)\\
&\qquad +p_k^{fi}(\chi-x,t-\tau)Q(\chi,\tau)\Big]\,d\chi\,d\tau\\
&+\int_0^t\int_\Omega p_k^{fi}(\chi-x,t-\tau)\zeta(\chi,0)\,d\chi
\end{aligned}
\tag{11}
$$

$$
\begin{aligned}
-\beta p(x,t) = &\int_0^t\int_\Gamma \Big[u_i^{li}(\chi-x,t-\tau)\sigma_{ij}(\chi,\tau)n_j(\chi)\\
&\qquad -\sigma_{ij}^{li}(\chi-x,t-\tau)n_j(\chi)u_i(\chi,\tau)\Big]\,d\chi\,d\tau\\
&-\int_0^t\int_\Gamma \Big[v_i^{li}(\chi-x,t-\tau)n_i(\chi)p(\chi,\tau)-p^{li}(\chi-x,t-\tau)v_i(\chi,\tau)n_i(\chi)\Big]\,d\chi\,d\tau\\
&+\int_0^t\int_\Omega \Big[u_i^{li}(\chi-x,t-\tau)F_i(\chi,\tau)+v_i^{li}(\chi-x,t-\tau)f_i(\chi,\tau)\\
&\qquad +p^{li}(\chi-x,t-\tau)Q(\chi,\tau)\Big]\,d\chi\,d\tau\\
&+\int_0^t\int_\Omega p^{li}(\chi-x,t-\tau)\zeta(\chi,0)\,d\chi
\end{aligned}
\tag{12}
$$

in which β is a constant determined by a Cauchy principal value integration and is equal to 0, 1 and 1/2 respectively for the base point x located outside, inside the domain Ω, and on the boundary Γ. The quantities denoted by superscripts are free-space Green's functions governed by equations (1) to (5). The conventions used to denote the singular solution types (force, source, continuous, instantaneous, etc.) are illustrated in Table 1. These solution have been compiled. They are not listed herein for space limitation, but will be published elsewhere.

	instantaneous	continuous
total force	*fi*	*fc*
fluid source	*si*	*sc*
fluid dipole	*pi*	*pc*
fluid dilatation	*li*	*lc*
displacement discontinuity	*di*	*dc*
edge dislocation	*ei*	*ec*

Table 1: Convention for naming singular solutions.

The above integral equations can be rewritten in terms of the fluid discharge q_i on the boundary and the source density γ in the domain, instead of the fluid displacement v_i and the injection volume Q, by performing integration by parts. We also carry out the time integration for the terms containing the initial condition $\zeta(\chi,0)$. The time integration introduces the continuous influence functions such as

$$\frac{\partial}{\partial \tau} p_k^{fc}(\boldsymbol{\chi}-\boldsymbol{x}, t-\tau) = -p_k^{fi}(\boldsymbol{\chi}-\boldsymbol{x}, t-\tau) \tag{13}$$

It can be shown that the following relations exist for the influence functions:

$$p^{lc} = p^{si} \tag{14a}$$
$$v_{ik}^{fi} = q_{ik}^{fc} \tag{14b}$$
$$v_{ik}^{li} = q_{ik}^{si} \tag{14c}$$

The following integral equations are obtained

$$\begin{aligned}
\beta u_k(\boldsymbol{x},t) = & \int_0^t \int_\Gamma \Big[u_{ik}^{fi}(\boldsymbol{\chi}-\boldsymbol{x},t-\tau)\sigma_{ij}(\boldsymbol{\chi},\tau)n_j(\boldsymbol{\chi}) \\
& \qquad -\sigma_{ijk}^{fi}(\boldsymbol{\chi}-\boldsymbol{x},t-\tau)n_j(\boldsymbol{\chi})u_i(\boldsymbol{\chi},\tau)\Big]\, d\boldsymbol{\chi}\, d\tau \\
& - \int_0^t \int_\Gamma \Big[q_{ik}^{fc}(\boldsymbol{\chi}-\boldsymbol{x},t-\tau)n_i(\boldsymbol{\chi})p(\boldsymbol{\chi},\tau) - p_k^{fc}(\boldsymbol{\chi}-\boldsymbol{x},t-\tau)q_i(\boldsymbol{\chi},\tau)n_i(\boldsymbol{\chi})\Big]\, d\boldsymbol{\chi}\, d\tau \\
& + \int_0^t \int_\Omega \Big[u_{ik}^{fi}(\boldsymbol{\chi}-\boldsymbol{x},t-\tau)F_i(\boldsymbol{\chi},\tau) + q_{ik}^{fc}(\boldsymbol{\chi}-\boldsymbol{x},t-\tau)f_i(\boldsymbol{\chi},\tau) \\
& \qquad +p_k^{fc}(\boldsymbol{\chi}-\boldsymbol{x},t-\tau)\gamma(\boldsymbol{\chi},\tau)\Big]\, d\boldsymbol{\chi}\, d\tau \\
& - \int_\Omega p_k^{fc}(\boldsymbol{\chi}-\boldsymbol{x},t)\zeta(\boldsymbol{\chi},0)\, d\boldsymbol{\chi}
\end{aligned} \tag{15}$$

$$\begin{aligned}
-\beta p(\boldsymbol{x},t) = & \int_0^t \int_\Gamma \Big[u_i^{li}(\boldsymbol{\chi}-\boldsymbol{x},t-\tau)\sigma_{ij}(\boldsymbol{\chi},\tau)n_j(\boldsymbol{\chi}) \\
& \qquad -\sigma_{ij}^{li}(\boldsymbol{\chi}-\boldsymbol{x},t-\tau)n_j(\boldsymbol{\chi})u_i(\boldsymbol{\chi},\tau)\Big]\, d\boldsymbol{\chi}\, d\tau \\
& - \int_0^t \int_\Gamma \Big[q_i^{si}(\boldsymbol{\chi}-\boldsymbol{x},t-\tau)n_i(\boldsymbol{\chi})p(\boldsymbol{\chi},\tau) - p^{si}(\boldsymbol{\chi}-\boldsymbol{x},t-\tau)q_i(\boldsymbol{\chi},\tau)n_i(\boldsymbol{\chi})\Big]\, d\boldsymbol{\chi}\, d\tau \\
& + \int_0^t \int_\Omega \Big[u_i^{li}(\boldsymbol{\chi}-\boldsymbol{x},t-\tau)F_i(\boldsymbol{\chi},\tau) + q_i^{si}(\boldsymbol{\chi}-\boldsymbol{x},t-\tau)f_i(\boldsymbol{\chi},\tau) \\
& \qquad +p^{si}(\boldsymbol{\chi}-\boldsymbol{x},t-\tau)\gamma(\boldsymbol{\chi},\tau)\Big]\, d\boldsymbol{\chi}\, d\tau \\
& - \int_\Omega p^{si}(\boldsymbol{\chi}-\boldsymbol{x},t)\zeta(\boldsymbol{\chi},0)\, d\boldsymbol{\chi}
\end{aligned} \tag{16}$$

In an initial/boundary value problem for poroelasticity, either the boundary traction $t_i = \sigma_{ij}n_j$ or the displacement u_i, and either the fluid pressure p or the normal flux $q = q_i n_i$, are prescribed on a given part of the boundary. Equations (15) and (16) are placed at a set of boundary nodes, and a collocation procedure is performed. Due to the transient nature of the integral equations, the discretization takes place both in time and in space. Through a time-stepping or convolutional integral process, the missing boundary data in terms of the physical parameters of traction, displacement, pressure or flux are directly solved.

Various non-transient versions of integral equations can be obtained by applying integral transform to (15) and (16). For example, applying Laplace transform and utilizing the convolutional theorem, the time integrals are eliminated and the following equations result[16–18]

$$\beta\tilde{u}_k(\boldsymbol{x},s) = \int_\Gamma \left[\tilde{u}^{fi}_{ik}(\chi-\boldsymbol{x},s)\tilde{\sigma}_{ij}(\chi,s)n_j(\chi) - \tilde{\sigma}^{fi}_{ijk}(\chi-\boldsymbol{x},s)n_j(\chi)\tilde{u}_i(\chi,s)\right] d\chi$$
$$-\int_\Gamma \frac{1}{s}\left[\tilde{q}^{fi}_{ik}(\chi-\boldsymbol{x},s)n_i(\chi)\tilde{p}(\chi,s) - \tilde{p}^{fi}_k(\chi-\boldsymbol{x},s)\tilde{q}_i(\chi,s)n_i(\chi)\right] d\chi$$
$$+\int_\Omega \left[\tilde{u}^{fi}_{ik}(\chi-\boldsymbol{x},s)\tilde{F}_i(\chi,s) + \frac{1}{s}\tilde{q}^{fi}_{ik}(\chi-\boldsymbol{x},s)\tilde{f}_i(\chi,s) + \frac{1}{s}\tilde{p}^{fi}_k(\chi-\boldsymbol{x},s)\tilde{\gamma}(\chi,s)\right] d\chi$$
$$-\int_\Omega \frac{1}{s}\tilde{p}^{fi}_k(\chi-\boldsymbol{x},s)\zeta(\chi,0)\, d\chi \qquad (17)$$

$$-\beta\tilde{p}(\boldsymbol{x},s) = \int_\Gamma s\left[\tilde{u}^{si}_i(\chi-\boldsymbol{x},s)\tilde{\sigma}_{ij}(\chi,s)n_j(\chi) - \tilde{\sigma}^{si}_{ij}(\chi-\boldsymbol{x},s)n_j(\chi)\tilde{u}_i(\chi,s)\right] d\chi$$
$$-\int_\Gamma \left[\tilde{q}^{si}_i(\chi-\boldsymbol{x},s)n_i(\chi)\tilde{p}(\chi,s) - \tilde{p}^{si}(\chi-\boldsymbol{x},s)\tilde{q}_i(\chi,s)n_i(\chi)\right] d\chi$$
$$+\int_\Omega \left[s\tilde{u}^{si}_i(\chi-\boldsymbol{x},s)\tilde{F}_i(\chi,s) + \tilde{q}^{si}_i(\chi-\boldsymbol{x},s)\tilde{f}_i(\chi,s) + \tilde{p}^{si}(\chi-\boldsymbol{x},s)\tilde{\gamma}(\chi,s)\right] d\chi$$
$$-\int_\Omega \tilde{p}^{si}(\chi-\boldsymbol{x},s)\zeta(\chi,0)\, d\chi \qquad (18)$$

where the tilde indicates a Laplace transform, and s the transform parameter. In writing the transformed integral equations, we have expressed the fluid dilatation Green's function by their time integral, the fluid source solutions and the continuous force influence functions in terms of their corresponding instantaneous solutions.

Integral equations of similar nature can also be derived in the frequency domain (or via Fourier transform),[19] and steadily moving coordinate system.[20]

Indirect Method

Consider a region Ω bounded by Γ. Its complementary region is denoted as Ω' as depicted in Figure 1. Equations (15) and (16) are now written for Ω', which is free from body force and fluid sources

$$0 = \int_0^t\int_\Gamma (u^{fi}_{ik}\sigma'_{ij}n'_j - \sigma^{fi}_{ijk}n'_ju'_i)\, d\chi\, d\tau - \int_0^t\int_\Gamma (q^{fc}_{ik}n'_ip' - p^{fc}_kq'_in'_i)\, d\chi\, d\tau$$
$$-\int_{\Omega'} (p^{fc}_k\zeta')\Big|_{\tau=0}\, d\chi \qquad (19)$$

$$0 = \int_0^t\int_\Gamma (u^{li}_i\sigma'_{ij}n'_j - \sigma^{li}_{ij}n'_ju'_i)\, d\chi\, d\tau - \int_0^t\int_\Gamma (q^{qi}_in'_ip' - p^{lc}q'_in'_i)\, d\chi\, d\tau$$
$$-\int_{\Omega'} (p^{si}\zeta')\Big|_{\tau=0}\, d\chi \qquad (20)$$

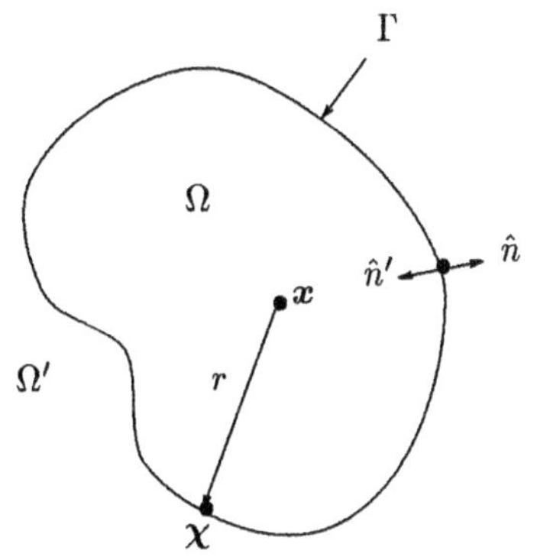

Figure 1: Two complementary regions Ω and Ω′ bounded by a common surface Γ.

In the above we have used prime to denote quantities associated with Ω'. The initial condition $\zeta'|_{t=0}$ is kept for reasons that will be explained later. We also note that $n'_i = -n_i$ because the outward normal of Ω and Ω' oppose each other. The left hand sides of (19) and (20) are zero because we require the base point to be located in Ω. Summing (19) and (20) with (15) and (16), we obtain the following

$$\begin{aligned}
\beta u_k &= \int_0^t\int_\Gamma \left[u_{ik}^{fi}(\sigma_{ij}-\sigma'_{ij})n_j - \sigma_{ijk}^{fi}n_j(u_i-u'_i)\right] d\chi\, d\tau \\
&- \int_0^t\int_\Gamma \left[q_{ik}^{fc}n_i(p-p') - p_k^{fc}(q_i-q'_i)n_i\right] d\chi\, d\tau \\
&+ \int_0^t\int_\Omega (u_{ik}^{fi}F_i + v_{ik}^{fi}f_i + p_k^{fc}\gamma)\, d\chi\, d\tau - \int_{\Omega+\Omega'} (p_k^{fc}\zeta)\Big|_{\tau=0} d\chi \qquad (21)\\
-\beta p &= \int_0^t\int_\Gamma \left[u_i^{li}(\sigma_{ij}-\sigma'_{ij})n_j - \sigma_{ij}^{li}n_j(u_i-u'_i)\right] d\chi\, d\tau \\
&- \int_0^t\int_\Gamma \left[q_i^{si}n_i(p-p') - p^{si}(q_i-q'_i)n_i\right] d\chi\, d\tau \\
&+ \int_0^t\int_\Omega (u_i^{li}F_i + q_i^{si}f_i + p^{si}\gamma)\, d\chi\, d\tau - \int_{\Omega+\Omega'} (p^{si}\zeta)\Big|_{\tau=0} d\chi \qquad (22)
\end{aligned}$$

Stress Discontinuity Method

For a problem defined in Ω, we can impose a complementary problem in Ω' with the displacement and pore pressure along the boundary identical to that of the primary problem. Equations (21) and (22) therefore reduce to

$$\beta u_k(\boldsymbol{x},t) = \int_0^t\int_\Gamma \left[u_{ki}^{fi}(\boldsymbol{x}-\chi,t-\tau)s_i(\chi,\tau) + p_k^{fc}(\boldsymbol{x}-\chi,t-\tau)d(\chi,\tau)\right] d\chi\, d\tau$$

$$+\int_0^t\int_\Omega \Big[u_{ki}^{fi}(\boldsymbol{x}-\boldsymbol{\chi},t-\tau)F_i(\boldsymbol{\chi},\tau)+q_{ki}^{fc}(\boldsymbol{x}-\boldsymbol{\chi},t-\tau)f_i(\boldsymbol{\chi},\tau)$$
$$-p_k^{fc}(\boldsymbol{x}-\boldsymbol{\chi},t-\tau)\gamma(\boldsymbol{\chi},\tau)\Big]\,d\boldsymbol{\chi}\,d\tau$$
$$+\int_{\Omega+\Omega'} p_k^{fc}(\boldsymbol{x}-\boldsymbol{\chi},t)\zeta(\boldsymbol{\chi},0)\,d\boldsymbol{\chi} \tag{23}$$

$$\beta p(\boldsymbol{x},t)=\int_0^t\int_\Gamma \Big[u_i^{li}(\boldsymbol{x}-\boldsymbol{\chi},t-\tau)s_i(\boldsymbol{\chi},\tau)+p^{si}(\boldsymbol{x}-\boldsymbol{\chi},t-\tau)d(\boldsymbol{\chi},\tau)\Big]\,d\boldsymbol{\chi}\,d\tau$$
$$+\int_0^t\int_\Omega \Big[u_i^{li}(\boldsymbol{x}-\boldsymbol{\chi},t-\tau)F_i(\boldsymbol{\chi},\tau)+q_i^{si}(\boldsymbol{x}-\boldsymbol{\chi},t-\tau)f_i(\boldsymbol{\chi},\tau)$$
$$-p^{si}(\boldsymbol{x}-\boldsymbol{\chi},t-\tau)\gamma(\boldsymbol{\chi},\tau)\Big]\,d\boldsymbol{\chi}\,d\tau$$
$$+\int_{\Omega+\Omega'} p^{si}(\boldsymbol{x}-\boldsymbol{\chi},t)\zeta(\boldsymbol{\chi},0)\,d\boldsymbol{\chi} \tag{24}$$

In the foregoing s_i and d are quantities related to the traction and normal flux jumps across the boundary Γ

$$s_i = (\sigma_{ij}-\sigma'_{ij})n_j \tag{25a}$$
$$d = -(q_i-q'_i)n_i \tag{25b}$$

Actually, s_i and d correspond to the boundary density of fictitious forces and fluid sources (the minus sign in the definition of d is introduced for consistency reasons, since the sources are defined positive for injection). In order to be consistent with the convention of indirect method, we have reversed the order of $\boldsymbol{x}$ and $\boldsymbol{\chi}$ in the Green's functions according to the following relations:

$$u_{ik}^{fi}(\boldsymbol{\chi}-\boldsymbol{x},t-\tau) = u_{ki}^{fi}(\boldsymbol{x}-\boldsymbol{\chi},t-\tau) \tag{26a}$$
$$p_k^{fc}(\boldsymbol{\chi}-\boldsymbol{x},t-\tau) = -p_k^{fc}(\boldsymbol{x}-\boldsymbol{\chi},t-\tau) \tag{26b}$$
$$u_i^{li}(\boldsymbol{\chi}-\boldsymbol{x},t-\tau) = -u_i^{li}(\boldsymbol{x}-\boldsymbol{\chi},t-\tau) \tag{26c}$$
$$p_k^{si}(\boldsymbol{\chi}-\boldsymbol{x},t-\tau) = p_k^{si}(\boldsymbol{x}-\boldsymbol{\chi},t-\tau) \tag{26d}$$

The above pair of equations (23) and (24) are equivalent to the 'simple-layer method' in the potential theory[1] in which weak singularities of the order $\ln r$ for 2-D and $1/r$ for 3-D are distributed. Equation (23) for example shows that for a problem free of body force, source and initial value, the displacement at a point $\boldsymbol{x}$ and time t can be evaluated by distributing along the boundary Γ influence functions of solid and fluid displacements created by instantaneous point force at 'fictitious densities'. Comparing equations (15) and (16) to (23) and (24), we observe that the distribution of singularities for the indirect method is along the integration contour $\boldsymbol{\chi}\in\Gamma$, while that for the direct method is at the fixed point $\boldsymbol{x}$. We also notice that the domain integration of the initial condition $\zeta|_{t=0}$ is performed for both Ω and Ω'. As pointed out by Sharp,[21] the need for integrating in the complementary domain is dependent on the numerical algorithm adopted. If the time integration in equations (23) and (24) is carried out in the convolutional sense, i.e., tracing back to the time origin for solution at any given time, one can simply assume that $\zeta|_{t=0} = 0$ for $\chi \in \Omega'$, which eliminates integration in the complementary domain. On the other hand, if a time stepping scheme is adopted, advancing solution to the next time level is based on information of the present time, not the time origin. In that case, the evolution of ζ in both Ω and Ω' is required for the domain integral, which is a serious drawback in terms of computational effort.

Equations (23) and (24) can be written in a different form to appeal to the physical intuition. The following relations exist for the Green's functions:

$$q_{ki}^{fc} = -\kappa u_{ki}^{pi} \tag{27a}$$
$$p_k^{fc} = u_k^{si} \tag{27b}$$
$$u_i^{li} = p_i^{fi} \tag{27c}$$
$$q_i^{si} = -\kappa p_i^{pi} \tag{27d}$$

We hence express (23) and (24) as

$$\begin{aligned}\beta u_k(\boldsymbol{x},t) = &\int_0^t\int_\Gamma \left[u_{ki}^{fi}(\boldsymbol{x}-\chi,t-\tau)s_i(\chi,\tau) + u_k^{si}(\boldsymbol{x}-\chi,t-\tau)d(\chi,\tau)\right]\,d\chi\,d\tau\\ &+\int_0^t\int_\Omega \left[u_{ki}^{fi}(\boldsymbol{x}-\chi,t-\tau)F_i(\chi,\tau) - \kappa u_{ki}^{pi}(\boldsymbol{x}-\chi,t-\tau)f_i(\chi,\tau)\right.\\ &\qquad\left. -u_k^{si}(\boldsymbol{x}-\chi,t-\tau)\gamma(\chi,\tau)\right]\,d\chi\,d\tau\\ &+\int_{\Omega+\Omega'} u_k^{si}(\boldsymbol{x}-\chi,t)\zeta(\chi,0)\,d\chi\end{aligned} \tag{28}$$

$$\begin{aligned}\beta p(\boldsymbol{x},t) = &\int_0^t\int_\Gamma \left[p_i^{fi}(\boldsymbol{x}-\chi,t-\tau)s_i(\chi,\tau) + p^{si}(\boldsymbol{x}-\chi,t-\tau)d(\chi,\tau)\right]\,d\chi\,d\tau\\ &+\int_0^t\int_\Omega \left[p_i^{fi}(\boldsymbol{x}-\chi,t-\tau)F_i(\chi,\tau) - \kappa p_i^{pi}(\boldsymbol{x}-\chi,t-\tau)f_i(\chi,\tau)\right.\\ &\qquad\left. - p^{si}(\boldsymbol{x}-\chi,t-\tau)\gamma(\chi,\tau)\right]\,d\chi\,d\tau\\ &+\int_{\Omega+\Omega'} p^{si}(\boldsymbol{x}-\chi,t)\zeta(\chi,0)\,d\chi\end{aligned} \tag{29}$$

The above equations state that the displacement and pressure in the field can be obtained by respectively distributing along Γ the influence functions of displacement and pressure due to instantaneous point force, fluid source and dipole.

In order to solve mixed boundary value problems, we also need integral expressions for boundary traction and fluid normal flux. They can be obtained by substituting (28) and (29) into (1) and (4), which yields

$$\begin{aligned}&\beta t_i(\boldsymbol{x},t) = \beta\sigma_{ij}(\boldsymbol{x},t)n_j(\boldsymbol{x}) =\\ &\quad\int_0^t\int_\Gamma \left[\sigma_{ijk}^{fi}(\boldsymbol{x}-\chi,t-\tau)s_k(\chi,\tau) + \sigma_{ij}^{si}(\boldsymbol{x}-\chi,t-\tau)d(\chi,\tau)\right]n_j(\boldsymbol{x})\,d\chi\,d\tau\end{aligned}$$

$$+\int_0^t\int_\Omega \Big[\sigma_{ijk}^{fi}(\boldsymbol{x}-\boldsymbol{\chi},t-\tau)F_k(\boldsymbol{\chi},\tau)-\kappa\sigma_{ijk}^{pi}(\boldsymbol{x}-\boldsymbol{\chi},t-\tau)f_k(\boldsymbol{\chi},\tau)$$
$$-\sigma_{ij}^{si}(\boldsymbol{x}-\boldsymbol{\chi},t-\tau)\gamma(\boldsymbol{\chi},\tau)\Big]\,n_j(\boldsymbol{x})\,d\boldsymbol{\chi}\,d\tau$$
$$+\int_{\Omega+\Omega'}\sigma_{ij}^{si}(\boldsymbol{x}-\boldsymbol{\chi},t)\zeta(\boldsymbol{\chi},0)n_j(\boldsymbol{x})\,d\boldsymbol{\chi} \tag{30}$$

$$\beta q(\boldsymbol{x},t)=\beta q_i(\boldsymbol{x},t)n_i(\boldsymbol{x})=$$
$$\int_0^t\int_\Gamma \Big[q_{ij}^{fi}(\boldsymbol{x}-\boldsymbol{\chi},t-\tau)s_j(\boldsymbol{\chi},\tau)+q_i^{si}(\boldsymbol{x}-\boldsymbol{\chi},t-\tau)d(\boldsymbol{\chi},\tau)\Big]\,n_i(\boldsymbol{x})\,d\boldsymbol{\chi}\,d\tau$$
$$+\int_0^t\int_\Omega \Big[q_{ij}^{fi}(\boldsymbol{x}-\boldsymbol{\chi},t-\tau)F_j(\boldsymbol{\chi},\tau)-\kappa q_{ij}^{pi}(\boldsymbol{x}-\boldsymbol{\chi},t-\tau)f_j(\boldsymbol{\chi},\tau)$$
$$-q_i^{si}(\boldsymbol{x}-\boldsymbol{\chi},t-\tau)\gamma(\boldsymbol{\chi},\tau)\Big]\,n_i(\boldsymbol{x})\,d\boldsymbol{\chi}\,d\tau$$
$$+\int_{\Omega+\Omega'} q_i^{si}(\boldsymbol{x}-\boldsymbol{\chi},t)\zeta(\boldsymbol{\chi},0)n_i(\boldsymbol{x})\,d\boldsymbol{\chi} \tag{31}$$

Equations (28) to (31) can then be exploited in a boundary collocation procedure to determine the discrete solutions of the distribution densities s_i and d.

Displacement Discontinuity Method

In contrast to the stress discontinuity method, we now consider the case where the boundary traction and the normal flux for the interior and exterior domain problems are set equal. The following set of integral equations are then deduced from (20) and (21):

$$\beta u_k(\boldsymbol{x},t)=\int_0^t\int_\Gamma \Big[\sigma_{ijk}^{fi}(\boldsymbol{x}-\boldsymbol{\chi},t-\tau)n_j(\boldsymbol{\chi})d_i(\boldsymbol{\chi},\tau)$$
$$-\frac{1}{\kappa}v_{ki}^{fi}(\boldsymbol{x}-\boldsymbol{\chi},t-\tau)n_i(\boldsymbol{\chi})s(\boldsymbol{\chi},\tau)\Big]\,d\boldsymbol{\chi}\,d\tau$$
$$+\int_0^t\int_\Omega \Big[u_{ki}^{fi}(\boldsymbol{x}-\boldsymbol{\chi},t-\tau)F_i(\boldsymbol{\chi},\tau)+v_{ki}^{fi}(\boldsymbol{x}-\boldsymbol{\chi},t-\tau)f_i(\boldsymbol{\chi},\tau)$$
$$+p_k^{fc}(\boldsymbol{x}-\boldsymbol{\chi},t-\tau)\gamma(\boldsymbol{\chi},\tau)\Big]\,d\boldsymbol{\chi}\,d\tau$$
$$+\int_{\Omega+\Omega'} p_k^{fc}(\boldsymbol{x}-\boldsymbol{\chi},t)\zeta(\boldsymbol{\chi},0)\,d\boldsymbol{\chi} \tag{32}$$

$$\beta p(\boldsymbol{x},t)=\int_0^t\int_\Gamma \Big[\sigma_{ij}^{li}(\boldsymbol{x}-\boldsymbol{\chi},t-\tau)n_j(\boldsymbol{\chi})d_i(\boldsymbol{\chi},\tau)$$
$$-\frac{1}{\kappa}v_i^{li}(\boldsymbol{x}-\boldsymbol{\chi},t-\tau)n_i(\boldsymbol{\chi})s(\boldsymbol{\chi},\tau)\Big]\,d\boldsymbol{\chi}\,d\tau$$
$$+\int_0^t\int_\Omega \Big[u_i^{li}(\boldsymbol{x}-\boldsymbol{\chi},t-\tau)F_i(\boldsymbol{\chi},\tau)+v_i^{li}(\boldsymbol{x}-\boldsymbol{\chi},t-\tau)f_i(\boldsymbol{\chi},\tau)$$
$$+p^{lc}(\boldsymbol{x}-\boldsymbol{\chi},t-\tau)\gamma(\boldsymbol{\chi},\tau)\Big]\,d\boldsymbol{\chi}\,d\tau$$
$$+\int_{\Omega+\Omega'} p^{lc}(\boldsymbol{x}-\boldsymbol{\chi},t)\zeta(\boldsymbol{\chi},0)\,d\boldsymbol{\chi} \tag{33}$$

In the above

$$d_i = u_i - u_i' \tag{34a}$$
$$s = \kappa(p - p') \tag{34b}$$

are respectively displacement and normal flux discontinuities. Noticing (14) and also

$$\sigma_{ijk}^{fi} = u_{kji}^{di} \tag{35a}$$
$$\sigma_{ij}^{li} = p_{ji}^{di} \tag{35b}$$

equations (32) and (33) can be rewritten as

$$\begin{aligned}\beta u_k(\boldsymbol{x},t) = &\int_0^t\int_\Gamma \left[u_{kji}^{di}(\boldsymbol{x}-\boldsymbol{\chi},t-\tau)n_j(\boldsymbol{\chi})d_i(\boldsymbol{\chi},\tau) + u_{ki}^{pi}(\boldsymbol{x}-\boldsymbol{\chi},t-\tau)n_i(\boldsymbol{\chi})s(\boldsymbol{\chi},\tau)\right] d\boldsymbol{\chi}\, d\tau \\ &+ \int_0^t\int_\Omega \left[u_{ki}^{fi}(\boldsymbol{x}-\boldsymbol{\chi},t-\tau)F_i(\boldsymbol{\chi},\tau) - \kappa u_{ki}^{pi}(\boldsymbol{x}-\boldsymbol{\chi},t-\tau)f_i(\boldsymbol{\chi},\tau)\right. \\ &\qquad \left. + u_k^{si}(\boldsymbol{x}-\boldsymbol{\chi},t-\tau)\gamma(\boldsymbol{\chi},\tau)\right] d\boldsymbol{\chi}\, d\tau \\ &+ \int_{\Omega+\Omega'} u_k^{si}(\boldsymbol{x}-\boldsymbol{\chi},t)\zeta(\boldsymbol{\chi},0)\, d\boldsymbol{\chi}\end{aligned} \tag{36}$$

$$\begin{aligned}\beta p(\boldsymbol{x},t) = &\int_0^t\int_\Gamma \left[p_{ji}^{di}(\boldsymbol{x}-\boldsymbol{\chi},t-\tau)n_j(\boldsymbol{\chi})d_i(\boldsymbol{\chi},\tau) + p_i^{pi}(\boldsymbol{x}-\boldsymbol{\chi},t-\tau)n_i(\boldsymbol{\chi})s(\boldsymbol{\chi},\tau)\right] d\boldsymbol{\chi}\, d\tau \\ &+ \int_0^t\int_\Omega \left[p_i^{fi}(\boldsymbol{x}-\boldsymbol{\chi},t-\tau)F_i(\boldsymbol{\chi},\tau) - \kappa p_i^{pi}(\boldsymbol{x}-\boldsymbol{\chi},t-\tau)f_i(\boldsymbol{\chi},\tau)\right. \\ &\qquad \left. + p^{si}(\boldsymbol{x}-\boldsymbol{\chi},t-\tau)\gamma(\boldsymbol{\chi},\tau)\right] d\boldsymbol{\chi}\, d\tau \\ &+ \int_{\Omega+\Omega'} p^{si}(\boldsymbol{x}-\boldsymbol{\chi},t)\zeta(\boldsymbol{\chi},0)\, d\boldsymbol{\chi}\end{aligned} \tag{37}$$

The above formulae hence define another indirect method in which displacement discontinuity and fluid dipole singularity are distributed.[22] It may be viewed as the equivalent of the double-layer method in potential theory.

In order to solve mixed boundary value problems, equations (36) and (37) are differentiated and combined according to the constitutive equation and Darcy's law to yield the following

$$\begin{aligned}\beta t_i(\boldsymbol{x},t) = \beta\sigma_{ij}(\boldsymbol{x},t)n_j(\boldsymbol{x}) = \\ &\int_0^t\int_\Gamma \left[\sigma_{ijkl}^{di}(\boldsymbol{x}-\boldsymbol{\chi},t-\tau)n_k(\boldsymbol{\chi})d_l(\boldsymbol{\chi},\tau) + \sigma_{ijk}^{pi}(\boldsymbol{x}-\boldsymbol{\chi},t-\tau)n_k(\boldsymbol{\chi})s(\boldsymbol{\chi},\tau)\right] n_j(\boldsymbol{x})\, d\boldsymbol{\chi}\, d\tau \\ &+ \int_0^t\int_\Omega \left[\sigma_{ijk}^{fi}(\boldsymbol{x}-\boldsymbol{\chi},t-\tau)F_k(\boldsymbol{\chi},\tau) - \kappa\sigma_{ijk}^{pi}(\boldsymbol{x}-\boldsymbol{\chi},t-\tau)f_k(\boldsymbol{\chi},\tau)\right. \\ &\qquad \left. + \sigma_{ij}^{si}(\boldsymbol{x}-\boldsymbol{\chi},t-\tau)\gamma(\boldsymbol{\chi},\tau)\right] n_j(\boldsymbol{x})\, d\boldsymbol{\chi}\, d\tau \\ &+ \int_{\Omega+\Omega'} \sigma_{ij}^{si}(\boldsymbol{x}-\boldsymbol{\chi},t)\zeta(\boldsymbol{\chi},0)n_j(\boldsymbol{x})\, d\boldsymbol{\chi}\end{aligned} \tag{38}$$

$$\beta q(\boldsymbol{x},t) = \beta q_i(\boldsymbol{x},t) n_i(\boldsymbol{x}) =$$

$$\int_0^t \int_\Gamma \left[q_{ijk}^{di}(\boldsymbol{x}-\boldsymbol{\chi},t-\tau) n_j(\boldsymbol{\chi}) d_k(\boldsymbol{\chi},\tau) + q_{ij}^{pi}(\boldsymbol{x}-\boldsymbol{\chi},t-\tau) n_j(\boldsymbol{\chi}) s(\boldsymbol{\chi},\tau) \right] n_i(\boldsymbol{x})\, d\boldsymbol{\chi}\, d\tau$$
$$+ \int_0^t \int_\Omega \left[q_{ij}^{fi}(\boldsymbol{x}-\boldsymbol{\chi},t-\tau) F_j(\boldsymbol{\chi},\tau) - \kappa q_{ij}^{pi}(\boldsymbol{x}-\boldsymbol{\chi},t-\tau) f_j(\boldsymbol{\chi},\tau) \right.$$
$$\left. + q_i^{si}(\boldsymbol{x}-\boldsymbol{\chi},t-\tau) \gamma(\boldsymbol{\chi},\tau) \right] n_i(\boldsymbol{x})\, d\boldsymbol{\chi}\, d\tau$$
$$+ \int_{\Omega+\Omega'} q_i^{lc}(\boldsymbol{x}-\boldsymbol{\chi},t) \zeta(\boldsymbol{\chi},0) n_i(\boldsymbol{x})\, d\boldsymbol{\chi} + \kappa f_i(\boldsymbol{x},t) n_i(\boldsymbol{x}) \tag{39}$$

It is of interest to denote that the transformation from (32) and (33) to (38) and (39) actually suggests that

$$\begin{aligned} \sigma_{ijkl}^{di} &= G\left(u_{ikl,j}^{di} + u_{jkl,i}^{di}\right) + \frac{2G\nu}{1-2\nu}\delta_{ij} u_{mkl,m}^{di} - \alpha \delta_{ij} p_{kl}^{di} \\ &= G\left(\sigma_{lki,j}^{fi} + \sigma_{lkj,i}^{fi}\right) + \frac{2G\nu}{1-2\nu}\delta_{ij} \sigma_{lkm,m}^{fi} - \alpha \delta_{ij} \sigma_{lk}^{li} \end{aligned} \tag{40}$$

The above can be integrated with respect to time to yield

$$\sigma_{ijkl}^{dc} = G\left(\sigma_{lki,j}^{fc} + \sigma_{lkj,i}^{fc}\right) + \frac{2G\nu}{1-2\nu}\delta_{ij} \sigma_{lkm,m}^{fc} - \alpha \delta_{ij} \sigma_{lk}^{si} \tag{41}$$

where we note that

$$\sigma_{lk}^{lc} = \sigma_{lk}^{si} \tag{42}$$

Equation (41) shows that the continuous displacement discontinuity solution is equal to the appropriate combination of continuous force dipole (also known as double force, which is the spatial derivative of force solution), quadrapole and hexapole (respectively two and three pairs of mutually orthogonal double forces for 2-D and 3-D problems), and also the instantaneous fluid source. The same statement has been made earlier in an effort to obtain displacement discontinuity solutions by physical intuition.[23,24] The rationale was based on the observation that such a combination exactly generates the constant displacement jump as required in the solution.

Dislocation Method

The displacement discontinuity equations (38) and (39) can also be applied to fracture problems. In that case, Γ represents the fracture locus, d_i is the actual displacement jump due to the opening and sliding of fracture walls, and d the fluid flux jump as the result of fluid injection. For example, for the geometry of a linear fracture under plane strain conditions (Figure 2), the integral equations can be written as

$$\sigma_n(x,t) = \int_0^t \int_{-L}^{L} \left[\sigma_{2222}^{di}(x-\chi,t-\tau) d_n(\chi,\tau) + \sigma_{2221}^{di}(x-\chi,t-\tau) d_s(\chi,\tau) \right)$$

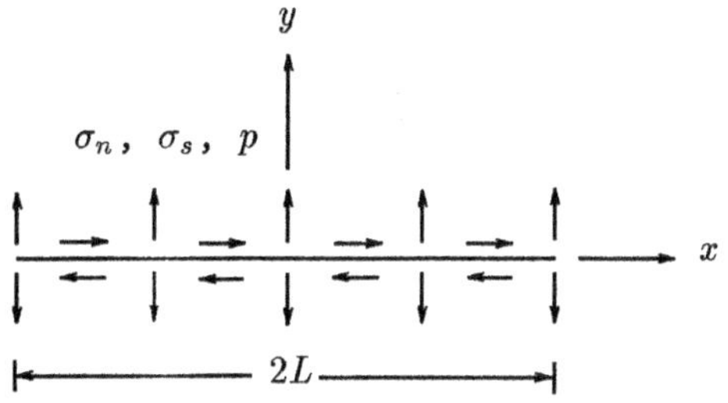

Figure 2: A linear fracture problem.

$$+ \sigma_{22}^{si}(x-\chi, t-\tau) d(\chi,\tau) \Big] \, d\chi \, d\tau \tag{43}$$

$$\sigma_s(x,t) = \int_0^t \int_{-L}^{L} \Big[\sigma_{1222}^{di}(x-\chi, t-\tau) d_n(\chi,\tau) + \sigma_{1221}^{di} d_s(\chi,\tau)(x-\chi, t-\tau)$$

$$+ \sigma_{12}^{si}(x-\chi, t-\tau) d(\chi,\tau) \Big] \, d\chi \, d\tau \tag{44}$$

$$p(x,t) = \int_0^t \int_{-L}^{L} \Big[p_{22}^{di}(x-\chi, t-\tau) d_n(\chi,\tau) + p_{21}^{di}(x-\chi, t-\tau) d_s(\chi,\tau)$$

$$+ p^{si}(x-\chi, t-\tau) d(\chi,\tau) \Big] \, d\chi \, d\tau \tag{45}$$

in which σ_n and σ_s are respectively the normal and shear stress on the fracture surface, d_n and d_s are the normal and shear displacement jumps across the fracture. These integral equations have been utilized for the numerical solution of fracture embedded in porous formation, opened by fluid pressurization (the hydraulic fracturing problem).[25,26]

An edge dislocation method, similar to that used in elasticity,[27] can also be devised. For example, performing integration by part on (43) produces the edge dislocation formula

$$\sigma_n(x,t) = \int_0^t \int_{-L}^{L} \Big[\sigma_{2222}^{ei}(x-\chi, t-\tau) d_n'(\chi,\tau) + \sigma_{2221}^{ei}(x-\chi, t-\tau) d_s'(\chi,\tau)$$

$$+ \sigma_{22}^{si}(x-\chi, t-\tau) d(\chi,\tau) \Big] \, d\chi \, d\tau \tag{46}$$

In the above, d_n' and d_s' are the derivatives with respect to χ of d_n and d_s. The influence functions denoted by the superscript e are edge dislocation solutions obtained as

$$\sigma_{2222}^{ei}(x-\chi, t-\tau) = \int_{-\infty}^{\chi} \sigma_{2222}^{di}(x-\chi', t-\tau) \, d\chi' \tag{47}$$

and so forth. Equation (46) has a singularity of $1/r$ (as compared with $1/r^2$ in (43)), which may be of some numerical advantage. In addition, the stress intensity factor can directly be solved in the edge dislocation method, which should be more accurate than that determined from the slope of displacement in the point dislocation method. The edge dislocation formulae have been numerically implemented.[28,29]

Conclusion

We have presented a unified formulation for the direct and indirect integral equations of linear quasi-static poroelasticity. The direct integral equations were derived based on the reciprocity relation. The indirect methods in the forms of displacement and stress discontinuity were normally presented following physical intuition of distributing singularities of actual or fictitious densities. Here a direct link between the direct and the indirect integral equations was established by rigorous mathematics. Throughout the derivation we also presented a number of interesting connections among the singular solutions. Some of the relations emerge as the by-product of the proof of the integral equations.

References

[1] Jaswon, M.A. and Symm, G.T., *Integral Equation Methods in Potential Theory and Elastostatics*, Academic Press, 1977.

[2] Brebbia, C.A. and Butterfield, R., "Formal equivalence of direct and indirect boundary element methods", Appl. Math. Modelling, **2**, 132-134, 1978.

[3] Banerjee, P.K., Butterfield, R. and Tomlin, G.R., "Boundary element methods for two-dimensional problems of transient ground water flow", Int. J. Num. Anal. Meth. Geomech., **5**, 15-31, 1981.

[4] Altiero, N.J. and Gavazza, S.D., "On a unified boundary-integral equation method", J. Elasticity, **10**, 1-9, 1980.

[5] Biot, M.A., "General theory of three-dimensional consolidation", J. Appl. Phys., **12**, 155-164, 1941.

[6] Biot, M.A., "Theory of elasticity and consolidation for a porous anisotropic solid", J. Appl. Phys., **26**, 182-185, 1955.

[7] Rice, J.R. and Cleary, M.P., "Some basic stress-diffusion solutions for fluid saturated elastic porous media with compressible constituents", Rev. Geophys. Space Phys., **14**, 227-241, 1976.

[8] Detournay, E. and Cheng, A.H-D., "Fundamentals of poroelasticity," Chap. 5 in *Comprehensive Rock Engineering: Principles, Practice & Projects, Vol. II, Analysis and Design Method*, ed. C. Fairhurst, Pergamon Press, 113-171, 1993.

[9] Crouch, S.L. and Starfield, A.M., *Boundary Element Methods in Solid Mechanics*, Allen and Unwin, 1983.

[10] Mura, T., *Micromechanics of Defects in Solids,* Martinus Nijhoff, Hague, 1982.

[11] Thompson, M. and Willis, J.R., "A reformulation of the equations of anisotropic poroelasticity", J. Appl. Mech., ASME., **58**, 612-616, 1991.

[12] Nowacki, W., *Thermoelasticity, 2nd ed.*, Pergamon, London, 1986.

[13] Predeleanu, M., "Reciprocal theorem in the consolidation theory of porous media", Anal. Univ. Bucuresti, Seria Stiintele Naturii, Matematica, Mecanica, **17**, 75-79, 1968.

[14] Cleary, M.P., "Fundamental solutions for a fluid-saturated porous solid", Int. J. Solids Struct., **13**, 785-806, 1977.

[15] Cheng, A.H-D. and Predeleanu, M., "Transient boundary element formulation for poroelasticity", Appl. Math. Modelling, **11**, 285-290, 1987.

[16] Cheng, A.H-D. and Liggett, J.A., "Boundary integral equation method for linear porous-elasticity with applications to soil consolidation", Int. J. Numer. Meth. Engng., **20**, 255-278, 1984.

[17] Cheng, A.H-D. and Detournay, E., "A direct boundary element method for plane strain poroelasticity", Int. J. Numer. Anal. Meth. Geomech., **12**, 551-572, 1988.

[18] Badmus, T., Cheng, A.H-D. and Grilli, S., "A Laplace-transform-based three-dimensional BEM for poroelasticity," Int. J. Numer. Meth. Engng., **36**,67-85, 1993.

[19] Cheng, A.H-D. and Liu, P.L-F., "Seepage force on a pipeline buried in a poroelastic seabed under wave loadings," Appl. Ocean Res., **8**, 22-32, 1986.

[20] Cheng, A.H-D. and Liggett, J.A., "Boundary integral equation method for linear porous-elasticity with applications to fracture propagation," Int. J. Numer. Meth. Eng., **20**, 279-296, 1984.

[21] Sharp, S., "A note on indirect boundary element methods for the diffusion equation", Int. J. Numer. Anal. Meth. Geomech., **7**, 269-272, 1983.

[22] Pan, E., "Dislocation in an infinite poroelastic medium", Acta Mechanica., **87**, 105-115, 1991.

[23] Curran, J.H. and Carvalho, J.L., "A displacement discontinuity model for fluid-saturated porous media," Proc. 6th Cong. Int. Soc. Rock Mech., Vol. 1, Montreal, 73-78, 1987.

[24] Carvalho, J.L., "Poroelastic effects and influence of material interfaces on hydraulic fracture behaviour", Ph.D. Dissertation, Univ. Toronto, 1990.

[25] Detournay, E. and Cheng, A.H-D., "Poroelastic solution of a plane strain point displacement discontinuity", J. Appl. Mech., **54**, 783-787, 1987.

[26] Vandamme, L., Detournay, E. and Cheng, A.H-D., "A 2-D poroelastic displacement discontinuity method for hydraulic fracture simulation," Int. J. Numer. Anal. Meth. Geomech., **13**, 215–224, 1989.

[27] Bilby, B.A. and Eshelby, J.D., "Dislocations and the theory of fracture", In: *Fracture*, (ed.) H. Liebowitz, Academic Press, 99-182, 1968.

[28] Cheng, A.H-D., Badmus, T. and Detournay, E., "Singular integral equation method for cracks embedded in poroelastic space," in *Boundary Elements X, Vol. 3 Stress Analysis*, Southampton, UK, ed. C.A. Brebbia, Springer-Verlag, 265-278, 1988.

[29] Detournay, E. and Cheng, A.H-D., "Plane strain analysis of a stationary hydraulic fracture in a poroelastic medium", Int. J. Solids Struct., **37**, 1645-1662, 1991.

MICROSTRUCTURE-BASED FINITE ELEMENT ANALYSIS OF HETEROGENEOUS MEDIA

J. ZHANG and N. KATSUBE
Department of Engineering Mechanics, The Ohio State University, Boyd Lab, 155 W. Woodruff Avenue, Columbus, OH 43210, U.S.A.

Abstract. A microstructure-based finite element analysis is presented in this paper for the mechanical response of two dimensional heterogeneous materials with randomly dispersed inclusions. A special n-sided polygonal element containing an elastic inclusion is developed on the basis of the Hellinger-Reissner principle and the hybrid finite element method. The element formulations are derived by decomposing the original problem into inclusion and matrix problems and relating each other through consistency conditions at their interface. For circular inclusions, the proposed method is verified against simple analytical solutions and shown to be suitable even for extreme cases such as porous materials and materials with rigid inclusions. It is also demonstrated that the number of degrees of freedom necessary to obtain accurate results can be significantly reduced by using the proposed method. The effect of random distributions of inclusions on local stress concentration factors can also be easily evaluated using the proposed method.

Key words: finite element method, microstructure, heterogeneous media, random distribution, polygonal mesh, variational principle, complex variable

1. Introduction

Historically, problems associated with heterogeneous media have been of great interest. ince the pioneering work of Eshelby [1, 2] in the late 1950s, extensive research [3, 4, 5,] has been done to study the elastic field of heterogeneous media induced by the presence f inclusions. The analytical approach based on the classical elasticity theory can provide xplicit solutions only for some ideal cases with particular loading and boundary onditions. In general cases, the equations formulated from the elasticity theory are nalytically intractable.

Gong and Meguid [7] have recently analyzed the problems of interacting circular iclusions by using the complex variable method of Muskhelishvili [8]. In their work, the omplex potentials governing the elastic field are expanded into a set of Laurent series nd general expressions for these potentials are derived. The coefficients of series are etermined by the perturbation technique. In this work, we seek a numerical approach by e finite element method to heterogeneous materials containing randomly dispersed clusions as shown in Figure 1(a).

In traditional finite element analysis, a complex mesh is required to represent one clusion. If all the inclusions are of the same geometry and periodically distributed, the me complex mesh representing the particular inclusion geometry can be repeatedly ed. In many cases, however, the distribution of the inclusions is random. While it is ll possible to use traditional methods with a very large number of elements and also grees of freedom as shown in Figure 1(b), it is certainly not a desirable approach.

P. S. Selvadurai (ed.), Mechanics of Poroelastic Media, 109–124.

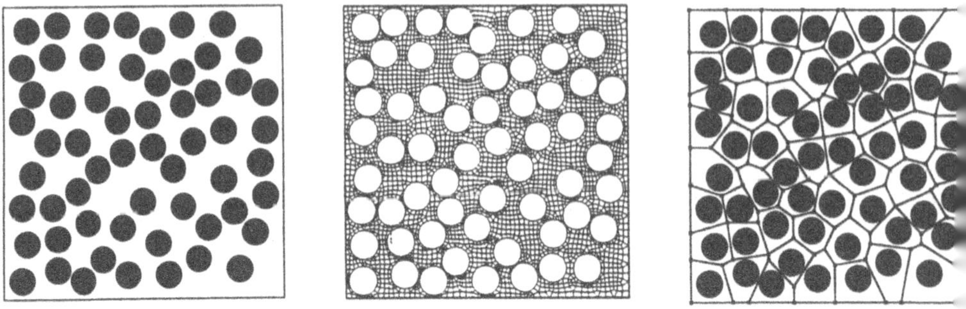

(a) (b) (c)

Figure 1- (a) A heterogeneous medium with randomly dispersed inclusions; (b) a comple finite element mesh generated by Nakamura and Suresh [9]; (c) an n-sided polygona network

Ghosh and Mukhopadhyay [10] recently proposed a finite element model by generatin an n-sided polygonal network for a two-dimensional heterogeneous medium. In thei model, each polygon contains an inclusion and serves as an element as shown in Figur 1(c). The complexity of generating a conventional mesh is thus avoided. The notion c transformation strain was introduced into their formulation as was proposed by Accor [11, 12]. However, it has been pointed out by Zhang and Katsube [13] that th transformation strain does not play any role in their applications. While generation of a n-sided network is indeed a notable contribution, stress irregularities surroundin inclusions are not well represented in their elements.

Even if there is no inclusion inside the n-sided polygonal element with $n > 4$, it difficult to approximate the displacement field by an interpolation polynomial and at th same time satisfy the inter-element displacement compatibility. It is possible t construct such an interpolation function by making use of a rational interpolation bas [14, 15]. However, the process is complicated. For these reasons, the convention displacement-based finite element approach is not suitable for formulating the n-side polygonal element.

Ghosh and Mukhopadhyay [10] used the hybrid method proposed by Pian [16, 17]. the conventional displacement-based approach, a displacement field over the enti element is assumed. In the hybrid method, however, a compatible displacement field assumed only along the element boundaries, and a self-equilibrating stress field assumed inside the element. Because of this, with the hybrid method, we can avoid th difficulties associated with the conventional approach in an n-sided polygonal elemen The hybrid method proposed by Pian was initially based upon the principle of minimu complementary energy. This idea was later extended, on the basis of the Hellinge Reissner principle, to allow another separate assumption of displacement field within th element [18].

Tong, Pian, and Lasry [19] successfully dealt with stress singularity at a crack tip applying the hybrid method to a special crack element where one crack tip is embedde Piltner [20] also employed the hybrid method to special elements with a circular elliptical hole within the element. In both cases, one special element with a crack or a hc replaces a large number of conventional elements. Their success comes from the fact th the numerical assumption of the elastic field in the element has a sound physical basis. their work, classical elasticity solutions were used in the element formulations represent the local stress irregularity surrounding a hole or a crack tip. Conform mapping techniques were employed to enforce the local stress boundary conditions.

Zhang and Katsube [21] recently focused their attention on the problems where randomly dispersed inclusions are rigid and circular. The n-sided polygonal element by Ghosh and Mukhopadhyay is employed in order to deal with randomness. In numerically approximating irregular stress distribution, however, their approach is similar to the works by Tong, Pian, and Lasry [19] and also by Piltner [20]. In the element formulation, classical elasticity solutions are directly used. Because of the simplification that the stress distribution only in the matrix needs to be analyzed, the resulting approach is very effective. Only one polygonal element with a rigid circular inclusion replaces many conventional elements in obtaining the same numerical accuracy.

This approach is extended to the cases where inclusions are also linearly elastic [22]. The difficulties associated with approximating irregular stress distribution both in the matrix and the inclusion within an element are overcome based on the notion of superposition. The element formulations for the matrix region and those for the inclusion region are performed separately and are later combined on the basis of consistency conditions at the interface. As observed in the limited case with rigid inclusions, only one polygonal element with an elastic circular inclusion (a hole or a rigid inclusion as extreme cases) is needed to replace many conventional elements in obtaining the same numerical accuracy. In this paper, the approach is summarized and presented with numerical examples.

2. Element Decomposition and Variational Formulations

Let us consider a special polygonal element embedded with an elastic inclusion of arbitrary geometry as shown in Figure 2(a). Our goal is to establish the relationship between the element's nodal forces and displacements, or simply, to formulate the element stiffness matrix. The key idea in formulating this special element is to decompose the original problem into two different boundary value problems as shown in Figures 2(b) and (c): (i) a specified boundary displacement problem in the matrix region Ω_1 with boundaries C_1 and C_2; (ii) a specified boundary traction problem in the inclusion region Ω_2 with boundary C_2. C_1 and C_2 are the element's outer boundary with neighboring elements and the inner interface between the matrix and inclusion, respectively. Following the Hellinger-Reissner principle, we define the following two separate hybrid functionals for these two problems:

$$\pi_m^e = \int_{\Omega_1} [\frac{1}{2}\sigma_m^T S_m \sigma_m - \sigma_m^T (D u_m)]\, d\Omega_1 - \oint_{C_1} t_m^T (\tilde{u}^{(1)} - u_m)\, dS_1 - \oint_{C_2} t_m^T (\tilde{u}^{(2)} - u_m)\, dS_2 \quad \text{for problem (i),} \qquad (1)$$

and

$$\pi_p^e = \int_{\Omega_2} [\frac{1}{2}\sigma_p^T S_p \sigma_p - \sigma_p^T (D u_p)]\, d\Omega_2 + \oint_{C_2} u_p^T \tilde{t}^{(2)}\, dS_2 \quad \text{for problem (ii),} \qquad (2)$$

where the subscripts 'm' and 'p' indicate the matrix and inclusion regions, respectively; the superscript 'T' denotes the transpose of a vector or a matrix. σ, u, t, and S are, respectively, the stress, displacement, and boundary traction vectors, and the elastic compliance matrix; D is the matrix differential operator relating strains to displacements; and the symbol '~' represents a specified quantity. Based on the hybrid finite element method, the boundary displacements $\tilde{u}^{(1)}$ are assumed separately from u_m and are expressed in terms of the nodal displacements of the element. The boundary displacements $\tilde{u}^{(2)}$ in problem (i) and the boundary traction $\tilde{t}^{(2)}$ in problem (ii) are assumed to be specified and are actually unknown.

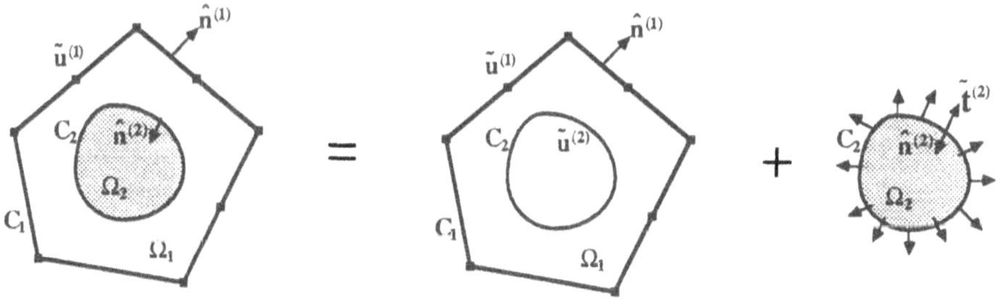

Figure 2- Element decomposition: (a) original element; (b) matrix region with specified boundary displacements; (c) inclusion region with specified boundary traction

The stationary values of the two functionals defined by Equations (1) and (2) yield the following two sets of equations:

problem (i)			and	problem (ii)		
$\mathbf{D}^T\sigma_m = 0$	in Ω_1,	(3)		$\mathbf{D}^T\sigma_p = 0$	in Ω_2,	(7)
$\mathbf{S}_m\sigma_m = \mathbf{D}\mathbf{u}_m$	in Ω_1,	(4)		$\mathbf{S}_p\sigma_p = \mathbf{D}\mathbf{u}_p$	in Ω_2,	(8)
$\mathbf{t}_m = \hat{\mathbf{n}}^{(i)}\sigma_m$	on C_i, i=1,2	(5)		$-\hat{\mathbf{n}}^{(2)}\sigma_p = \tilde{\mathbf{t}}^{(2)}$	on C_2,	(9)
$\mathbf{u}_m = \tilde{\mathbf{u}}^{(i)}$	on C_i, i=1,2	(6)				

where $\hat{\mathbf{n}}^{(i)}$ is a 2 x 3 matrix of the unit normal to boundary C_i (i=1,2). It is seen from the above equations that the true solutions of the two problems minimize the two hybrid functionals. As indicated by Tong, Pian, and Lasry [19], each quantity in Equations (1) and (2) may be individually approximated (or assumed) in principle. But in practice, the stress σ, displacement $\mathbf{u}$, and traction $\mathbf{t}$ fields can be constructed in such a manner that Equations (3), (4), (5), (7) and (8) are automatically satisfied. By doing this and using the divergence theorem over Ω_1 and Ω_2, the two functionals are reduced to

$$\pi_m^e = \frac{1}{2}\left(\oint_{C_1} \mathbf{t}_m^T\mathbf{u}_m \, dS_1 + \oint_{C_2} \mathbf{t}_m^T\mathbf{u}_m \, dS_2\right) - \oint_{C_1} \mathbf{t}_m^T\tilde{\mathbf{u}}^{(1)} \, dS_1 - \oint_{C_2} \mathbf{t}_m^T\tilde{\mathbf{u}}^{(2)} \, dS_2, \tag{10}$$

and

$$\pi_p^e = -\frac{1}{2}\oint_{C_2} \mathbf{t}_p^T\mathbf{u}_p \, dS_2 + \oint_{C_2} \mathbf{u}_p^T\tilde{\mathbf{t}}^{(2)} \, dS_2, \tag{11}$$

where

$$\mathbf{t}_p = -\hat{\mathbf{n}}^{(2)}\sigma_p. \tag{12}$$

In order to recover the solution for the original problem from those of the two decomposed problems, the following consistency conditions need to be imposed:

$$\tilde{\mathbf{u}}^{(2)} = \mathbf{u}_p \qquad \text{on } C_2 \tag{13}$$

and

$$\tilde{\mathbf{t}}^{(2)} = -\mathbf{t}_m \qquad \text{on } C_2. \tag{14}$$

Thus, Equations (10), (11), (12), (13), and (14) consist of the fundamental variational formulations for the special elastic inclusion-embedded polygonal element.

3. Construction of Special Approximating Functions

The performance of an element largely depends on the selection of approximating functions for the element. In the conventional displacement-based or hybrid finite element methods, approximating functions for displacements and stresses are constructed in the form of a complete polynomial of a certain degree. As long as the displacement and stress fields are regular (smooth) in the element and the size of the element is sufficiently small, a good approximation can be achieved by assuming such a polynomial. In the present case, however, the presence of the inclusion severely perturbs an otherwise regular distribution of the stresses and displacements in the matrix. Moreover, the element mesh is not allowed to be refined. Thus, one has to search for special approximating functions such that the actual stress and displacement fields in the element can be properly characterized. In addition, in order to satisfy the requirements for using the two hybrid functionals of Equations (10) and (11) mentioned in the previous section, special approximating functions are needed such that the assumed stresses and displacements are able to satisfy *a priori* both the equilibrium equations (3) and (7) and the constitutive equations (4) and (8).

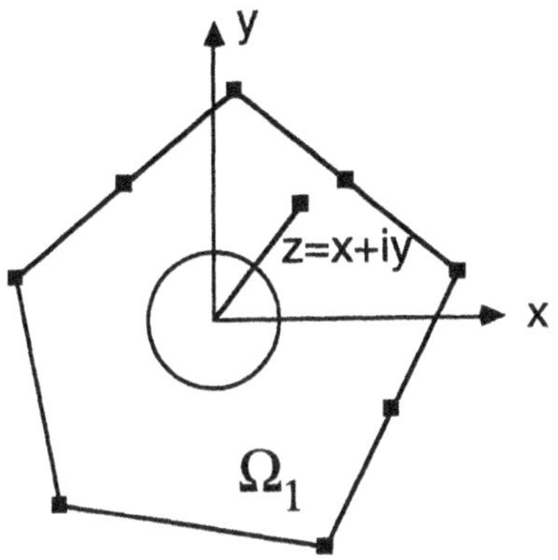

Figure 3- Matrix region of an n-sided polygonal element with a circular inclusion

The analytical solutions based on the elasticity theory will be utilized to seek a better approximation of the actual stress and displacement fields than using polynomials. Let us first consider the matrix region Ω_1 as shown in Figure 3. For a two dimensional multiply-connected region, the components of stress and displacement can be expressed in terms of two analytical functions of a complex variable z (z=x+iy) as follows [8]:

$$\sigma_{11} + i\sigma_{12} = \varphi'(z) + \overline{\varphi'(z)} - \overline{z\varphi''(z)} - \overline{\psi'(z)}, \tag{15}$$

$$\sigma_{22} - i\sigma_{12} = \varphi'(z) + \overline{\varphi'(z)} + \overline{z\varphi''(z)} + \overline{\psi'(z)}, \tag{16}$$

and

$$2\mu(u_1 + iu_2) = \kappa\varphi(z) - z\overline{\varphi'(z)} - \overline{\psi(z)}, \tag{17}$$

where $\varphi(z)$ and $\psi(z)$ are the two analytical functions of z; $(\ )'$ denotes differentiation with respect to the complex variable z and $\overline{(\ )}$ is the complex conjugate; $\kappa = (3-\nu)/(1+\nu)$ for plane stress and $\kappa = 3-4\nu$ for plane strain; and μ and ν are the shear modulus and Poisson's ratio, respectively, of the material in the region.

For the case with only one internal contour, $\varphi(z)$ and $\psi(z)$ are expressed as follows:

$$\varphi(z) = -\frac{1}{2\pi(1+\kappa)}(F_x + i\,F_y)\log(z - z_o) + \varphi_o(z), \tag{18}$$

and

$$\psi(z) = \frac{\kappa}{2\pi(1+\kappa)}(F_x - i\,F_y)\log(z - z_o) + \psi_o(z), \tag{19}$$

where F_x and F_y are the x and y components of the resultant force acting on the internal contour; z_o represents a fixed but arbitrarily chosen point inside the internal contour;

and $\varphi_o(z)$ and $\psi_o(z)$ are two single-valued holomorphic functions. It is assumed that there is no external force directly acting on the inclusions. This leads to

$$F_x = F_y = 0. \tag{20}$$

If the internal contour is circular, the two holomorphic functions $\varphi_o(z)$ and $\psi_o(z)$ can be represented by the Laurent series of z. Thus, Equations (18) and (19) are reduced to

$$\varphi(z) = \varphi_o(z) = \sum_{k=m_b}^{m_u} a_k z^k, \tag{21}$$

and

$$\psi(z) = \psi_o(z) = \sum_{k=m_b}^{m_u} b_k z^k, \tag{22}$$

where m_u and m_b are upper and low limits of the series and m_b is generally set to be $-m_u$ for symmetry; a_k and b_k are complex coefficients with the forms of $a_k = \tilde{a}_k + i\,\hat{a}_k$ and $b_k = \tilde{b}_k + i\,\hat{b}_k$; and $\tilde{a}_k$, $\hat{a}_k$, $\tilde{b}_k$, and $\hat{b}_k$ are real numbers.

By substituting Equations (21) and (22) into Equations (15), (16), and (17) and separately equating the real and imaginary parts of both sides of the resulting equations, we obtain the analytical expressions for the stresses and displacements in terms of these unknown real coefficients as follows:

$$\begin{Bmatrix} \sigma_{11}^m \\ \sigma_{22}^m \\ \sigma_{12}^m \end{Bmatrix} = \sum_{k=m_b}^{m_u} \begin{bmatrix} 2A_1 - A_3 & -2A_2 + A_4 & -A_1 & A_2 \\ 2A_1 + A_3 & -2A_2 - A_4 & A_1 & -A_2 \\ A_4 & A_3 & A_2 & A_1 \end{bmatrix} \begin{Bmatrix} \tilde{a}_k^m \\ \hat{a}_k^m \\ \tilde{b}_k^m \\ \hat{b}_k^m \end{Bmatrix}, \quad \text{or} \quad \boldsymbol{\sigma}_m = \mathbf{P}_m \boldsymbol{\beta}_m \tag{23}$$

and

$$\begin{Bmatrix} u_1^m \\ u_2^m \end{Bmatrix} = \frac{1}{2\mu_m} \sum_{k=m_b}^{m_u} \begin{bmatrix} \kappa B_1 - B_3 & -\kappa B_2 + B_4 & -B_1 & B_2 \\ \kappa B_2 + B_4 & \kappa B_1 + B_3 & B_2 & B_1 \end{bmatrix} \begin{Bmatrix} \tilde{a}_k^m \\ \hat{a}_k^m \\ \tilde{b}_k^m \\ \hat{b}_k^m \end{Bmatrix}, \quad \text{or} \quad \mathbf{u}_m = \mathbf{U}_m \boldsymbol{\beta}_m \tag{24}$$

where

$$A_1(k,r,\theta) = k\,r^{k-1}\cos(k-1)\theta \tag{25}$$
$$A_2(k,r,\theta) = k\,r^{k-1}\sin(k-1)\theta, \tag{26}$$
$$A_3(k,r,\theta) = k(k-1)\,r^{k-1}\cos(k-3)\theta, \tag{27}$$
$$A_4(k,r,\theta) = k(k-1)\,r^{k-1}\sin(k-3)\theta, \tag{28}$$
$$B_1(k,r,\theta) = r^k \cos k\theta, \tag{29}$$
$$B_2(k,r,\theta) = r^k \sin k\theta, \tag{30}$$
$$B_3(k,r,\theta) = k\,r^k\cos(k-2)\theta, \tag{31}$$
$$B_4(k,r,\theta) = k\,r^k\sin(k-2)\theta. \tag{32}$$

In the above expressions, the two variables r and θ are the polar coordinates. The superscripts and subscripts 'm' in Equations (23) and (24) indicate the matrix.

The expressions of Equations (23) and (24) can also be used as approximating functions for the elastic circular inclusion region Ω_2 if the terms involving negative k are excluded. Thus, by setting the value of m_b equal to 1, the following approximating functions similar to those for the matrix region are obtained for the inclusion region:

$$\boldsymbol{\sigma}_p = \mathbf{P}_p \boldsymbol{\beta}_p, \tag{33}$$

and

$$\mathbf{u}_p = \mathbf{U}_p \boldsymbol{\beta}_p, \tag{34}$$

where the subscript 'p' indicates the inclusion.

The five coefficients $\tilde{a}_o$, $\hat{a}_o$, $\tilde{b}_o$, $\hat{b}_o$, and $\hat{a}_1$ in Equations (23) and (24) make no contribution to the stresses because they represent the rigid body movement. Thus, all these five coefficients may be set as zero. In addition, the coefficient $\hat{b}_{-1}$ also has to be zero to satisfy the condition of zero resultant moment exerted on the inclusion. This will be shown as follows.

The magnitude of the resultant moment on the inclusion can be expressed in indicial notation as follows:

$$M_o = -\oint_{C_2} e_{ij3} t_i x_j \, ds = -\oint_{C_2} e_{ij3} \sigma_{ir} (-\hat{n}_r^{(2)}) x_j \, ds, \qquad i,j,r = 1,2. \tag{35}$$

where x_j are the coordinate components with $x_1 = x$ and $x_2 = y$; and e_{ij3} are the components of the permutation symbol. With the substitution of Equation (23) and Equations (25) through (28) into Equation (35), it can be shown that the expression for the resultant moment is reduced to

$$M_o = 2\pi \hat{b}_{-1}. \tag{36}$$

Therefore, $\hat{b}_{-1}$ has to be zero to guarantee the zero resultant moment on the inclusion. All of the six zero coefficients mentioned above are removed from the vector β to reduce the dimensions of the matrices **P** and **U**.

4. Element Stiffness Matrix

In this section, the stiffness matrix of the special polygonal element is derived by incorporating the results in the previous two sections. With Equations (5), (12), (23), and (33), the boundary traction on C_1 and C_2 can be expressed as follows:

$$\mathbf{t}_m = (\hat{\mathbf{n}}^{(i)} \mathbf{P}_m) \beta_m = \mathbf{R}_m^{(i)} \beta_m \quad \text{on } C_i, \; i = 1,2 \tag{37}$$

and

$$\mathbf{t}_p = (-\hat{\mathbf{n}}^{(2)} \mathbf{P}_p) \beta_p = \mathbf{R}_p^{(2)} \beta_p \quad \text{on } C_2. \tag{38}$$

By substituting Equations (24), (34), (37), and (38) into Equations (10) and (11), the two hybrid functionals become

$$\pi_m^e = \frac{1}{2} \beta_m^T (\mathbf{H}_m^{(1)} + \mathbf{H}_m^{(2)}) \beta_m - \beta_m^T \mathbf{G}_m \mathbf{q} - \beta_m^T \mathbf{D}_m, \tag{39}$$

and

$$\pi_p^e = -\frac{1}{2} \beta_p^T \mathbf{H}_p \beta_p + \beta_p^T \mathbf{D}_p, \tag{40}$$

where

$$\mathbf{H}_m^{(1)} = \frac{1}{2} \oint_{C_1} [(\mathbf{R}_m^{(1)})^T \mathbf{U}_m + \mathbf{U}_m^T \mathbf{R}_m^{(1)}] \, dS_1, \tag{41}$$

$$\mathbf{H}_m^{(2)} = \frac{1}{2} \oint_{C_2} [(\mathbf{R}_m^{(2)})^T \mathbf{U}_m + \mathbf{U}_m^T \mathbf{R}_m^{(2)}] \, dS_2, \tag{42}$$

$$\mathbf{H}_p = \frac{1}{2} \oint_{C_2} [(\mathbf{R}_p^{(2)})^T \mathbf{U}_p + \mathbf{U}_p^T \mathbf{R}_p^{(2)}] \, dS_2, \tag{43}$$

$$\mathbf{G}_m = \oint_{C_1} (\mathbf{R}_m^{(1)})^T \mathbf{L} \, dS_1, \tag{44}$$

$$\mathbf{D}_m = \oint_{C_2} (\mathbf{R}_m^{(2)})^T \tilde{\mathbf{u}}^{(2)} dS_2, \tag{45}$$

$$\mathbf{D}_p = \oint_{C_2} \mathbf{U}_p^T \tilde{\mathbf{t}}^{(2)} dS_2. \tag{46}$$

L is the matrix relating the boundary displacements $\tilde{\mathbf{u}}^{(1)}$ to the nodal displacements **q**. According to the hybrid finite element method, **L** is assumed separately from the approximating functions inside the element in order to satisfy the inter-element displacement compatibility. Therefore, the matrix **L** is determined by the number of nodal points on each side of the element along with the mesh pattern of how neighboring elements connect to the sides of the element. The interpolation may be linear if there are only two end nodes on the side line, quadratic or doubly linear if there exists one additional mid-node, and 4th order or doubly quadratic if there are 5 nodes on the side.

For the specified boundary displacement problem (i), the stationary value of π_m^e of Equation (39) with respect to β_m gives

$$(\mathbf{H}_m^{(1)} + \mathbf{H}_m^{(2)})\beta_m - \mathbf{G}_m\mathbf{q} - \mathbf{D}_m = 0. \tag{47}$$

Similarly, for the specified boundary traction problem (ii), we have

$$-\mathbf{H}_p\beta_p + \mathbf{D}_p = 0. \tag{48}$$

Using the consistency conditions of Equations (13) and (14), we may solve Equation (47) and (48) simultaneously for the unknown coefficients β_m and β_p in terms of the nodal displacements $\mathbf{q}$ as follows:

$$\beta_m = (\mathbf{H}_m^{(1)} + \mathbf{H}_m^{(2)} + \mathbf{H}_{mp})^{-1}\mathbf{G}_m\mathbf{q}, \tag{49}$$

and

$$\beta_p = -\mathbf{H}_p^{-1}\mathbf{G}_{mp}^T\beta_m, \tag{50}$$

where

$$\mathbf{H}_{mp} = \mathbf{G}_{mp}\mathbf{H}_p^{-1}\mathbf{G}_{mp}^T, \tag{51}$$

and

$$\mathbf{G}_{mp} = \oint_{C_2} (\mathbf{R}_m^{(2)})^T \mathbf{U}_p \, dS_2. \tag{52}$$

As we know, within the linearly elastic range, an element stiffness matrix can b obtained by expressing the strain energy of the element in terms of the element's nod displacements. For the special polygonal element, the strain energy V_{st} is expressed as

$$V_{st} = \frac{1}{2}\beta_m^T(\mathbf{H}_m^{(1)} + \mathbf{H}_m^{(2)})\beta_m + \frac{1}{2}\beta_p^T\mathbf{H}_p\beta_p. \tag{53}$$

The substitution of Equations (49) and (50) into Equation (53) leads to

$$V_{st} = \frac{1}{2}\beta_m^T(\mathbf{H}_m^{(1)} + \mathbf{H}_m^{(2)} + \mathbf{H}_{mp})\beta_m = \frac{1}{2}\mathbf{q}^T\mathbf{G}_m^T(\mathbf{H}_m^{(1)} + \mathbf{H}_m^{(2)} + \mathbf{H}_{mp})^{-1}\mathbf{G}_m\mathbf{q}. \tag{54}$$

Then, the stiffness matrix of the special element is

$$\mathbf{K}_e = \mathbf{G}_m^T(\mathbf{H}_m^{(1)} + \mathbf{H}_m^{(2)} + \mathbf{H}_{mp})^{-1}\mathbf{G}_m. \tag{55}$$

It is seen from the above equations that the element stiffness matrix $\mathbf{K}_e$ is symmet and can be evaluated by simple matrix operations and numerical integration only alo two boundaries. Because of the elimination of area integration, the proposed method h great advantage in dealing with high order approximating functions. The stresses a displacements in the element can be calculated from Equations (23), (24), (35), and (3 once the unknown coefficients β_m and β_p are determined.

5. Numerical Examples

In this section, four numerical examples are presented to demonstrate the validity a versatility of the proposed method for the analysis of heterogeneous media with randon dispersed inclusions. In the first two examples, the accuracy of the proposed hyb method is verified against simple analytical solutions and the results computed by conventional finite element method. The computer program ABAQUS [25] is used for conventional method. In the third example, the hybrid method is applied to examine effects of random packing of fibers on stress concentration factors for a unidirectio fibrous composite. Finally, a cantilever beam problem with randomly dispersed inclusi of various sizes is analyzed using the proposed method.

5.1 VERIFICATION AGAINST SIMPLE ANALYTICAL SOLUTIONS

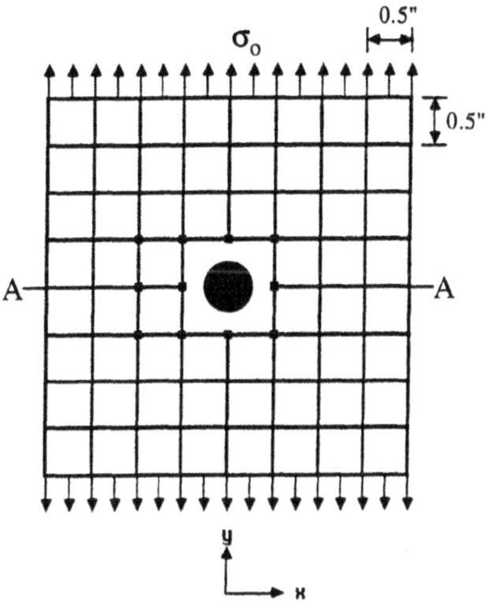

Figure 4- A large plate with a small circular inclusion subjected to uniform tension

As shown in Figure 4, an elastic circular inclusion of radius 0.25" is embedded at the center of a 4" x 4" square plate (plane stress). The plate is subjected to a remote uniform tension σ_o in the vertical direction. The material properties of the plate and the inclusion, respectively, are: Young's moduli $E_m = 2.5 \times 10^5$ psi and $E_p = k^*E_m$; and shear moduli $G_m = 0.97 \times 10^5$ psi and $G_p = k^*G_m$. k is the parameter representing the ratio of the elastic moduli of the plate and the inclusion materials. In the finite element model, the plate is partitioned into 60 conventional 4-node and one special 8-node (4-sided) elements. Doubly linear interpolation is used for the boundary displacement approximation on each side of the special element to satisfy the inter-element displacement compatibility. The 4-node elements can either be conventional displacement-based quadrilateral elements or homogeneous hybrid ones.

In evaluating the hybrid element's stiffness matrix, the values of the upper and low limits m_u and m_b of the Laurent series in Equations (23) and (24) are determined such that the resulting matrix is rank-sufficient. The guideline of this stability condition for the element's stiffness matrix is as follows [17, 23, 24]:

$$n_\beta \geq n_q - 3 \tag{56}$$

where n_β is the number of the unknown coefficients in vector β, and n_q is the total number of degrees of freedom of the element. The above criterion applies to homogeneous elements. Our special lement, however, ontains an elastic nclusion. In pproximating the tress field urrounding the nclusion, the special lement requires erms r^k where the nteger k is both egative and positive, hile a homogeneous lement requires rms r^k with positive only. Due to the mmetry of the quired integers in e special inclusion-nbedded element,

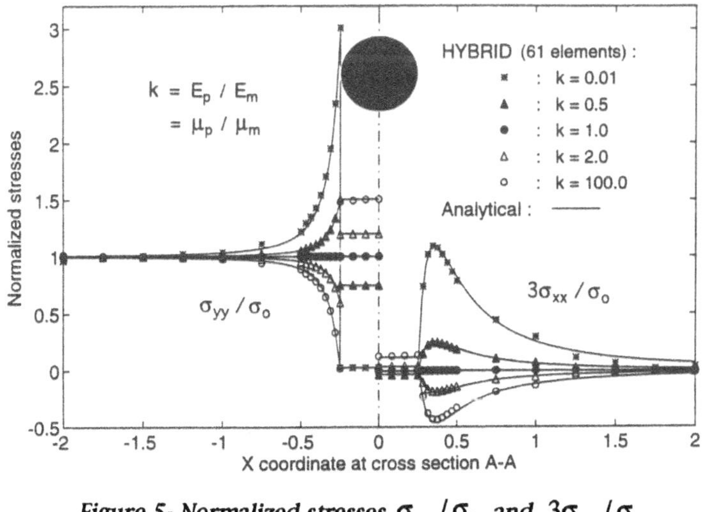

Figure 5- Normalized stresses σ_{yy}/σ_o and $3\sigma_{xx}/\sigma_o$ at cross-section A-A for various cases

the actual value of n_β is chosen to be twice the one needed for the homogeneous case. In this example, the values of m_u and m_b are set to be 5 and -5 for the 8-node inclusion-embedded element.

Figure 5 shows the normalized stresses σ_{yy}/σ_o and $3\sigma_{xx}/\sigma_o$ at cross section A-A for five cases with different values of k. The analytical results are shown in a solid line for each case. A small value of k represents the case where the inclusion approaches a hole, while a large value of k represents the case where the inclusion approaches a rigid one. In all five cases, excellent agreement is observed for both σ_{yy} and σ_{xx} between the numerical results obtained by the proposed method and the analytical predictions. This implies that the specially formulated element can be used for any inhomogeneous material including porous materials and materials with rigid inclusions.

5.2 VERIFICATION AGAINST THE CONVENTIONAL METHOD

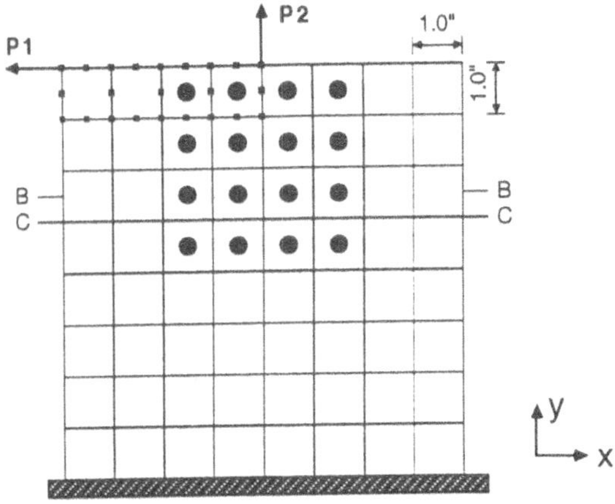

Figure 6- A plate with 16 rigid inclusions and the mesh used by the hybrid method

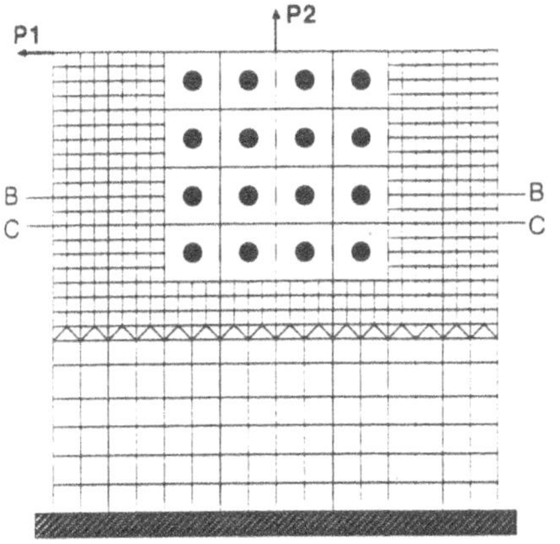

Figure 7- Complex mesh used by ABAQUS

As shown in Figure 6, an 8" x 8" square plate is subjected to a 100 lb horizontal load (P1) and an 800 lb vertical load (P2) at the top end. Sixteen small circular rigid inclusions with radius 0.15" are periodically arranged in the upper half of the plate. The material properties of the plate are the same as those used in the previous example. Again, a plane stress deformation is assumed in the analysis. For the proposed hybrid method, the plate is uniformly discretized into 64 8-node (4-sided) hybrid elements. Quadratic interpolation is used for the displacement field along the inter-element boundaries. The values of the two limits m_u and m_b are set to be 5 and 1 for the homogeneous elements and 5 and -5 for the special inclusion-embedded ones. The complex mesh used by ABAQUS consists of 2928 conventional displacement-based elements as shown in Figure 7. The region around each rigid inclusion is discretized into 152 elements as shown in Figure 8.

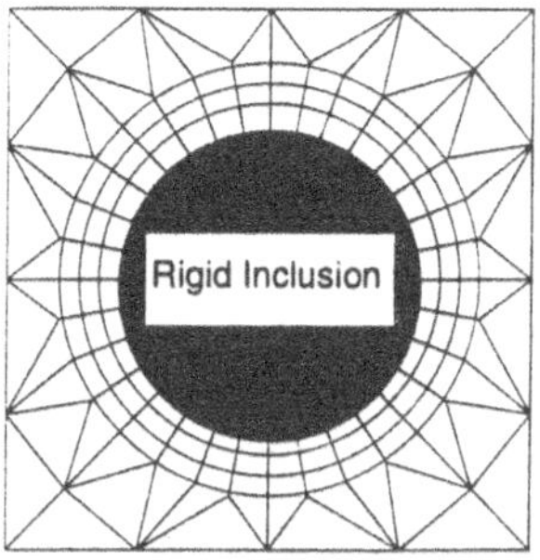

Figure 8- Complex mes discretizing the regio around each rigid inclusion

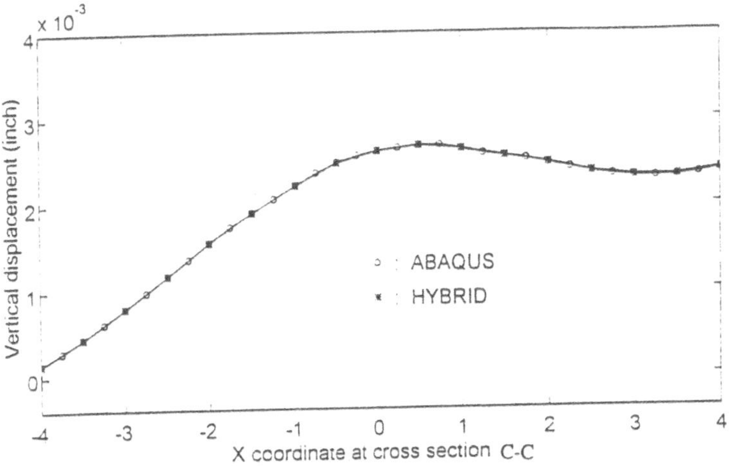

Figure 9- Vertical displacement δ_y *at cross section C-C*

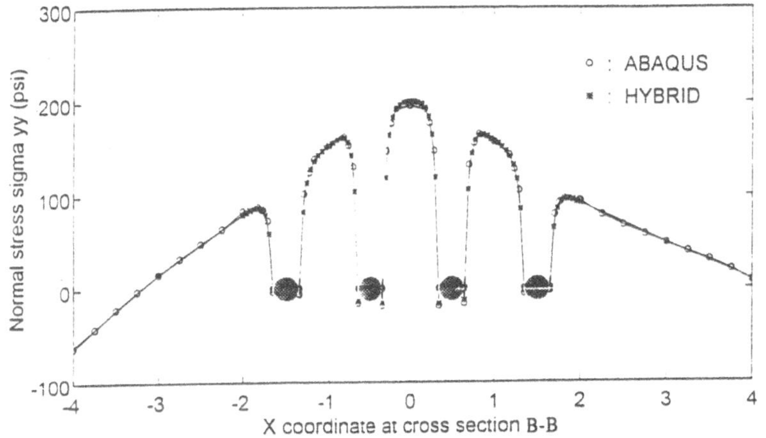

Figure 10- Normal stress σ_{yy} *at cross section B-B*

Figures 9 to 11 show the vertical displacements at cross sections C-C and the normal tresses at cross sections B-B and C-C, respectively. As clearly seen from these figures, he results obtained from the two methods agree very well. Using the coarse mesh of 64 ybrid elements, one can obtain as good results as using the complex mesh of 2928 onventional displacement-based elements. The number of degrees of freedom in this xample is reduced from 5526 in the conventional method to 450 in the proposed hybrid ethod. This clearly demonstrate the efficiency of the proposed method.

3 EFFECT OF RANDOM PACKING ON STRESS CONCENTRATION FACTORS

unit cell model is often used in the micromechanical analysis of fibrous composites. is concept is based on the assumption that the fibers are perfectly spaced and riodically arranged in the matrix such as square in Figure 12(a) and hexagon in Figure (b). The unit cell methods work reasonably well for predicting some properties such as stic moduli in the axial direction. However, many other physical quantities including stic moduli in transverse directions are greatly influenced by the distribution of

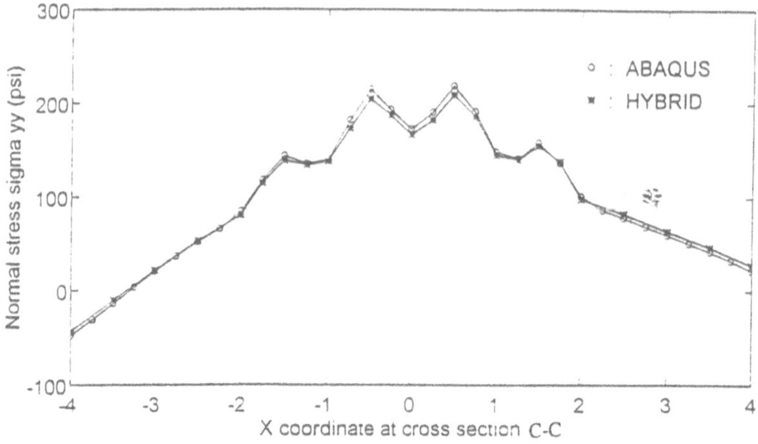

Figure 11- Normal stress σ_{yy} *at cross section C-C*

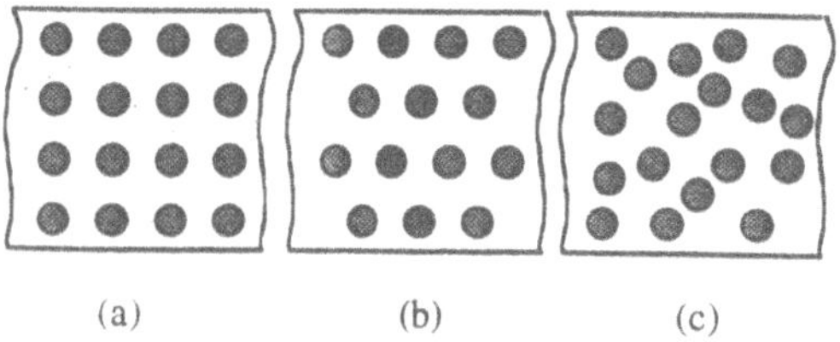

Figure 12- Three different packing patterns: (a) square; (b) hexagonal; (c) random

fibers. Failure mechanisms such a microcrack initiation and propagatio also depend on the microstructure Since the distribution of fibers in a rea composite is often random as shown i Figure 12(c), it is important to examin the effect of random packing on th local stress concentration factors.

In this example, a unidirectiona Ti/SCS composite specimen [26] i analyzed using the proposed finit element method. A portion of the cross section (0.15" x 0.04") is shown in Figure 13 base on a photomicrograph of the specimen. The specimen is subjected to uniformly transvers tension σ_o, and a plane strain deformation model is used in the analysis. The materia properties of the composite are: E_m= 12.3 x 10^6 psi; ν_m= 0.32; E_p= 62.0 x 10^6 ps and ν_p= 0.30. Based on the microstructure of the cross section, an n-sided polygonal mes is generated. It consists of 24 4-sided and 41 5-sided elements. There are 3 nodal poin on each side of the elements and quadratic interpolation is used for the displacemer approximation on each side. The values of m_u and m_b are set to be 5 and -5 for the 4-side elements and 6 and -6 for the 5-sided ones.

Figure 14 shows the normalized radial, circumferential, maximum principal, an maximum shear stresses around inclusion #A. These numerical results are useful i determining the initiation of matrix cracking and matrix-fiber debonding. In Figure 15,

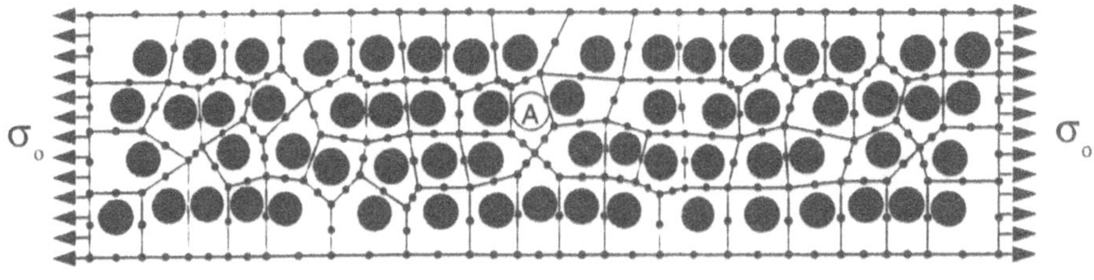

Figure 13- Cross section of Ti/SCS composite specimen with its n-sided polygonal mesh

comparison is made among the maximum principal stresses for three different fiber packing patterns: (a) random distribution (inclusion #A), (b) square distribution, and (c) hexagonal distribution. The peak value of the principal stresses obtained from the random (real) distribution is about 42% higher than those based on periodic distribution assumptions.

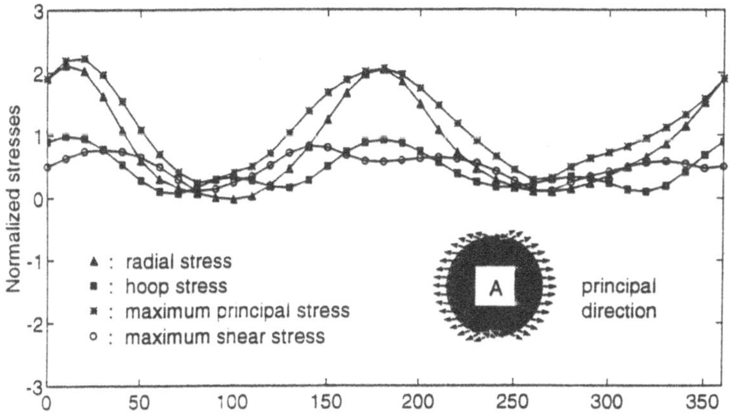

Figure 14- Normalized stresses around inclusion #A

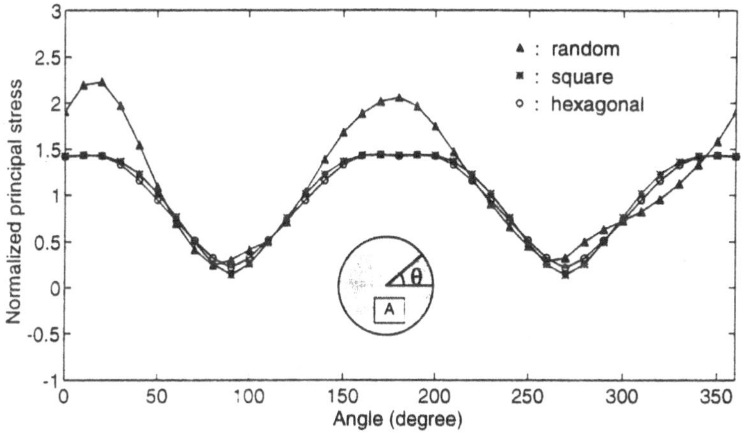

Figure 15- Comparison of maximum principal stresses for three different distributions

4 A CANTILEVER BEAM WITH RANDOMLY DISPERSED RIGID INCLUSIONS

this example, a cantilever beam (8" x 4") with 79 rigid inclusions of various sizes is bjected to a concentrated load 500 lbs at the free end as shown in Figure 16. The clusions are randomly scattered in the beam and their radii range from 0.07" to 0.12". e generated n-sided polygonal network consists of 64 4-sided (8-node) hybrid elements d 15 5-sided (10-node) ones. Doubly linear displacement interpolation is used along all element boundaries. Again, the values of m_u and m_b are set to be 5 and -5 for the 4-ied elements and 6 and -6 for the 5-sided ones. Figure 16 illustrates in dotted line the formed configuration of the polygonal mesh along with its initial position. The

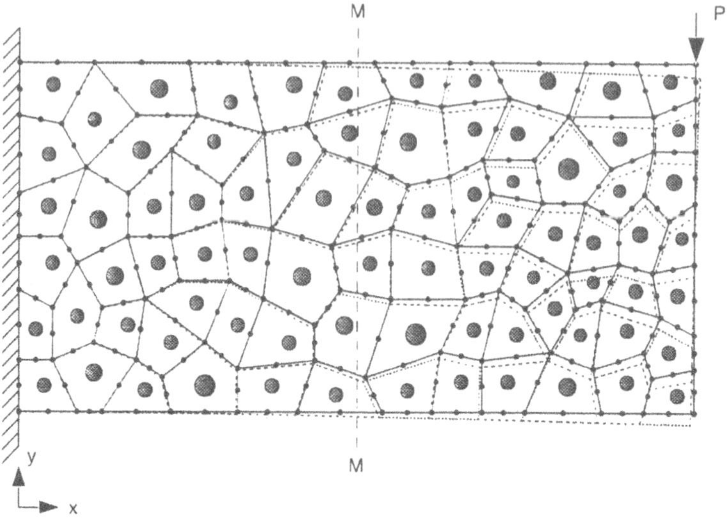

Figure 16- Polygonal mesh of a heterogeneous cantilever beam and its deformed configuration (in dotted line)

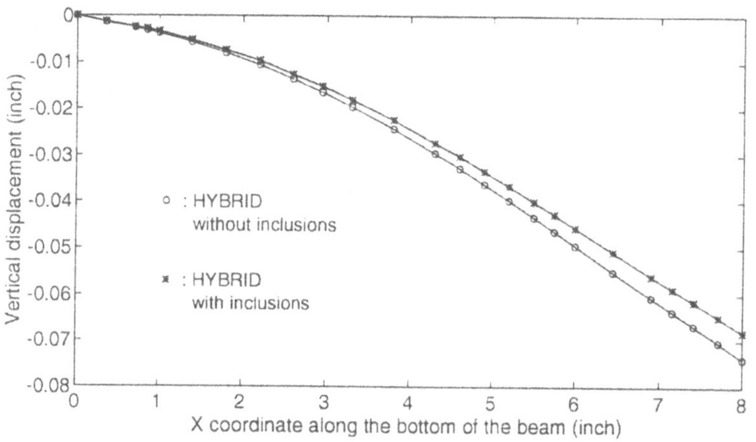

Figure 17- Vertical displacement δ_y along the bottom of the beam

deformation of the mesh shown in Figure 16 is exaggerated proportionally at each no point.

The vertical displacements δ_y along the bottom of the beam are plotted in Figure It is seen from Figure 17 that there is about an 8 percent increase in the global stiffn of the cantilever beam with the addition of the rigid inclusions compared to homogeneous one. The normal stress σ_{xx} along the middle cross section M-M of the be is plotted in Figure 18. The results without inclusions based on the proposed meth agree very well with the analytical solution without inclusions. The stress distribution the beam with inclusions is shown to be extremely irregular.

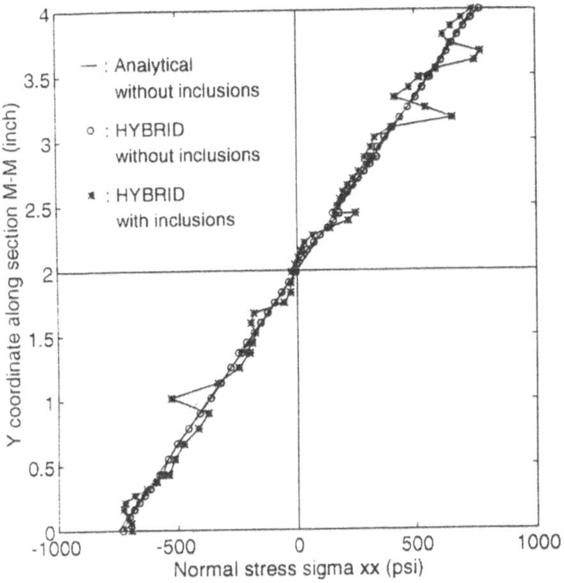

Figure 18- Normal stress σ_{xx} along section M-M

6. Conclusions

In this work, a new microstructure-based hybrid finite element method is presented for the mechanical analysis of heterogeneous media with elastic matrix and randomly dispersed inclusions. An n-sided polygonal mesh is used to model the heterogeneity, and a special polygonal element containing an elastic inclusion is developed. The element formulations are derived by separating the inclusion from the matrix and using two different hybrid functionals for the matrix and the inclusion respectively. Consistency conditions are imposed at the interface of the matrix and inclusion to recover the solution for the original element. For circular inclusions, special approximating functions are constructed based on the classical elasticity solutions for a doubly-connected circular region. The resulting element stiffness matrix only involves integration over the element's outer and inner boundaries. This makes numerical evaluation of the element stiffness matrix relatively easy.

The proposed method is verified against analytical solutions of problems with one circular inclusion in a large plate. By using one special element with an inclusion, the analytical stress distribution surrounding the inclusion is reproduced based on the proposed method. Excellent agreement is obtained even in extreme cases where the inclusion becomes rigid or a hole. Similarly, the proposed method is verified against the conventional finite element solutions of a problem with inclusion interaction. Versatility of the proposed method is demonstrated by examining the effect of random packing on the local stress concentration factors and also by solving a heterogeneous cantilever beam problem.

Acknowledgment: This work was supported by the National Science Foundation under the grant No. MSS-9009347. The numerical computations by ABAQUS were performed on CRAY Y-MP8/864 at Ohio Supercomputer Center under the grant No. PAS805-1.

References

1. J. D. Eshelby, 'The determination of the elastic field of an ellipsoidal inclusion, and related problems', *Proc. Roy. Soc.* A, **241**, 376-396 (1957).
2. J. D. Eshelby, 'The elastic field outside an ellipsoidal inclusion', *Proc. Roy. Soc. A*, **252**, 561-569 (1959).
3. M. A. Jaswon and R. D. Bhargava, 'Two dimensional elastic inclusion problems', *Proc. Camb. Phil. Soc. Math. Phys. Sci.*, **57**, 669-680 (1961).
4. R. D. List and J. P. O. Silberstein, 'Two dimensional elastic inclusion problems', *Proc. Camb. Phil. Soc. Math. Phys. Sci.*, **62**, 303-311 (1966).
5. P. S. Theocaris and N. I. Ioakimidis, 'The inclusion problem in plane elasticity', *Q. J. Mech. Appl. Math.*, **30**, 437-448 (1977).
6. T. Honein and G. Hermann, "On the bonded inclusions with circular or straight boundaries in plane elastostatics', *J. Appl. Mech.*, **57**, 850-856 (1990).
7. S. X. Gong and S. A. Meguid, 'Interacting circular inhomogeneities in plane elastostatics', *Acta Mechanica*, **99**, 49-60 (1993).
8. N. I. Muskhelishvili, 'Some basic Problems of the mathematical theory of elasticity', P. Noordhoff Ltd., Groningen, Holland, pp. 120-129 (1953).
9. T. Nakamura and S. Suresh, 'Effects of thermal residual stresses and fiber packing on deformation of metal-matrix composites', *Acta Metall. Mater.*, **41**, No. **6**, pp. 1665-1681 [1993].
10. S. Ghosh and S. N. Mukhopadhyay, 'A material based finite element analysis of heterogeneous media involving Dirichlet tessellations', *Computer Methods in Applied Mechanics and Engineering*, **104**, pp. 211-247 (1993).
11. M. L. Accorsi, 'A method for modeling microstructural material discontinuities in the finite element analysis', *Int. J. Numer. Methods Eng.*, **26**, pp. 2187-2197 (1988).
12. M. L. Accorsi and R. Chamarajanagar, 'Numerical validation of a hybrid finite element method using eigenstrain', *Computers & Structures*, **41**, pp. 1065-1071 (1990).
13. J. Zhang and N. Katsube, 'Problems related to application of transformation strains in a finite element analysis', *Int. J. Numer. Methods Eng.*, **37**, pp. 3185-3193 (1994).
14. E. Wachspress, 'A rational finite element basis', Academic Press, New York, 1975.
15. G. F. Carey and J. T. Oden, 'Finite elements: a second course', II, Prentice-Hall, Inc., New Jersey, 1983.
16. T. H. H. Pian, 'Derivation of element stiffness matrices by assumed stress distributions', *AIAA J.*, **2**, 1333-1336 (1964).
17. P. Tong and T. H. H. Pian, 'A variational principle and the convergence of a finite-element method based on assumed stress distribution', *Int. J. Solids Structures*, **5**, 463-472 (1969).
18. T. H. H. Pian and D. P. Chen, 'Alternative ways for formulation of hybrid stress elements', *Int. J Numer. Methods Eng.*, **18**, 1679-1684 (1982).
19. P. Tong, T. H. H. Pian and S. J. Lasry, 'A hybrid-element approach to crack problems in plane elasticity', *Int. J. Numer. Methods Eng.*, **7**, pp. 297-308 (1973).
20. R. Piltner, 'Special finite elements with holes and internal cracks', *Int. J. Numer. Methods Eng.* **21**, pp. 1471-1485 (1985).
21. J. Zhang and N. Katsube, 'A hybrid finite element method for heterogeneous materials with randomly dispersed rigid inclusions', to appear in *Int. J. Numer. Methods Eng.*
22. J. Zhang and N. Katsube, 'A hybrid finite element method for heterogeneous materials with randomly dispersed elastic inclusions', to appear in *Int. J. Finite Element Method in Design and Analysis.*
23. T. H. H. Pian and D. P. Chen, 'On the suppression of zero energy deformation modes', *Int. J. Numer. Methods Eng.*, **19**, 1741-1752 (1983).
24. T. H. H. Pian and C. C. Wu, 'A rational approach for choosing stress terms for hybrid finite element formulations', *Int. J. Numer. Methods Eng.*, **26**, 2331-2343 (1988).
25. Hibbitt, Karlsson & Sorensen, Inc., 'ABAQUS user's manual', Version **4.8**, 1989.
26. B. M. Rabeeh and W. O. Soboyejo, unpublished research work.

BOUNDARY ELEMENT APPROACH TO COUPLED POROELASTODYNAMIC PROBLEMS

J. DOMÍNGUEZ AND R. GALLEGO
Escuela Superior de Ingenieros, Universidad de Sevilla
Av. Reina Mercedes s/n, 41012-Sevilla, Spain

Abstract: In this paper a Boundary Element (BE) model for the dynamic analysis of multiregion domains including viscoelastic, porous-viscoelastic and water regions is presented. The integral equation formulation based on Biot's theory is used to represent the porous regions. The propagation of acoustic waves due to a point source in a homogeneous water layer overlying a homogeneous porous viscoelastic bottom half-space is studied using the proposed BE model. The far-field results show a good agreement between the BE solution and that obtained by combination of a few normal modes taken from Yamamoto (1983). The number of modes required decreases with the distance to the source. The effects of the shear modulus of the porous sediment are also studied.

1 Introduction

The seabed and the bottom of water reservoirs and lakes are covered by sediments which are a mixture of solid grains and pore water and gases. There are important applications in ocean acoustics and earthquake engineering which require the study of the dynamic behavior of those sediments and in general of coupled systems where dynamic interaction effects between the sediments, the water layer, and the underlying solid region take place.

The theory of wave propagation in saturated poroelastic sediments was developed by Biot (1941, 1956a, 1956b, 1962). This theory was extended by Stoll (1977) to include the viscoelastic behavior of the solid skeleton using complex elastic moduli. Studies of acoustic wave propagation in water domains with poroelastic bottom have been carried out by several authors. Stoll and Kan (1981) and Huang (1992) studied the general problem of reflection and transmition of plane acoustic waves at water-sediment interfaces. A two-dimensional model closer to the actual ocean geometry consisting of a homogeneous water layer overlying a homogeneous poroelastic bottom half-space was analyzed by Yamamoto (1983). He obtained the normal modes of wave propagation in closed form.

Models for dynamic soil-structure or soil-water-structure interaction problems including poroelastic sediments have been presented by Halpern and Christiano (1986 a, b), Cheng (1986), Bougacha and Tassoulas (1991 a, b) and Bougacha et al. (1993 a, b).

A. P. S. Selvadurai (ed.), Mechanics of Poroelastic Media, 125–142.

Mei and Foda (1981) studied waves in a poroelastic domain with a free surface and showed the existence of a Stokes' type boundary layer.

In recent years boundary integral formulations for dynamic poroelasticity have been presented by several investigators, e.g. Predeleanu (1984), Manolis and Beskos (1989), Domínguez (1991), Cheng et al. (1991) and Wiebe and Antes (1991). Based on those integral equation formulations Cheng et al. (1991) and Domínguez (1992) developed Boundary Element Method (BEM) implementations for the analysis of time harmonic poroelastic problems. Details of the BEM approach can be found in Domínguez (1993).

In the present paper a BE model for the analysis of multi region domains including solid viscoelastic zones, porous viscoelastic zones and water zones is briefly presented for the first time. The dynamic interaction along the interfaces between the different media is taken into account rigorously. The model is used for the analysis of a realistic problem. A homogeneous water layer overlying a homogeneous poroelastic half-space is considered. An acoustic point source is located at mid-depth in the water layer. The pressure distribution along the water layer is analyzed. The results are compared with a synthesis of the plane wave normal modes obtained by Yamamoto (1983) for the same geometry and properties. The effects of the rigidity of the sediments on the acoustic propagation are studied.

2 Basic theory

Following Biot (1956a), the constitutive equations for linear dynamic poroelasticit are expressed as

$$\tau_{ij} = (\lambda + \frac{Q^2}{R})\, \delta_{ij}\, e + 2\mu e_{ij} + Q\delta_{ij}\, \varepsilon \tag{1}$$

$$\tau = Q\, e + R\, \varepsilon \tag{2}$$

where τ_{ij} are the stresses in the solid, $\tau = -\beta$ p the stress in the fluid, β the porosity, p th pore pressure, δ_{ij} is the Kronecker delta function, $e_{ij} = 0.5\,(u_{i,j} + u_{j,i})$ are the soli strains, $\varepsilon = U_{i,i}$ and $e = u_{i,i}$ are the fluid and solid dilation, respectively, λ, μ, Q ar R are material constants, and u_i and U_i are the solid and fluid displacement, respectivel

Assuming that the excitation and the solution are harmonic in time with a tin dependence $\exp(i\omega t)$, the equilibrium equations in terms of displacements can be written

$$\mu \, \Delta \, u_i + (\lambda + \mu + \frac{Q^2}{R}) \, e_{,\,i} + Q \, \varepsilon_{,\,i} + X_i \tag{3a}$$

$$= -\omega^2 \, (\rho_{11} \, u_i + \rho_{12} \, U_i) + i\omega b \, (u_i - U_i)$$

$$(Qe + R\varepsilon)_{,\,i} + X'_i = -\omega^2 \, (\rho_{12} \, u_i + \rho_{22} \, U_i) - i\omega b \, (u_i - U_i) \tag{3b}$$

here b is the dissipation constant ρ_{11}, ρ_{12}, ρ_{22} are the mass densities and, X_i an X'_i are body forces acting in the solid and in the fluid respectively.

The six displacements in equations (3) are not independent and can be expressed in terms of only four variables, namely the solid displacements u_i and the fluid stress τ.

Using equations (2) and (3b) the displacements in the fluid are written as

$$U_i = \frac{\tau_{,\,i} + X'_i + (i\omega b + \omega^2 \, \rho_{12}) \, u_i}{i\omega b - \omega^2 \, \rho_{22}} \tag{4}$$

By substituting equations (3b) and (4) in equation (3a) the following is obtained:

$$\mu \Delta u_i + (\lambda + \mu) \, e_{,\,i} + \tau_{,\,i} \, (\frac{Q}{R} + \frac{i\omega b + \rho_{12} \, \omega^2}{i\omega b - \omega^2 \, \rho_{22}})$$

$$+ u_i \, \omega^2 \, (\frac{\omega^2 \, (-\rho_{11} \, \rho_{22} + \rho_{12}^2) + i\omega b \, (\rho_{11} + \rho_{22} + 2\rho_{12})}{i\omega b - \omega^2 \, \rho_{22}}) \tag{5}$$

$$+ X_i + \frac{i\omega b + \rho_{12} \, \omega^2}{i\omega b - \omega^2 \, \rho_{22}} \, X'_i = 0$$

Equation (5) depends only on the three (or two) components of the solid isplacement and on the fluid stress. The fourth equation, which completes the system, is otained by taking the divergences of equation (3b) and substituting equation (2).

$$\Delta\tau + \frac{\tau}{R} \, (-i\omega b + \omega^2 \, \rho_{22})$$

$$+ e \, [i\omega b \, (1 + \frac{Q}{R}) + \omega^2 \, (\rho_{12} - \rho_{22} \, \frac{Q}{R})] + X'_{i,\,i} = 0 \tag{6}$$

3 Boundary integral formulation

The boundary integral formulation emerges from a reciprocal relation established in terms of solid displacements, boundary tractions on the solid, fluid stresses, fluid displacements normal to the boundary and body forces in both phases. The fundamental solution corresponds to unit point loads in the solid phase and also to a field of body forces in the solid and in the fluid, chosen in such a way that it gives rise to a simple integral representation of the fluid stress. The required fundamental solution can be obtained with the help of the analogy between poroelasticity and thermoelasticity, well known for the quasi-static case, that can be extended to dynamics, as shown by Bonnet (1987).

The reciprocal relation can be obtained as usual starting from the equilibrium equations for time-harmonic behavior, weighting them with displacement functions and integrating by parts

$$\int_\Gamma (t_i\, u_i^* + \tau\, U_n^*)\, d\Gamma + \int_\Omega (X_i\, u_i^* + X'_i\, U_i^*)\, d\Omega = \int_\Gamma (t_i^*\, u_i + \tau^*\, U_n)\, d\Gamma + \int_\Omega (X_i^*\, u_i + X'^*_i\, U_i)\, d\Omega \qquad (7)$$

where Γ is the boundary of the body Ω, $t_i = \tau_{ij}\, n_j$ and $U_n = U_i\, n_i$, , n being the unit normal to the boundary. The variables denoted by superscript * are those associated with the weighting displacement fields.

The integral representation of the solid displacement u_i is obtained by substitution of the fundamental solution derived from the following body forces:

$$X_i^* = \delta(x - \xi)\, \delta_{ij}$$

$$X'^*_i = 0 \qquad (8)$$

in which $\delta(x - \xi)$ is the Dirac delta function, ξ indicates the point of application of the delta function and δ_{ij} is the Kronecker delta.

The reciprocal relation gives

$$c_{ij}\, u_i + \int_\Gamma t_{ij}^*\, u_i\, d\Gamma + \int_\Gamma \tau_j^*\, U_n\, d\Gamma = \int_\Gamma u_{ij}^*\, t_i\, d\Gamma + \int_\Gamma \tau\, U_{nj}^*\, d\Gamma + \int_\Omega (X_i\, u_{ij}^* + X'_i\, U_{ij}^*)\, d\Omega \qquad (9)$$

Equations (8) and (9) contain three (two for plane problems) individual substitutions, i.e. $j = 1, 2$ and 3. The j subscript in equation (9) is used to indicate the jth set of the fundamental solution. The coefficient c_{ij} has a value δ_{ij} for points inside Ω and zero outside Ω. The value of c_{ij} for points on the boundary Γ can be determined following the same approach as in the case of scalar or elastic problems.

To obtain the integral representation of the fluid stress, the fundamental solution corresponding to the following body forces is used:

$$X'^*_i = [\frac{1}{2\pi} \ln r]_{,i} \qquad \textit{for 2-D}$$

$$X'^*_i = [\frac{-1}{4\pi r}]_{,i} \qquad \textit{for 3-D}$$

$$X^*_i = \frac{i\omega b + \omega^2 \rho_{12}}{-i\omega b + \omega^2 \rho_{22}} X'^*_i \tag{10}$$

where r is the distance to the point ξ at which the fluid stress is going to be represented. Notice that in this case, forces in all directions exist simultaneously in the two phases. Using these body forces, equation (4), and taking into account that $X'^*_{i,i} = \delta(x - \xi)$, the integral representation for the fluid stress is obtained:

$$\frac{c\,\tau}{-i\omega b + \omega^2 \rho_{22}} + \int_\Gamma (t^*_{ij}\, u_i + \tau^*_j\, U_n)\, d\Gamma$$

$$= \int_\Gamma [t_i\, u^*_{ij} + \tau\, (U^*_{nj} + \frac{X'^*_i\, n_i}{-i\omega b + \omega^2 \rho_{22}})]\, d\Gamma \tag{11}$$

$$+ \int_\Omega [X_i\, u^*_{ij} + X'_i\, (U^*_{ij} + \frac{X'^*_i}{-i\omega b + \omega^2 \rho_{22}})]\, d\Omega$$

where the j subscript corresponds in this case to the fundamental solution due to the second type of body forces, the integrals are in the sense of Cauchy's principal value, $c = 1$ for points inside Ω and $c = 0$ for points outside Ω. The value of c for points on the boundary is computed as usual. It is equal to 0.5 for points on Γ where the boundary is smooth.

In the two dimensional case, a vector notation for the two components of solid displacement and the fluid traction is used

$$V = \begin{bmatrix} u_1 \\ u_2 \\ \tau \end{bmatrix} \tag{12}$$

The two-dimensional solution can be obtained with the help of the thermoelastic analogy (Bonnet, 1987). The fundamental solution matrix is

$$\psi_{kj} = \sum_{m=1}^{3} \{ A_{kjm}\, K_0(i\lambda_m\, r) + (B_{kjm}\, r_{,j} + C_{kjm}\, r_{,k} + D_{kjm} \frac{1}{r})\, K_1(i\lambda_m\, r)$$
$$+ E_{kjm}\, r_{,k}\, r_{,j}\, K_2\, (i\lambda_m\, r) \} \tag{13}$$

where j indicates the set of body forces and k the component in the vector (12), K_1 and K_2 are the modified Bessel functions of the second kind and first and second order, respectively and

$$A_{kjm} = (1 - \delta_{k3})\, (1 - \delta_{j3})\, \frac{\delta_{kj}\, \delta_{3m}}{2\pi\mu} + c_m\, \delta_{k3}\, \delta_{j3}$$

$$B_{kjm} = -\, \lambda_m b_m \omega \eta \delta_{k3}\, (1 - \delta_{j3})$$

$$C_{kjm} = i\lambda_m b_m \gamma \delta_{j3}\, (1 - \delta_{k3}) \tag{14}$$

$$D_{kjm} = i\lambda_m \delta_{kj}\, a_m\, (1 - \delta_{k3})\, (1 - \delta_{j3})$$

$$E_{kjm} = \lambda_m^2 a_m\, (1 - \delta_{k3})\, (1 - \delta_{j3})$$

The coefficients λ_1 and λ_2 are the wave numbers of the coupled two dilatio waves. The coefficient λ_3 is the wave number of the rotational wave.

Once the displacements in the solid and the pressure in the fluid corresponding the fundamental solution are known, the boundary tractions can be computed fro equations (1). Using the subscript "j" to indicate the jth set of fundamental solution, α 1, 2 and β = 1, 2

$$t^*_{\alpha j} = \tau^*_{\alpha\beta j}\, n_\beta = (\lambda + \frac{Q^2}{R})\, \delta_{\alpha\beta}\, e_j^*\, n_\beta + 2\mu\, e^*_{\alpha\beta j}\, n_\beta + Q\, \delta_{\alpha\beta}\, \varepsilon_j^*\, n_\beta \tag{15}$$

where

$$e^*_{\alpha\beta j} = \frac{1}{2}\, (\psi_{\alpha j,\, \beta} + \psi_{\beta j,\, \alpha})$$

$$e_j^* = e^*_{\alpha\alpha j} \tag{16}$$

By substitution of $\varepsilon_j^* = \tau_j^*/R - (Q/R)e_j^*$ and $e_{\alpha\beta j}^*\, n_\beta = e_{\alpha\, j}^*$, the following expressions for the tractions is obtained:

$$t^*_{\alpha j} = (\lambda e_j^* + \frac{Q}{R}\, \psi_{3j})\, n_\alpha + 2\mu e^*_{\alpha j} \tag{17}$$

The fluid normal displacement for each set of fundamental solutions can be obtained by substitution of equations (4) and (10) into

$$U^*_{nj} = U^*_{\alpha j}\, n_\alpha \tag{18}$$

to yield

$$U^*_{nj} = J\, (\psi_{3j,\, \alpha} + \frac{1}{2\pi}\, \frac{r_{,\,\alpha}}{r}\, \delta_{3j})\, n_\alpha + Z\, \psi_{\alpha j}\, n_\alpha \tag{19}$$

where

$$J = \frac{1}{i\omega b - \omega^2\, \rho_{22}}$$

$$Z = J\, (i\omega b + \omega^2\, \rho_{12}) \tag{20}$$

4 Boundary element representation

The integral representations in equations (9) and (11) can be written for two-dimensional domains under zero body forces conditions as

$$c_{\alpha\beta}\, u_\alpha + \int_\Gamma t^*_{\alpha\beta}\, u_\alpha\, d\Gamma + \int_\Gamma \tau^*_\beta\, U_n\, d\Gamma = \int_\Gamma u^*_{\alpha\beta}\, t_\alpha\, d\Gamma + \int_\Gamma \tau\, U^*_{n\beta}\, d\Gamma \tag{21}$$

$$\int_\Gamma t^*_{\alpha 3}\, u_\alpha\, d\Gamma + \int_\Gamma \tau^*_3\, U_n\, d\Gamma = \int_\Gamma u^*_{\alpha 3}\, t_\alpha\, d\Gamma + \int_\Gamma \tau\, (U^*_{n3} - J\, X'^*_\alpha\, n_\alpha)\, d\Gamma + J\, c_{33}\, \tau \tag{22}$$

where J is as defined by equation (20).

Using vector notation, the integral representation for a point "i" can be written as

$$c^i\, u^i + \int_\Gamma p^*\, u\, d\Gamma = \int_\Gamma u^*\, p\, d\Gamma \tag{23}$$

where **u** and **p** are the field variables vectors for displacements and stresses, respectively

$$u = \begin{bmatrix} u_1 \\ u_2 \\ \tau \end{bmatrix} \quad \textit{and} \quad p = \begin{bmatrix} t_1 \\ t_2 \\ U_n \end{bmatrix} \tag{24}$$

$\mathbf{p}^*$ and $\mathbf{u}^*$ are the fundamental solution tensors

$$p^* = \begin{bmatrix} t^*_{11} & t^*_{21} & -U^*_{n1} \\ t^*_{12} & t^*_{22} & -U^*_{n2} \\ t^*_{13} & t^*_{23} & -\hat{U}^*_{n3} \end{bmatrix}$$

$$u^* = \begin{bmatrix} u_{11}^* & u_{21}^* & -\tau_1^* \\ u_{12}^* & u_{22}^* & -\tau_2^* \\ u_{13}^* & u_{23}^* & -\tau_3^* \end{bmatrix} \tag{25}$$

where $\hat{U}_{n3}{}^* = U_{n3}{}^* - J\, X'_{\alpha}{}^* n_\alpha = (J\tau_{3,\alpha}{}^* + Zu_{3\alpha}{}^*)\, n_\alpha$

$$c^i = \frac{1}{2}\begin{bmatrix} 1 & 0 & 0 \\ 0 & 1 & 0 \\ 0 & 0 & -J \end{bmatrix} \tag{26}$$

for a smooth boundary point.

Consider now that the boundary of the domain under study is discretized into boundary elements. Equation (23) can be written as

$$c^i\, u^i + \sum_{j=1}^{N} \{ \int_{\Gamma_j} p^* \;\Phi\; d\Gamma\}\; u^j = \sum_{j=1}^{N}\{ \int_{\Gamma_j} u^* \;\Phi\; d\Gamma\}\; p^j \tag{27}$$

where each integral extends over element "j" and Φ is the shape functions matrix. Using the traditional BE notation, equation (27) can be written as

$$H\; u = G\; p \tag{28}$$

5 Water layer on a porous elastic bottom

A homogeneous water layer overlying a homogeneous half-space of porous elastic sediment is assumed. (Figure 1).

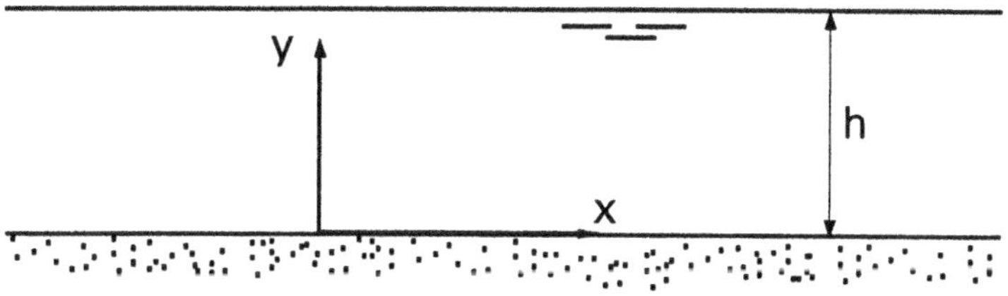

Figure 1. Water layer on poroelastic half-space.

The water pressure p_w in the layer for time-harmonic behavior is governed by the scalar wave equation

$$\nabla^2 p_w + \frac{\omega^2}{c^2} p_w = 0 \tag{29}$$

An integral equation formulation and a Boundary Element implementation can be done for this region to obtain a relation of the type (see, for instance, Domínguez, 1993)

$$\boldsymbol{H}\ \boldsymbol{U}_w = \boldsymbol{G}\ \boldsymbol{P}_w \tag{30}$$

where $\boldsymbol{U}_w$ is a vector containing the water normal displacement at the boundary nodes, $\boldsymbol{P}$ contains the nodal values of the pressure and $\boldsymbol{H}$ and $\boldsymbol{G}$ are the system matrices obtained from the fundamental solution. Using the half-space fundamental solution only the lower boundary of the water layer has to be discretized. Therefore, this is the only boundary which requires discretization in the problem at hand. Equation (30) is written for the elements along this boundary as part of the water region. Equation (28) is written for the same elements as part of the poroelastic half-space. In case of existence of internal boundaries within the water region or the porous half-space those boundaries must also be discretized and the corresponding integral representation (for the fluid or the porous medium) written for the nodes along the internal boundaries.

The boundary conditions to be set along the interface between the water and the poroelastic half-space ($y = 0$) are:

(1) Equilibrium of normal traction; i.e., zere effective traction on the porous medium

$$t'_n = t_n + (1-\beta)\ p = 0 \tag{31}$$

where p is the pore pressure, t'_n is the effective traction and t_n the normal traction on the solid skeleton.

(2) Equilibrium of fluid pressure.

$$p = p_w \tag{32}$$

(3) Equilibrium of tangential traction.

$$t_t = 0 \tag{33}$$

(4) Continuity of the fluid movement in and out of the skeletal frame in a direction normal to the interface

$$U_w = u_n\,(1-\beta) + U_n\,\beta \tag{34}$$

where U_w is the normal displacement of the water as part of the layer, u_n the normal displacement of the solid and U_n the normal displacement of the pore water.

6 Results and discussion

Using the above BEM formulation, a problem with practical interest is studied. A horizontal layer with a water depth h = 100 m is assumed, the water density ρ_f = 1000 Kg/m^3 and the bulk modulus of water $K_f = 2.3 \times 10^9$ N/m^2 which corresponds to a wave velocity c = 1500 m/s. The properties of the bottom material correspond to a fine sand and are the same assumed by Yamamoto (1983): Porosity, $\beta = 0.3$; shear modulus, $G = 5 \times 10^7 (1 + 0.02\,i)$ N/m^2; Poisson's ratio, $\upsilon = 0.333$; density of the grain material, $\rho_s = 2.65 \times 10^3$ Kg/m^3; added mass, $\rho_a = 75$ Kg/m^3 = $-\rho_{12}$; Biot's dissipation coefficient, b = 0.09×10^7 Ns/m^4; $Q = 1.395 \times 10^9$ N/m^2 and $R = 6.01 \times 10^8$ N/m^2. The densities in Biot's equations are $\rho_{11} = (1-\beta)\,\rho_s + \rho_a = 1930$ Kg/m^3 and $\rho_{22} = \beta\,\rho_f + \rho_a = 375$ Kg/m^3. A time harmonic point source, represented by a circular cavity of radius r = 1 cm, is located at mid-depth (y = 50 m) within the water layer. The boundary of the circular cavity is discretized into 4 quadratic elements. A length of the bottom equal to 15 h = 1500 m is discretized into 50 equal length quadratic elements (Figure 2).

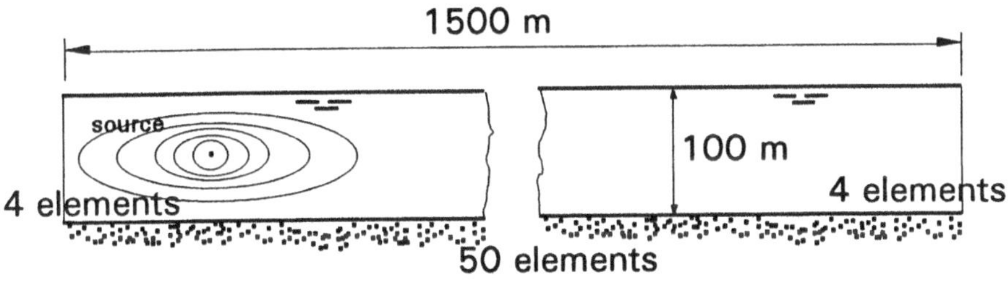

Figure 2. Problem geometry

Because of the absorption by the poroelastic bottom, the water waves damp as they progress along the channel; however, in order to avoid possible spurious reflections a mechanism of wave absorption is located at both ends of the model. To do so, four boundary elements are located at each end and boundary conditions are established assuming that the water layer away from the model is on top of a rigid bottom. By doing so a relation between the pressure and the normal displacement can be obtained using the analytical solution for that problem which consist of simple normal modes of wave propagation. Only those modes with a frequency under the value of the excitation frequency have to be considered. Using a number of nodes at the vertical boundary equal to the number of normal modes, a matrix relating the hydrodynamic pressure and the normal displacement can be obtained and used as boundary condition for those nodes. Further details about this absorbing boundary can be found in the book of Domínguez (1993).

Using the BEM model of Figure 2 and assuming an excitation frequency f = 20 Hz, the problem is solved to obtain the pressure in the water layer and the pore pressure, the displacements and the tractions in the half-space. The real and imaginary parts of the pressure in the water along a horizontal line y = 50 m are shown in Figures 3 and 4, respectively. Yamamoto (1983) obtained the analytical expressions of the normal modes for plane harmonic waves in the same water layer on the same poroelastic half-space. The complex wave numbers for the first three modes at f = 20 Hz are: $k_1\, h = 8.061 - i\, 0.00233$, $k_2\, h = 6.949 - i\, 0.0932$ and $k_3\, h = 3.147 - i\, 1.032$. A comparison of the solution obtained with the BEM and the water pressure obtained using the normal modes is also shown in Figures 3 and 4. It can be seen from these figures that as soon as one gets at a certain distance from the source the solution can be accurately represented using a small number of normal modes. The complex contribution of each mode to the far field solution has been obtained by a minimum square optimization of the difference between the BEM solution and the normal mode solution along 200 points going from 360 m to 1360 m from the source.

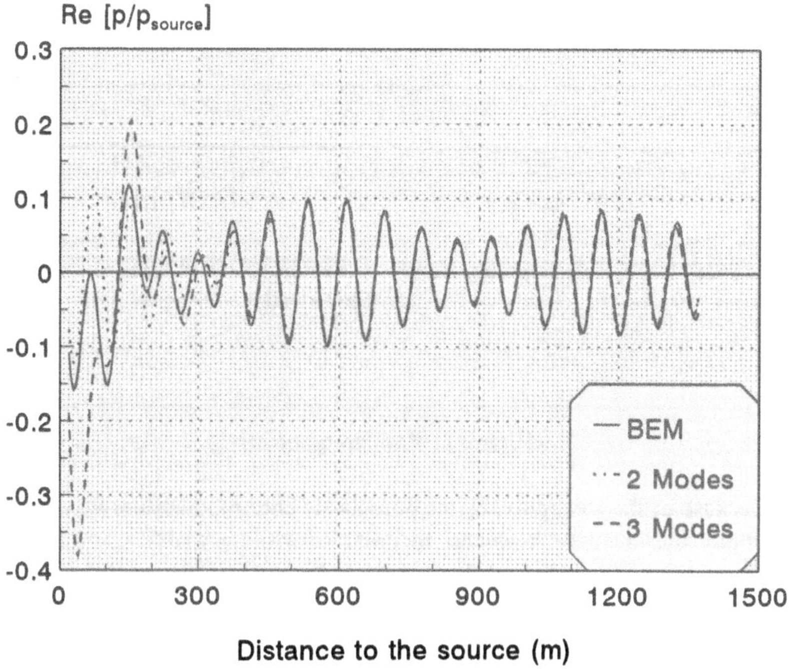

Figure 3. Pressure distribution at mid-depth. Real part.

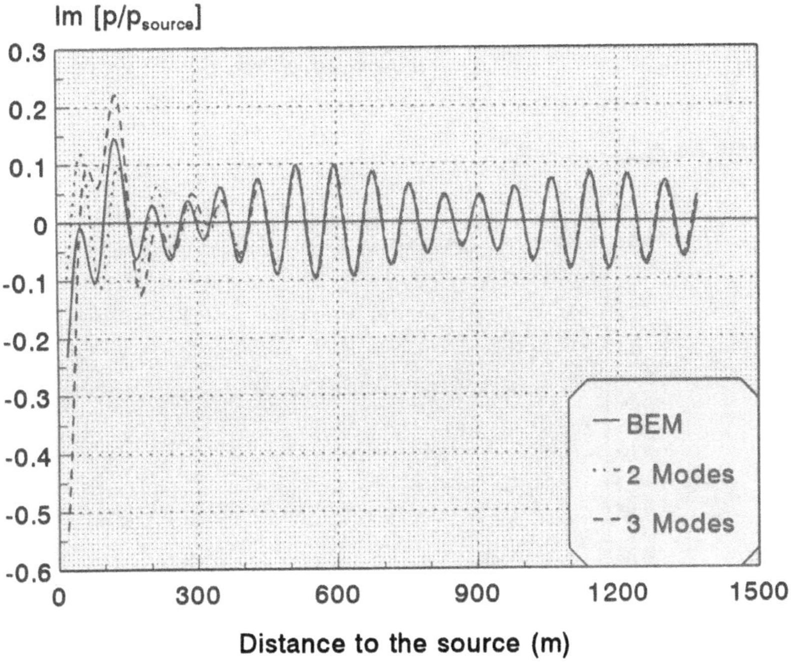

Figure 4. Pressure distribution at mid-depth. Imaginary part.

Figures 5 and 6 show the water pressure distribution at several vertical lines ahead of the source. Again the pressure distribution is compared with the approximation in terms of the first few modes. The number of modes required to obtain a good approximation is smaller as one gets furthermost from the source. In order to show the effect of the shear modulus of the sediment the problem was solved for a range of the shear modulus between 10^5 and 3×10^8 N/m^2. The values of other physical properties are kept constant. The far-field solution is again approximated using the first two, three and four normal modes. The real and imaginary parts of the contribution of the first two modes to the far-field solution obtained using two normal modes are shown in Figure 7 for the abovementioned Shear Modulus range. Table I shows the value of the residues obtained comparing the BEM solution and the normal modes solution obtained with two and three modes. Those residues are computed as

$$e = \sqrt{e_{real}^2 + e_{imag}^2} \tag{35}$$

where

$$e_{real} = \frac{\sqrt{\sum (f_{1real} - f_{2real})}}{number\ of\ points} \tag{36}$$

and the same for e_{imag} .

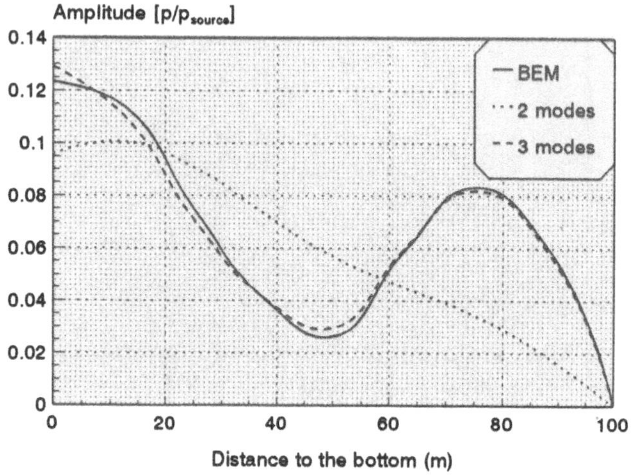

x=2H=200 m

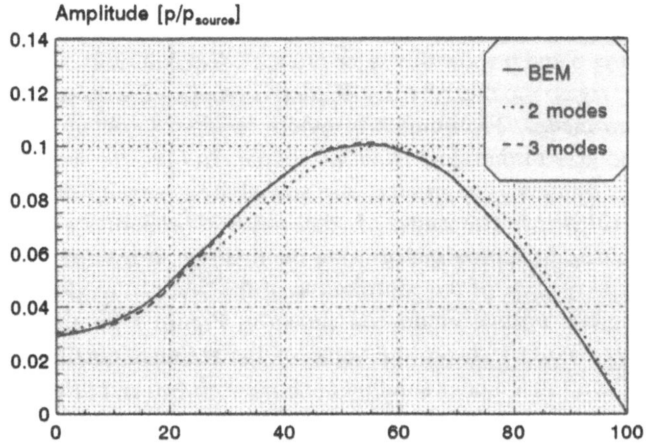

x=5H=500 m

Figure 5. Pressure distribution along two vertical lines. Distance to the source; a) x = 200 m, b) x = 500 m

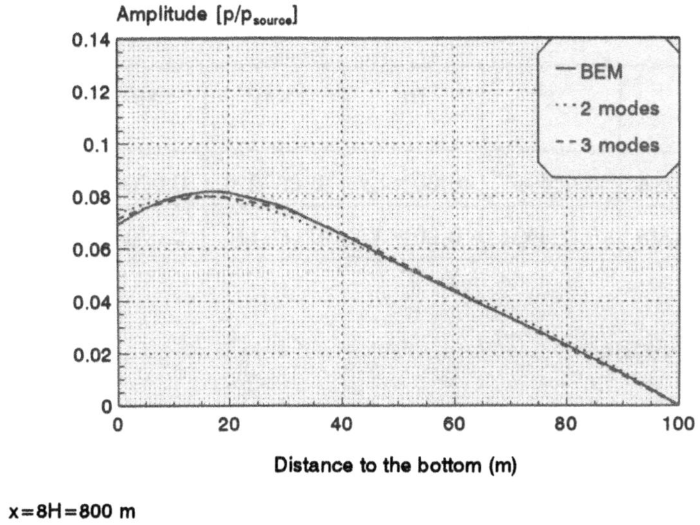

Figure 6. Pressure distribution along a vertical line. Distance to the source; x = 800 m

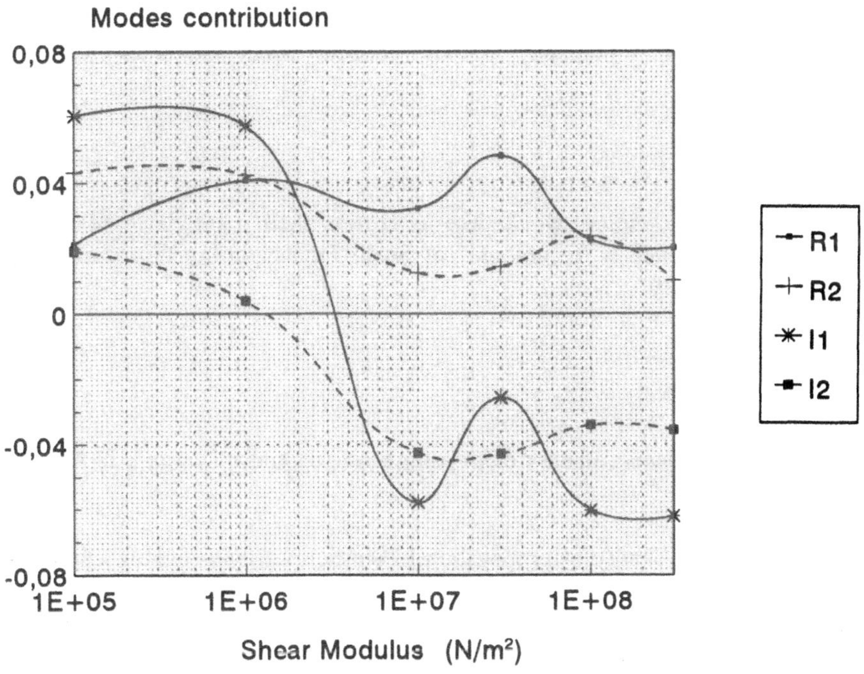

Figure 7. Contribution of the first two modes.

Table I

G	10^5	10^6	10^7	3×10^7	10^8	3×10^8
2 modes	0.005294	0.004967	0.001645	0.007214	0.000444	0.000294
3 modes	0.005246	0.004924	0.001637	0.007146	0.000444	0.000280

7 Conclusions

A BEM model for the dynamic analysis of multiregion domains including wate regions, viscoelastic regions and porous viscoelastic regions has been presented. Th problem of a sound source in a homogeneous ocean overlying a porous viscoelastic botto has been analyzed. The effects of the shear modulus has been examined numerically. Th computed far-field pressure distribution is compared with the approximation obtained fro the series expansion in terms of the known close form complex normal modes. Th agreement between the BEM solution and the expansion in terms of only a few modes very good.

The model presented in this paper can be used without any further developme for non-uniform bottom topography, porous sediment layers and other more realist geometries.

Acknowledgments: The authors would like to express their gratitude to the "Comisi Interministerial de Ciencia y Tecnología" of Spain for supporting this work under t research grant PB93-1191.

8 References

Biot, M.A. (1941). "General theory of three-dimensional consolidation". J. Appl. Phy Vol. 12, pp. 155-164.

Biot, M.A. (1956a). "Theory of propagation of elastic in a fluid-saturated porous so Part I: Low frequency range". J. Acoust. Soc. Am., Vol. 28, pp. 168-178.

Biot, M.A. (1956b). "Theory of propagation of elastic waves in a fluid-saturated por solid. Part II: Higher frequency range. "J. Acoust. Soc. Am., Vol. 28, pp. 179-191.

Biot, M.A. (1962). "Mechanics of deformation and acoustic propagation in porous media". J. Appl. Phys., Vol. 33, pp. 1482-1498.

Bonnet, G. (1987). "Basic singular solutions for a poroelastic medium in the dynamic range", J. Acoust. Soc. Am., Vol. 82, pp. 1758-1762.

Bougacha. S. and Tassoulas, J.L. (1991a). "Seismic analysis of gravity dams.I: modeling of sediments". J. Engrg. Mech., ASCE, Vol. 117, pp. 1826-1838.

Bougacha. S. and Tassoulas, J.L. (1991b). "Seismic analysis of gravity dams.II: effects of sediments". J. Engrg. Mech., ASCE, Vol. 117, pp. 1839-1850.

Bougacha S., Tassoulas, J.L. and Roësset, J. M. (1993a). "Analysis of foundations on fluid-filled poroelastic stratum". J. Engrg. Mech., ASCE, Vol. 119, pp. 1632-1648.

Bougacha. S., Roësset, J.M. and Tassoulas, J.L. (1993b). "Dynamic stiffness of foundations on fluid-filled poroelastic stratum". J. Engrg. Mech., ASCE, Vol. 119, pp. 1649-1662.

Cheng, A.H.-D. (1986). "Effect of sediment on earthquake-induced reservoir hydrodynamic response. "J. Engrg. Mech., ASCE, Vol. 112, pp. 654-665.

Cheng, A.H.D., Badmus, T. and Beskos, D.E., (1991), "Integral equation for dynamic poroelasticity in frequency domain with boundary element solution", J.Eng. Mech., ASCE, Vol. 117, pp. 1136-1157.

Domínguez, J. (1991), "An integral formulation for dynamic poroelasticity", J. Appl. Mech. ASME, Vol. 58, pp. 588-591.

Domínguez, J. (1992), "Boundary element approach for dynamic poroelastic problems", Int. J. Num. Meth. Eng., Vol. 35, pp. 307-324.

Domínguez, J. (1993), "Boundary elements in dynamics", Computational Mechanics Publications, Southampton and Elsevier, London.

Halpern, M.R., and Christiano, P. (1986a). "Response of poroelastic halfspace to steady-state harmonic surface tractions". Int. J. Numer. and Analytical Methods in Geomechanics, Vol. 10, pp. 609-632.

Halpern, M.R., and Christiano, P. (1986b). "Steady-state harmonic response of a rigid plate bearing on a liquid-saturated poroelastic halfspace". Earthquake Engrg. and Struct. Dyn., Vol. 14, pp. 439-454.

uang, L.H. (1992). "Influence of seafloor on acoustic plane wave", J. Eng. Mech., SCE, Vol. 118, pp. 1987-2004.

Manolis, G., and Beskos, D.E. (1989). "Integral formulation and fundamental solutions of dynamic poroelasticity and thermoelasticity". Acta Mech., Vol. 76, pp. 89-104, also "Errata". (1990). Acta Mech., Vol. 83, pp. 223-226.

Mei, C.C., and Foda, M.A. (1981). "Wave-induced responses in a fluid-filled poroelastic solid with a free surface -A boundary layer theory". Geophys. J. Roy. Astr. Soc., Vol. 66, pp. 597-631.

Predeleanu, M. (1984). "Development of boundary element method to dynamic problems for porous media". Appl. Math. Modeling, Vol. 8, pp. 378-382.

Stoll, R.D. (1977). "Acoustic waves in ocean sediments", Geophys., Vol. 42, pp. 715-725.

Stoll, R.D. and Kan, T.K. (1981). "Reflection of acoustic waves at a water-sediment interface", J. Acoust. Soc. Am., Vol. 70, pp. 149-164.

Wiebe, Th. and Antes, H. (1991). "A time domain integral formulation of dynamic poroelasticity", Acta Mech., Vol. 90, pp. 125-137.

Yamamoto, T. (1983). "Acoustic propagation in the ocean with a poro-elastic bottom", J. Acoust. Soc. Am., Vol. 73, pp. 1587-1596.

NUMERICAL MODELLING OF SATURATED POROUS MEDIA SUBJECTED TO DYNAMIC LOADING

H. MODARESSI
Bureau de Recherches Géologiques et Minières,
Department of Geotechnical Engineering and Mineral Technology,
Avenue de Concyr, B.P. 6009, Orléans CEDEX 2, FRANCE

Abstract. With the increase in the computing power, the approximation of the equations governing the behaviour of saturated and partially saturated porous materials such as soil, rock and concrete is emerging as an active field of research. The Finite Element method provides a convenient tool for handling non-linear coupled formulations involved in the context of soil mechanics. This paper deals with major aspects of the analysis of dynamic behaviour of soils as porous media. The mathematical formulation of saturated porous media under dynamic conditions established following the works of Biot, the treatment of problems with respect to numerical modelling of spatially unbounded domains and finally, the numerical algorithms and associated difficulties are discussed. Absorbing boundaries are developed using the paraxial approximation. The seismic analysis of soil sediments is considered as a special case and the introduction of seismic motion is presented.

Key words: two-phase porous media, spectral impedance, numerical algorithms, absorbing boundaries, paraxial approximation, seismic analysis.

1. Introduction

The analysis of seismic wave propagation in soils can be considered as one of the important fields of research in Civil Engineering in the past three or four decades. Besides the non-linear aspects of the soil behaviour the role of pore water pressure has been of a major interest for geotechnical engineers.

The mathematical formulation of saturated poroelastic media under dynamic conditions can be considered as well established following the original works of Biot. With respect to the frequency content of the dynamic process, simplified versions, in which the effective stress principle of Terzaghi is included, have been proposed (Zienkiewicz and co-workers, 1980, 1984).

For poroelastoplastic materials recent theoretical investigations (Coussy 1991) confirm the applicability of effective stress principle for soil but suggest a total stress approach for materials with small porosity as rock and concrete. In fact, in this case the proportionality coefficient of the pore fluid pressure (also called Biot's α parameter) is no more constant and evolves with plastic state of the material.

Various numerical techniques can be used to handle the complex initial-boundary value problems using either the Finite Element method (Ghaboussi and Wilson, 1972, Zienkiewicz and Shiomi, 1984, Aubry and Modaressi 1988) or the Boundary Element method (Bonnet, 1987, Chen, 1992, Kaynia and Banerjee 1993, Banerjee *et al.*, 1994). The

A. P. S. Selvadurai (ed.), Mechanics of Poroelastic Media, 143–156.

major difference between these techniques comes from the spatial and time discretization process. The stability and accuracy of the selected numerical schemes has to be established in order to guarantee the quality of obtained results. Under certain circumstances the discretized formulations, may alter the computed solutions, leading to dispersion and amplitude modulations, even though the stability requirements are satisfied.

The transient Finite Element application is well-adapted to non-linear coupled problems. However, in spatially unbounded problems, the numerically finite outer boundaries are problematic as undesirable spurious reflections are generated. They affect the computed results and have to be removed from the calculations. Several techniques have been proposed in the literature for modelling unbounded domains, assuming a linear behaviour of the studied domain. For poroelastodynamic stationary problems, transmitting boundaries permit the obtaining of a local-in-space impedance operator (Degrande and Roeck, 1994). Unfortunately, these boundaries are frequency dependent and lead to very heavy calculations in the transient domain, unless a recursive evaluation method as the one proposed for one-phase material by Mohasseb and Wolf (1989) is generalised to two-phase media.

For transient analyses, it is possible to develop frequency independent local absorbing boundaries with respect to time and space. Amongst different existing techniques developed for absorbing boundaries, the paraxial approximation represents an elegant framework. In order to obtain a local transient impedance, a limited development of the wave number is performed on the elastodynamic equations expressed in Fourier domain. For high-frequency waves and for waves hitting the boundary at a low angle this approximation is very accurate. The performance of such boundary elements is proportional to the approximation order. Modaressi and Benzenati (1994), extended this approach to the case of saturated porous media for low-frequency phenomenon modelled by the so-called *u-p* formulation.

In this paper, some new results regarding spectral impedance and absorbing boundaries for a more general formulation of poroelastic material are given. The seismic motion can be introduced into numerical computations by these absorbing boundary elements and reveal that far field pore pressure is also requested in seismic analysis of saturated soils. This result, to the knowledge of the author, has not been reported in computational seismology.

2. Governing Equations

If n is the porosity of the porous medium, the relative velocity of the fluid phase v_{rf}, given by $v_{rf} = n\,(\partial_t u_f - \partial_t u_s)$, the displacement of the solid skeleton u_s and the pore-water pressure p, will be kept as the primary unknowns. The momentum conservation for a saturated two-phase medium in terms of effective stress is given by

$$\mathcal{D}iv\ \sigma' - \mathcal{g}rad\ p + \rho\, g = \rho\, \partial_{tt} u_s + \rho_f\, \partial_t v_{rf}$$

In above:

- σ': is the effective stress tensor,
- n: is the porosity of porous medium,
- u_s: is the skeleton displacement vector,
- u_f: is the fluid displacement vector,
- v_{rf}: is the fluid relative velocity vector $(=n\,(\partial_t u_f - \partial_t u_s))$,
- p: is the pore-fluid pressure,
- g: is the body force vector per unit mass,
- ρ: is the bulk density of the overall medium $(\rho = (1-n)\,\rho_s + n\,\rho_f)$,

ρ_s: is the density of the solid phase,
ρ_f: is the density of the fluid phase,

The divergence operator is defined as:

$$\{Div\ \sigma\}_i = \Sigma_j\ \partial_j \sigma_{ij}$$

The mass conservation for a saturated two-phase medium may be written as:

$$div\ v_{rf} + div\ \partial_t u_s + \partial_t p/Q = 0$$

with the compressibility parameter Q given by

$$\frac{1}{Q} = \frac{n}{K_f} + \frac{1-n}{K_s}$$

in which K_s and K_f are compressibility moduli of, respectively, the solid grains and the fluid phase.

The conservation equation must, of course, be augmented with an equation that describes the movement of one phase with respect to the other. This equation is given by a generalised Darcy's law:

$$v_{rf} = \mathbb{K}\left\{-grad\ p + \rho_f(g - \partial_{tt}u_s - \partial_t v_{rf}/n)\right\}$$

where $\mathbb{K}$ is the permeability tensor.

The above set of equations results in formulation called henceforth *full model* [also u_s-v_{rf}-p model]:

$$\begin{cases} Div\ \sigma' - grad p + \rho\ g = \rho\ \partial_{tt}u_s + \rho_f\ \partial_t v_{rf} \\ \mathbb{K}^{-1} v_{rf} + grad\ p - \rho_f(-g + \partial_{tt}u_s + \partial_t v_{rf}/n) = 0 \\ div v_{rf} + div \partial_t u_s + Q^{-1}\partial_t p = 0 \end{cases}$$

The presented model is interesting for any frequency range. It is also applicable when the solid skeleton and pore fluid are both nearly incompressible. Furthermore, this formulation has given slightly more accurate results for pore pressure in calculations performed by Simon *et al.* (1986). However, for the frequency range suitable for earthquake engineering a simpler version can be used. In this model, so-called *simplified model* [also u_s-p model] in the next sections, the relative fluid acceleration is neglected so that the relative fluid velocity is eliminated and displacement and pore pressure become the only unknowns:

$$\begin{cases} Div\ \sigma' - grad\ p + \rho g = \rho\ \partial_{tt}u_s \\ div\ \partial_t u_s - div\ (\mathbb{K}\ grad(p - \rho_f g x)) - div\ (\mathbb{K}\rho_f\ \partial_{tt}u_s) + \partial_t p/Q = 0 \end{cases}$$

In some cases when the influence of the acceleration vector in Darcy's relationship i small, it is possible to neglect the third term in the second equation as it i computationally undesirable (Park, 1983). Nevertheless, it should be noted that if this term is neglected then the above system does not represent the limiting case of *true* drained dynamic behaviour without coupling.

3. Boundary and Interface Conditions

The boundary conditions will obviously be more complex than for one-phase material an must be decomposed into boundary conditions relative to the bulk material and to the flui flow. For the first equation, the boundary Γ of the domain Ω is thus decomposed into tw first parts: Γ_σ and Γ_{us}. On Γ_{us} a displacement boundary condition is prescribed, while o Γ_σ a traction boundary condition is applied. For the second equation only an essentia boundary condition exists on Γ_{vrf} on which a (no)drainage type boundary condition i selected. For the third equation, a second partition of the boundary is introduced to deal wit the fluid flow: Γ_p and Γ_φ. On Γ_φ there is a normal velocity (or flux) condition, and on Γ a pressure boundary condition is considered. We note that boundary conditions of the secon and third equations are coupled. On the interface between a saturated porous material and one-phase medium, the continuity of absolute solid displacements as well as the total stres vector has to be verified and the relative velocity of the fluid phase must vanish.

The one-phase medium symbolises *either* a two-phase material with a very sma permeability for which a one-phase analysis is more attractive, hence:

$$\begin{cases} u_s^{(two\text{-}phase)} = u_s^{(one\text{-}phase)} \\ v_{rf}^{(two\text{-}phase)} = 0 \\ t^{(two\text{-}phase)} + t^{(one\text{-}phase)} = 0 \end{cases}$$

or a two-phase material with a very high permeability so that the pore pressure on t interface could be prescribed:

$$\begin{cases} u_s^{(two\text{-}phase)} = u_s^{(one\text{-}phase)} \\ p^{(two\text{-}phase)} = prescribed \\ t^{(two\text{-}phase)} + t^{(one\text{-}phase)} = 0 \end{cases}$$

In *simplified model*, for the fluid flow there is only a normal velocity condition, th weaker than in the *full model* :

$$\text{Either: } \begin{cases} u_s^{(two\text{-}phase)} = u_s^{(one\text{-}phase)} \\ v_{rf}^{(two\text{-}phase)} \cdot n = 0 \\ t^{(two\text{-}phase)} + t^{(one\text{-}phase)} = 0 \end{cases} \quad \text{or: } \begin{cases} u_s^{(two\text{-}phase)} = u_s^{(one\text{-}phase)} \\ p^{(two\text{-}phase)} = prescribed \\ t^{(two\text{-}phase)} + t^{(one\text{-}phase)} = 0 \end{cases}$$

In dynamic problems, especially in those of seismic analysis, radiation bound conditions are required to avoid artificial reflections on the mesh boundary. Only in so conditions when the contrast of impedance between the bedrock and soil is high enough that the bedrock can be assumed undeformable, the studied domain can be limited deformable soil and numerical analysis is simplified, otherwise the influence of deforma bedrock must be included.

To clarify this purpose, let us consider a semi-infinite domain composed of two sub-domains; Ω_E (outer domain, unbounded except on the ground) and Ω_I (inner domain, bounded). On the interface Σ between these two domains, the continuity equations between displacement (u_s and u_E), fluid relative velocity (v_{rf} and v_E,) pore pressure (p and p_E), stress vectors (t and t_E) and the hydraulic flux (φ and φ_E) must be verified:

$$\begin{cases} u_s = u_E \\ v_{rf} = v_E \\ p = p_E \\ t + t_E(u_E) = 0 \\ \varphi = \varphi_E \end{cases}$$

At the vicinity of the boundary Σ, a linear elastic behaviour for the constituting materials is assumed so that the notation $t_E(u_E)$ has a meaning. To replace the outer domain in calculations a dynamic impedance on boundary Σ has to be introduced. The spectral impedance on the interface represents the spectral action exerted by the outer domain Ω_E on the inner domain Ω_I, when Ω_E is submitted to scattering waves coming from Σ and going into the infinity. To obtain the spectral impedance on interface Σ, we carry out a Fourier transformation of the total stress vector and the hydraulic flux on this interface with respect to time and space vector components in a plane *tangent to the boundary* (Σ_P (x',x_3)) with normal e_3. This impedance is not local.

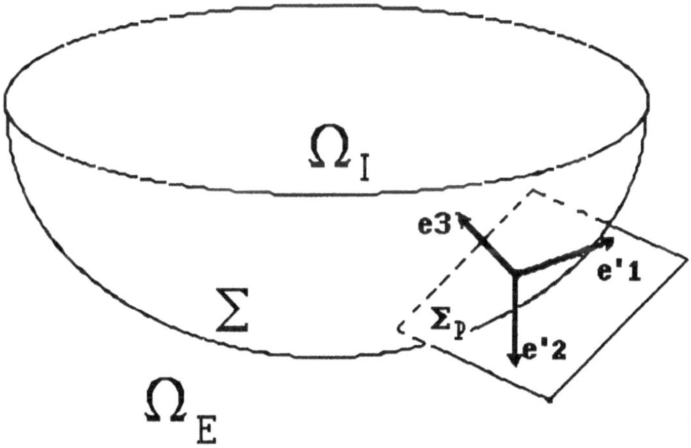

The total stress vector and the hydraulic flux on this boundary in transient domain are given as:

$$\begin{cases} t\,(x',x_3,t) = \{[\lambda\, tr\, \varepsilon(u_s) - p\,]\, I + 2\mu\, \varepsilon(u_s)\}\, e_3 \\ \varphi(x',x_3,t) = v_{rf}\, e_3 \end{cases}$$

where λ and μ are Lamé coefficients, I is the unity tensor, and ε is the strain tensor.

The Fourier Transform of above expressions with respect to x' and t, when x_3 is assumed to be zero on Σ_p is:

$$\begin{cases} \hat{t}\,(\xi',0,\omega) = [i\lambda\,\xi'\hat{u}_0'+(\lambda+2\mu)\partial_3\hat{u}_{30} - \hat{p}_0]\,e_3 + \mu[\partial_3\hat{u}_0'+ i\,\xi'\hat{u}_{30}] \\ \hat{\phi}_0\,(\xi',0,\omega) = \hat{v}_{rf30} \end{cases}$$

where ξ' and ω, respectively, are the wave vector and the angular frequency.

For Finite Element transient analysis the impedance has to be local with respect to time and space, for which purpose the paraxial approximation is used. This technique is very accurate for high-frequency waves and for waves impinging on the boundary at a low angle. The paraxial approximation is performed by limited development of wave numbers but before, $\hat{u}_0'$, $\hat{u}_{30}$, $\hat{p}_0$ and $\hat{v}_{rf30}$ solutions of differential equations governing poroelastodynamic media (i.e. *full* or *simplified model*) projected on Σ_p plane must be computed.

The existence of a shear wave and two dilatational waves (resp. V_s, V_{p1} and V_{p2}) are standard features of porous media. The solution of the differential equations with plane-wave assumption is obtained in terms of the velocities of these waves. For the *full model* these wave velocities are complex numbers. Therefore, the dilatational as well as shear waves are dispersive and attenuated. The solid skeleton and the pore-fluid are in-phase in the first dilatational wave and out-of-phase for the second one. For *simplified model* only dilatational wave velocities contain imaginary parts and the shear wave velocity is equivalent to that of a one-phase medium.

In practical engineering problems, the following assumptions are usually made regarding the value of wave velocities:

$$\begin{cases} V_{p1}=V_c = \sqrt{\dfrac{\lambda+2\mu+Q}{\rho}} \\ V_{p2}=0 \\ V_s = C_s \end{cases}$$

where V_c corresponds to the dilatational wave velocity in a *dynamically compatible* condition. It represents the case in which the waves move through the porous media without any relative motion between the solid skeleton and the fluid phase.

The permeability being assumed isotropic, the general solutions for *full model* are calculated as (Benzenati, 1993):

$$\begin{aligned} \xi'\hat{u}' &= |\xi'|\,[A_s\,exp(-i\xi_s x_3) + A_{p1}\,exp(-i\,\xi_{p1}x_3) + A_{p2}\,exp(-i\,\xi_{p2}x_3)] \\ \hat{u}_3 &= A_s\frac{|\xi'|}{\xi_s}\,exp(-i\,\xi_s x_3) - A_{p1}\frac{\xi_{p1}}{|\xi'|}\,exp(-i\,\xi_{p1}x_3) - A_{p2}\frac{\xi_{p2}}{|\xi'|}\,exp(-i\,\xi_{p2}\,x_3) \\ \hat{p} &= \frac{-i\,\rho\omega^2}{|\xi'|(1-Y)}\Big[(1-\frac{C_s^2}{V_s^2}-\rho_f\omega^2 Y)\,A_s\,exp(-i\xi_s x_3)\,+ \\ &(1-\frac{C_p^2}{V_{p1}^2}-\rho_f\omega^2 Y)\,A_{p1}\,exp(-i\xi_{p1}x_3)+(1-\frac{C_p^2}{V_{p2}^2}-\rho_f\omega^2 Y)\,A_{p2}\,exp(-i\xi_{p2}x_3)\Big] \\ \hat{v}_{rf3} &= ZK\,(\,\partial_3\hat{p}\, - \rho_f\,\omega^2\,\hat{u}_3) \end{aligned}$$

with for *simplified model*: $Z=-1$ and $Y = 0$

and for *full model:* $Z = \dfrac{n(n+i\omega K\rho_f)}{n^2\omega^2+K^2\rho_f^2\omega^4}$ *and* $Y = K\rho_f\omega^2 Z$

C_s and C_p are, respectively, the shear and dilatational wave velocities in the medium and in the absence of pore-fluid, given by

$$\begin{cases} C_s^2 = \dfrac{\mu}{\rho} \\ C_p^2 = \dfrac{(\lambda+2\mu)}{\rho} \end{cases}$$

and:

$$\xi_\Theta^2 = \frac{\omega^2}{V_\Theta^2} - |\xi'|^2 \quad ; \quad \Theta = p1, p2 \text{ or } s$$

We note that for the *simplified model* the pore pressure is not influenced by shear waves.

Now, to form a local operator the above terms should be approximated either by using a Taylor approximation or a continued fraction expansion. The former, in the case of a one-phase medium, can result in strongly unstable boundary conditions if higher order paraxial elements are necessary (Engquist and Majda, 1977). For *zeroth* and *first order* paraxial approximation both techniques result in the same expression.

The finite continued fraction expansions for *jth order* are recursively given by

$$\xi_\Theta^j = \frac{\omega}{V_\Theta}\left[1 - \frac{\dfrac{V_\Theta^2|\xi'|^2}{\omega^2}}{1+\dfrac{V_\Theta\,\xi_\Theta^{j-1}}{\omega}}\right] \quad ; \quad \Theta = p1, p2 \text{ or } s;\ \xi_\Theta^0 = 1$$

It is easily seen that for small $|\xi'|/\omega$ values the waves propagate in directions close to e_3.

In the previous work on *simplified model* with *zeroth order* paraxial approximation we obtained the following relationships for the total stress vector and the hydraulic flux on Σ (Modaressi and Benzenati, 1994).

$$\begin{cases} t_0(t) = -\left[\rho\dfrac{C_p^2}{V_{p1}}\partial_t u_{30} + p_0\right] e_3 - \rho\, C_s\, \partial_t u_0' \\ \varphi_0(t) = K\left[\rho\left(1 - \dfrac{C_p^2}{V_{p1}^2}\right) - \rho_f\right]\partial_{tt} u_{30} \end{cases}$$

The above expressions were obtained after some simplifying hypotheses, i.e. the amplitude of second dilatational wave was assumed to be negligible. This assumption results in a physically meaningless expression for the stress vector in the case of highly incompressible pore fluid (i.e. zero impedance for normal component). The general form for the total stress vector and the hydraulic flux on Σ without such simplifying assumptions and at the *zeroth order* in the frequency domain becomes:

$$\begin{cases} \hat{t}_0(\xi',0,\omega) = -\left[i\omega\rho \dfrac{C_p^2 C_s^2 (V_{p1}+V_{p2})}{C_p^2 V_s^2 + C_s^2 V_{p1} V_{p2}} \hat{u}_{30} + \hat{p}_0 \right] e_3 - \\ \qquad i\omega\rho \dfrac{C_s^2}{V_s} \left[\dfrac{\xi' \hat{u}_0'}{|\xi'|^2} \xi' + \dfrac{(\hat{u}_0', \xi', e_3)}{|\xi'|^2} \xi \wedge e_3 \right] \\ \hat{\varphi}_0(\xi',0,\omega) = K\omega^2 \left[\rho\, f(V_{p1}, V_{p2}, V_s, C_p, C_s, Y) + \rho_f \right] \hat{u}_{30} \end{cases}$$

The expression of $f(V_{p1}, V_{p2}, V_s, C_p, C_s, Y)$ is too long and is not given here.

In time domain and for *simplified model*, when taking into account only the real part of dilatational wave velocities we obtain:

$$\begin{cases} t_0(t) = -\left[\rho \dfrac{C_p^2 (V_{p1}+V_{p2})}{C_p^2 + V_{p1} V_{p2}} \partial_t u_{30} + p_0 \right] e_3 - \rho\, C_s\, \partial_t u_0' \\ \varphi_0(t) = -K \left[\rho \dfrac{(V_{p1}^2 - C_p^2)(V_{p2}^2 - C_p^2)}{V_{p1} V_{p2} (C_p^2 + V_{p1} V_{p2})} + \rho_f \right] \partial_{tt} u_{30} \end{cases}$$

For *full model*, some preliminary assumptions should be made on the expression of wave velocity and coupling term (Y) before the inverse Fourier Transform.

4. Variational Formulation

The weak form of the governing equations may be obtained using a variational formulation. A finite-element approximation of $u_s(t)$, $v_{rf}(t)$ and $p(t)$, defined by $u_{sh}(t)$, $v_{rfh}(t)$ and $p_h(t)$, should be computed by restricting the shape functions to finite dimensional spaces for displacement of the solid skeleton, relative pore-fluid velocity and pore pressure.

Let υ and ϖ be admissible virtual displacement and velocity fields and π an admissible virtual pressure field over Ω_I with boundary Γ given as:

$$\begin{cases} \Gamma = \Gamma_\sigma \cup \Gamma_u \cup \Sigma \\ \Gamma = \Gamma_\varphi \cup \Gamma_p \cup \Sigma \end{cases}$$

From the governing equations and following the virtual work principle for every υ, ϖ and π, the following system of equations can be written:

$$(\rho \partial_{tt} u_s, \upsilon)_{\Omega I} + (\rho \partial_t v_{rf}, \upsilon)_{\Omega I} + (\sigma', \varepsilon(\upsilon))_{\Omega I} + (p, div(\upsilon))_{\Omega I} - \langle t, \upsilon \rangle_\Sigma = (\rho g, \upsilon,)_{\Omega I} + \langle t, \upsilon, \rangle_{\Gamma\sigma}$$

$$(\rho_f \partial_{tt} u_s, \varpi)_{\Omega I} + (\rho_f \partial_t v_{rf}/n, \varpi)_{\Omega I} + (K^{-1} v_{rf}, \varpi)_{\Omega I} + (grad p, \varpi)_{\Omega I} = (\rho_f g, \varpi)_{\Omega I}$$

$$(div\,\partial_t u_s,\pi)_{\Omega I} - (v_{rf}, grad\pi,)_{\Omega I} + (\partial_t p/Q,\pi)_{\Omega I} + <\varphi,\pi>_{\Sigma} = - <\varphi,\pi>_{\Gamma\varphi}$$

in which $(\ ,\ )_{\Omega I}$ denotes the integration over the domain Ω_I and $<\ ,\ >_{\Gamma}$ designates the integration along its boundary Γ.

The above equations for the *zeroth order* paraxial approximation become:

$$(\rho\partial_{tt}u_s, \upsilon)_{\Omega I}+(\rho\partial_t v_{rf}, \upsilon)_{\Omega I}+(\sigma',\varepsilon(\upsilon))_{\Omega I}+(p,div(\upsilon))_{\Omega I}-<A(\partial_t u_s),\upsilon>_{\Sigma}-<A'(p),\upsilon>_{\Sigma} = (\rho g,\upsilon,)_{\Omega I}+<t,\upsilon,>_{\Gamma\sigma}$$

$$(\rho_f\partial_{tt}u_s,\varpi)_{\Omega I} + (\rho_f\partial_t v_{rf}/n,\varpi)_{\Omega I} + (K^{-1}v_{rf},\varpi)_{\Omega I} + (grad p,\varpi)_{\Omega I} = (\rho_f g,\varpi)_{\Omega I}$$

$$(div\,\partial_t u_s,\pi)_{\Omega I} - (v_{rf}, grad\pi,)_{\Omega I} + (\partial_t p/Q,\pi)_{\Omega I} + <B(\partial_{tt}u_s),\pi>_{\Sigma} = - <\varphi,\pi>_{\Gamma\varphi}$$

The Finite Element spaces have the usual Finite Element shape functions $w_{sJ}(x)\cdot e_j$, $w_{rfJ}(x)\cdot e_j$ and $w_{pJ}(x)$, where J stands for a node number and e_j is the unit basis vector of Euclidean space. In this manner the following expansion may be written:

$$\begin{cases} u_{sh} = \sum_{Ii} u_{sIi}\cdot w_{sI}\, e_i \\ v_{rfh} = \sum_{Ii} v_{rfIi}\cdot w_{rfI}\, e_i \\ p_h = \sum_I p_I\cdot w_{pI} \end{cases}$$

Correspondingly, the test functions will be successively chosen as equal to each member of the unit basis, e.g.:

$$\begin{cases} \upsilon_{sh} = w_{sJ}\cdot e_j \\ \varpi_{rfh} = w_{rfJ}\cdot e_j \\ \pi_{ph} = w_{pJ} \end{cases}$$

Obviously, the shape functions for skeleton displacement, pore fluid relative velocity and pore pressure may be chosen differently. The above equations written in the matrix form become:

$$\begin{cases} m_{ss}\,\partial_{tt}u_s + p_{ss}\,\partial_t u_s + B\,\sigma' + c_{sf}\,\partial_t v_{rf} + k_{sp}\,p + p_{sp}\,p = f_s \\ m_{fs}\,\partial_{tt}u_s + c_{ff}\,\partial_t v_{rf} + k_{ff}\,v_{rf} + k_{fp}\,p = f_f \\ c_{ps}\,\partial_t u_s + k_{pf}\,v_{rf} + p_{pf}\,v_{rf} + c_{pp}\,\partial_t p = f_p \end{cases}$$

with:

$$
\begin{aligned}
(m_{ss})_{IiJj} &= \delta_{ij}\ (\rho\, w_{sI}, w_{sJ})_{\Omega I} \\
(p_{ss})_{IiJj} &= \delta_{ij} <A(w_{sI}), w_{sJ}>_{\Sigma} \\
(B\sigma')_{Jj} &= (\sigma', \varepsilon(w_{sJ} \cdot e_j))_{\Omega I} \\
(c_{sf})_{IiJj} &= (\rho_f w_{rfI}, w_{sJ})_{\Omega I} \\
(k_{sp})_{IJj} &= -(w_{pI}, div\, w_{sJ}\, e_j)_{\Omega I} \\
(p_{sp})_{IiJj} &= \delta_{ij} <A'(w_{pI}), w_{sJ}>_{\Sigma} \\
(m_{fs})_{IiJj} &= \delta_{ij}\ (\rho\, w_{sI}, w_{rfJ})_{\Omega I} \\
(c_{ff})_{IiJj} &= \delta_{ij}\ (\rho\, w_{rfI}, w_{rfJ})_{\Omega I} \\
(k_{ff})_{IiJj} &= \delta_{ij}\ (K^{-1} w_{rfI}, w_{rfJ})_{\Omega I} \\
(k_{fp})_{IiJ} &= (grad w_{pJ}, w_{rfI}\, e_i)_{\Omega I} \\
(c_{ps})_{IiJ} &= (div\, w_{sI}\, e_i, w_{pJ})_{\Omega I} \\
(k_{pf})_{IiJ} &= (w_{rfI}\, e_i, grad w_{pJ})_{\Omega I} \\
(p_{pf})_{IiJj} &= \delta_{ij} <B(w_{rfI}), w_{pJ}>_{\Sigma} \\
(c_{pp})_{IJ} &= (1/Q\, w_{pI}, w_{pJ})_{\Omega I} \\
(f_s)_{Jj} &= (\rho\, g, w_{sJ} \cdot e_j)_{\Omega I} + <t, w_{sJ} \cdot e_j>_{\Gamma\sigma} \\
(f_f)_{Jj} &= (\rho_f g, w_{rfJ} \cdot e_j)_{\Omega I} \\
(f_p)_J &= -<\varphi, w_{pJ}>_{\Gamma\varphi}
\end{aligned}
$$

We note that with a suitable choice of numerical quadrature the coefficient matrices i the coupled discretized equilibrium equations are symmetric, unless for dynamic bounda terms.

5. Remarks on Time and Space Discretization

The numerical treatment of porous media in presence of small permeability ar incompressibility constraints is problematic. Additionally, in earthquake engineeri problems it is interesting to dispose numerical algorithms able to follow a rapidly changi event (i.e. earthquake) as well as a steady-state situation (i.e. post-seismic consolidatior Many papers have been dedicated to this topic (Park, 1983, Zienkiewicz and co-worke 1984, 1988, 1991, 1994)

In the coupled problems, the incompressibility constraints may largely affect the quali of numerical results. In general, with only certain types of approximate functions of t displacement u_s, velocity v_{rf} and pore pressure p it is possible to obtain stable resu governed by so-called *B-B* (Babuska-Brezzi) conditions. To avoid such restrictions on t spatial approximation Zienkiewicz and Wu (1991) applied the splitting-operator procedu to poroelastodynamic problems formulated by the *simplified model.* Then, both transie and steady state solutions may be obtained by the equal interpolation functions f displacement and pore pressure variables. More recently, Zienkiewicz *et al.* (1994) propos a new algorithm for drained and undrained conditions for saturated porous media unc dynamic loading. This method seems to be very efficient as it avoids the difficulti associated to mixed interpolation and compared to the previous one represents a signific improvement for non-uniform meshes.

In our work, with respect to spatial discretization different interpolation functions displacement/velocity and pore pressure are used. With respect to time, a standard predict corrector algorithm close to the concept of implicit/explicit partition, introduced some ye ago by Hughes and Liu (1978), especially with respect to finite-element partitioning opposed to nodal partitioning, is implemented. Obviously the main reason for develop such schemes is to circumvent the shortcomings of both approaches, mainly the hard stability requirements of explicit schemes difficult to satisfy for two-phase sediments, and the higher cost of implicit schemes.

Either the skeleton displacement, pore-fluid velocity and pore-fluid pressure or their second derivatives with respect to time can be considered as principal unknowns. In linear dynamic problems these two approaches are identical, but for non-linear problems the second choice may be preferable. For one-phase problems, Xie and Steven (1994) proposed an energy-conserving scheme in which, the increment of acceleration is kept as unknown. Their method seems to be very interesting as it removes chaotic behaviour produced by the Newmark scheme in their non-linear dynamic application.

It should be noted that if the system obtained after the discretization with respect to space is not symmetrical, following the strategy adopted for time discretization, it is possible to render the systems symmetrical. The easiest way to do such operations is to treat the non-symmetrical terms explicitly. Therefore, the paraxial elements may be treated as totally explicit elements. However, the critical time step is then smaller than the one required without paraxial elements.

6. Introduction of Seismic Incident Field

The definition of seismic incident field is one of the most difficult aspects in earthquake engineering problems. In almost every case, the seismic loading (acceleration vector) is supposed to be known on the ground surface at so-called control points (usually one). For modelling purposes, due to theoretical limitations, this control point has to be as far as possible from the surface or buried irregularities.

Now, if the outer domain can be assumed as a homogeneous linear elastic domain, the seismic incident field can be evaluated at paraxial-element nodes directly by a simple deconvolution of the acceleration at the control point for any incident angle. In any other cases (i.e. a multilayer substratum), the deconvolution must use other techniques (i.e. the Thomson-Haskell method for a multilayer substratum), after which, the time history of the stress and velocity fields can be entered for each paraxial-element node.

At infinity, the total displacement field tends to the incident displacement field. In one-phase medium, the total displacement vector is split into the (free field) incident and radiant displacement fields on the boundary Σ (Aubry *et al.*, 1985, Modaressi *et al.* 1991):

$$u_E = u_{in} + u_r$$

Moreover, the displacement field must satisfy the Sommerfeld radiation condition:

$$u_E = u_{in} \qquad \text{for } x \longrightarrow \infty$$

and the free field incident motion must be compatible with the free surface conditions.

For a saturated two-phase medium and *zeroth order* paraxial approximation, the total stress vector and flux on the boundaries of the problem depend not only on wave velocities and mechanical properties, but also on pore pressure. This dependency implies that the incident pore pressure must be also known near the boundary. Such an aspect has not been discussed in earlier work on this subject and to our knowledge, in practice, the information on the pore-pressure field is not available. We use the same procedure for the pore-pressure field:.

$$p_E = p_{in} + p_r$$

Now, we will write the radiant fields through Ω_E due to the diffraction over Σ at zeroth order approximation:

$$\begin{cases} t_E(u_r, p_r) = A_0(\partial_t u_r) + A'_0(p_r)\, e_3 \\ \varphi_E(\partial_{tt} u_r, p_r) = B_0(\partial_{tt} u_r, p_r) \end{cases}$$

This radiant field obeys the paraxial equation and approximately verifies the radiation condition. The continuity condition for the stress vector over Σ and the linearity hypotheses in its vicinity enable us to write for the *zeroth order* approximation:

$$\begin{cases} t = - t_E(u_E) = -t_E(u_{in}) - t_E(u_r) = -t'_E(u_{in}) - p_{in} e_3 - A_0(\partial_t u - \partial_t u_{in}) - A'_0(p - p_{in}) e_3 \\ \varphi = - \varphi_E = \varphi_{in} + \varphi_r = - K\,[\partial_3 p_{in} + \rho_f \partial_{tt} u_{3in}] - B(\partial_{tt} u_3) + B(\partial_{tt} u_{3in}) \end{cases}$$

Therefore, to introduce the seismic input motion into calculations, the (free field) incident displacement and pore pressure fields are required.

7. Conclusion

The dynamic formulation of saturated porous media and its numerical implementation in the framework of Finite Element method is presented. For the *full model* suitable for any frequency range an absorbing boundary element is developed using the paraxial approximation. This boundary is local in time and space and, therefore can be easily integrated in standard Finite Element programs for transient analysis of unbounded porous media. The seismic motion can be introduced using these elements. It has been shown that far field pore pressure is also requested in seismic analysis of saturated soils. Unfortunately, there is lack of instrumental data in this field and it is difficult to verify the influence of such a factor in real cases. The above developments have been implemented in a three-dimensional Finite Element code for non-linear transient analysis.

References

Aubry D., Chouvet D., Modaressi H. and Mouroux P., 'Local amplification of a seismic incident field through an elastoplastic sedimentary valley.', *Num. Meth Geomech.* (Eds. Kawamoto and Ichikawa), Balkema, 421-428, 1985.

Aubry D. and Modaressi H., 'Numerical Modelling of the Dynamics of Saturated Anelastic Soils.', *Seismic Hazard in Mediterranean Regions.* (Eds. Bonnin, Cara Cisternas and Frantechi), Kluwer Academic Publishers, 151-171, 1988.

Banerjee P.K., Guin J. and Raveendra, S.T., 'Application of BEM to some dynamic problems of saturated porous media.', *Comput. Meth. Adv. Geomech.* (Eds. Siriwardanne and Zaman), Balkema, 79-89, 1994.

Benzenati, I., '*Modélisation du comportement dynamique des sols saturés appliquée aux barrages et effets de site.*', Doctoral Thesis, Ecole Centrale de Paris, 1993.

Bonnet, G., 'Basic singular solutions for a poroelastic medium in the dynamic range *J. Acoust. Soc. Am.*, 82, 1758-1762, 1987.

Chen, J., 'Fundamental solutions and boundary element formulations for dynamic problems of poroelasticity and thermoelasticity.', Ph.D. dissertation, State University of New York at Buffalo, Buffalo, NY, 1992.

Coussy, O., *Mécanique des Milieux Poreux*, Editions Technip, Paris, 1991

Degrande G. and De Roeck, 'An absorbing boundary condition for wave propagation in saturated poroelastic media, part I: Formulation and efficiency evaluation.', *Soil Dyn. Earthquake Eng.*, 12, 411-421, 1994.

Engquist, B. and Majda, A., 'Absorbing boundary conditions for the numerical simulation of waves.', *Math. Comput.* 31:139, 625-651, 1977.

Ghaboussi, J. and Wilson E.L., 'Variational formulation of dynamics of fluid saturated porous elastic solids.', *Proc. ASCE*, 98(EM4), 947-963, 1972.

Hughes T.J.R. and Liu W.K., 'Implicit-Explicit Finite Elements in transient analysis: Implementation and numerical examples.', *ASME, J. Appl. Mech.*, 45, 375-378, 1978.

Kaynia, A.M. and Banerjee, P.K., 'Fundamental solutions of Biot's equation of dynamic poroelasticity.', *Int. J. Eng. Sci.*, 31, 817-830, 1993.

Modaressi H., Aubry, D. and Bour, M., 'A finite element approach for prediction of site effects.', *Proc. 4th Int. Conf. Seism Zonation*, Stanford, 261-268, 1991.

Modaressi H. and Benzenati I., 'Paraxial Approximation for Poroelastic Media.', *Soil Dyn. Earthquake Eng.*, 13, 117-129, 1994.

Mohasseb S.K. and Wolf J., 'Recursive evaluation of interaction forces of unbounded soil in frequency domain.', *Soil Dyn. Earthquake Eng.*, 8(4), 176-188, 1989.

Park K.C., 'Stabilization of partitionned solution procedures for pore-fluid interaction analysis.' *Int. J. Num. Eng.*,19, 1669-1673, 1983.

Simon, B.R., Wu, J.S.S., Zienkiewicz O.C. and Paul D.K., 'Evaluation of u-w and u-π finite element methods for the dynamic response of saturated porous media using one-dimensional models.', *Int. J. Num. Anal. Meth. in Geomech.*, 10, 461-482, 1986.

Xie Y.M. and Steven G.P., 'Instability, chaos, and growth and decay of energy of time-stepping schemes for non-linear dynamic equations.', *Comm. Num. Meth. Eng.*, 10, 393-401, 1994.

Zienkiewicz O.C., Chang C.T. and Bettess P., 'Drained, undrained, consolidating and dynamic behaviour assumptions in soils. Limits of validity.', *Geotechnique*, 30, 385-395, 1980.

Zienkiewicz O.C. and Shiomi T., 'Dynamic behaviour of saturated porous media; The generalized Biot formulation and its numerical solution, *Int. J. Num. Anal. Meth. in Geomech.*, 8, 71-96, 1984.

Zienkiewicz O.C., Paul, D.K. and Chan A.H.C., 'Unconditionally stable staggered solution procedure for soil-pore fluid interaction problems.', *Int. J. Num. Meth. Eng.*, 26, 1039-1055, 1988.

Zienkiewicz O.C., Chan A.H.C., Pastor, M., Paul, D.K. and Shiomi, T., 'Static and dynamic behaviour of soil: a rational approach to quantitative solutions. I. Fully saturated problems.', *Proc. R. Soc. Lond.*, A 429, 285-309, 1991.

Zienkiewicz O.C. and Wu, J., 'Incompressibility without tears- How to avoid restrictions of mixed formulation.', *Int. J. Num. Meth. Eng.*, 32, 1189-1203, 1991.

Zienkiewicz O.C., Huang, M. and Pastor, M., 'Computational soil dynamics- A new algorithm for drained and undrained conditions.', *Comput. Meth. Adv. Geomech.* (Eds. Siriwardanne and Zaman), Balkema, 47-59, 1994.

SECTION 3

POROELASTICITY:

APPLICATIONS IN GEOMECHANICS

Modelling of thermal consolidation of sparsely fractured rock in the context of nuclear waste management

T.S. Nguyen, Atomic Energy Control Board, Ottawa, Canada
A.P.S. Selvadurai, McGill University, Montreal, Canada

Abstract. This paper[1] summarizes the results of research activities which focus on thermal-mechanical-hydrological (T-M-H) coupling and its implications on the safety issues related to the Canadian concept for the disposal of nuclear fuel wastes. These research activities include the development of a finite element computer code, FRACON, its verification against analytical solutions and its calibration against laboratory and field experiments. The FRACON code was used in the preliminary analysis of the coupled T-M-H response of a sparsely fractured rock mass around a hypothetical nuclear fuel waste repository. It was found that the flow and structural characteristics of this rock mass could be significantly perturbed by the thermal pulse generated by the wastes. The implications of these perturbations on the safety of the disposal system have to be further quantified.

Keywords: Consolidation, Thermal, Coupled Processes, Joint, Finite Element, Nuclear Waste, Fractured Rock, Heater experiments, Hydraulic jacking Tests, Pulse Tests, Constant pressure tests.

1 Introduction

Many countries which utilize nuclear energy for electricity production are investigating permanent disposal solutions for the nuclear fuel wastes (NFW) produced by their reactors. The disposal concept currently being studied by these countries consists of burying the wastes deep into geological media such as granite, salt or clay. In Canada, Atomic Energy of Canada Ltd (AECL) is studying the concept of NFW disposal in granites of the Canadian Shield at depths of 500-1000 m. Although the granitic rock formations of the Canadian Shield have estimated ages of 500 millions to more than 2.5 billions years they have remained mostly stable for the last hundreds of millions of years under different geological and geochemical processes. According to AECL's current conceptual design (AECL, 1992), a repository consisting of a series of disposal rooms

[1] This paper reflects the authors' opinion and not necessarily that of the Atomic Energy Control Board or of McGill University

A. P. S. Selvadurai (ed.), *Mechanics of Poroelastic Media*, 159–180.

would be excavated at a depth of 500 to 1000 m in a granitic rock formation of the Canadian Shield. Three rows of boreholes would be drilled along the length of each room. Corrosion resistant containers containing the wastes would be emplaced in the boreholes. A buffer material, consisting of a mixture of sand and bentonite would be packed around the containers in the boreholes; after all boreholes have been filled , each disposal room would be backfilled with a mixture of clay and crushed granite. When such a repository is completely filled and sealed, it is generally recognized that with the eventual decay of the containers, groundwater is the most important agent that could transport contaminants from the wastes to the surface . The groundwater migration of contaminants would be retarded by several barriers. The waste forms have very low dissolution rates and constitute the first barrier. The waste containers if properly designed should be structurally stable, and have a very slow rate of corrosion. The buffer and backfill by virtue of their low hydraulic conductivities and high chemical sorption potential would ensure slow rates of contaminant migration through them. The combined effects of these multiple barriers should be such that by the time contaminants reach the ground surface, radioactive decay would reduce their concentrations to levels harmless to humans and the environment.

In the current Canadian concept for NFW disposal, the natural geological barrier is probably the most important one (Nguyen, 1992). The performance of the geological barrier ultimately depends on its groundwater regime. Since the wastes will generate heat for hundreds to thousands of years, the stability , deformation and water flow characteristics of the rock mass will be influenced by the introduced thermal pulse. These coupled Thermal-Hydrological-Mechanical (THM) processes can be interpreted by generalizing the theory of consolidation proposed by Biot (1941) to include thermal effects. To develop plausible models of THM processes in geological media, it is desirable to consider the characteristics of both the intact rock and discontinuities such as fracture zones that are usually present in most hard rocks. An extensive literature review has shown that the study of coupled THM processes in porous , sparsely fractured geological media, such as granitic rock of the Canadian Shield is still at a relatively early stage of theoretical and experimental development (Cook, 1987; Noorishad and Tsang, 1987; Onishi et al, 1987; Witherspoon, 1987).

The Atomic Energy Control Board (AECB) is the Canadian agency responsible for the regulation of the nuclear industry in Canada. Presently the AECB is actively reviewing the concept for NFW disposal proposed by AECL. In order to gain an independent view on the implications of THM coupling on the safety of the concept, the AECB has initiated a research program, in cooperation with Carleton University and McGill University which consists of: i) the development of a finite element code, FRACON, with a variety of element capabilities which can examine THM processes in both solid media and in discontinuities; ii) verification of the code against analytical solutions for both isothermal and non-isothermal cases; iii) calibration of the code against laboratory and field experiments performed at Carleton University and McGill University, and by other researchers involved in experimental THM modelling; vi) simulation of the coupled THM behaviour of a hypothetical NFW repository and perform parametric studies to examine the influences of jointing, stress- dependence of permeability, nonlinear stress-strain behaviour (particularly for fractures and joints), etc. on the performance of the hypothetical NFW repository. In this paper, we will present the progress achieved to date in the above research program.

2 Governing equations of coupled T-H-M processes in saturated geological media and finite element formulation

By generalizing Biot's (1941) theory of consolidation to include thermal effects, we can obtain the following equations governing the non-isothermal consolidation of saturated porous media (Selvadurai and Nguyen, 1993):

$$\frac{\partial}{\partial x_i}\left(\kappa_{ij}\frac{\partial T}{\partial x_j}\right) = \rho C\frac{\partial T}{\partial t} \quad (1)$$

$$G\frac{\partial^2 u_i}{\partial x_j \partial x_j} + (G+\lambda)\frac{\partial^2 u_j}{\partial x_i \partial x_j} + \alpha\frac{\partial p}{\partial x_i} - \beta K_D\frac{\partial T}{\partial x_i} + F_i = 0 \quad (2)$$

$$\frac{\partial}{\partial x_i}\left(\frac{k_{ij}}{\mu}\left(\frac{\partial p}{\partial x_j} + \rho_f g_j\right)\right) - \left(\frac{n}{K_f} - \frac{n}{K_s} + \frac{\alpha}{K_s}\right)\frac{\partial p}{\partial t} +$$
$$\alpha\frac{\partial}{\partial t}\left(\frac{\partial u_i}{\partial x_i}\right) + \left((1-\alpha)\beta - (1-n)\beta_s - n\beta_f\right)\frac{\partial T}{\partial t} = 0 \quad (3)$$

where:
the unknowns are the displacement u_i [m], temperature T [°C] and pore pressure p [Pa]
κ_{ij} is the thermal conductivity tensor of the bulk medium [W/m/°C]
ρ is the density of the bulk medium [kg/m^3]
C is the specific heat per unit mass of the bulk medium[J/kg/°C]
G and λ are the Lame's constants [Pa]
$\alpha = 1 - K_D/K_s$
K_D,K_s and K_f are respectively the bulk moduli of the drained material, the solid phase and the fluid phase [Pa]
F_i is the volumetric body force [N/m^3]
β,β_s and β_f are respectively the coefficient of thermal expansion of the drained material,

the solid phase and the fluid phase [°C^{-1}]
n : is the porosity of the medium [dimensonless]
ρ_f is the density of the fluid [kg/m^3]
μ is the viscosity of the fluid [kg/m/s]
k_{ij} is the intrinsic permeability tensor [m^2]
g_i is the ith component corresponding to the acceleration due to gravity [m/s^2]
(In the above parameter definitions the bracketed designations are in typical SI units)

In the above equations, the Cartesian tensor notation, with Einstein's convention of summation on repeated indices was adopted. In addition, the sign convention for stress and fluid pressure is considered positive for tension fields. The governing equations are approximated by matrix equations via a standard Galerkin finite element procedure (Huyakorn and Pinder, 1983). Two types of elements are considered:

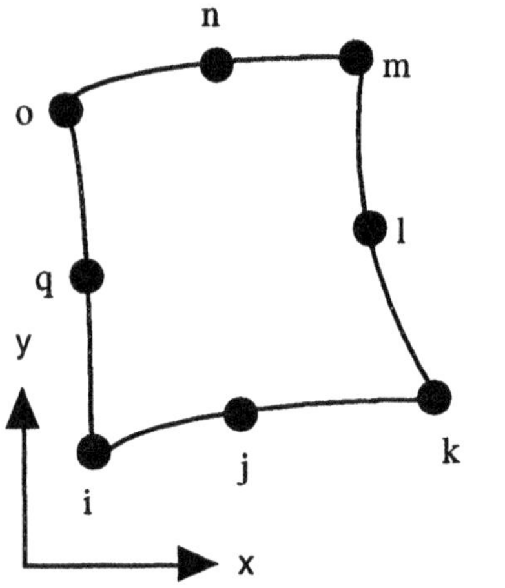

Figure 1 Plane isoparametric element

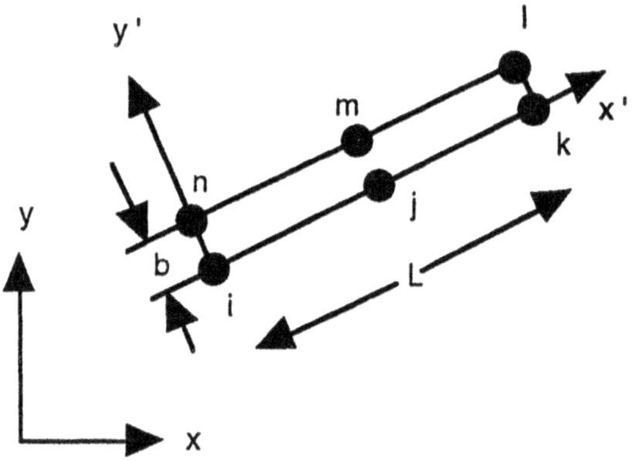

Figure 2 Joint Element

i) Plane isoparametric elements: This element (Figure 1) is used to represent the unfractured rock mass. Displacements within the element are interpolated as functions of the displacements at all 8 nodes, while the pore pressure and temperature are interpolated as functions of the same values at only the four corner nodes i, k, m and o. A detailed description of this element is given , for example, by Smith and Griffiths(1988).

ii) Joint Element: This element (Figure 2) is used to simulate discontinuities in the roc mass such as joints, fracture zones, and fault zones. In finite element terminology (see e.g. Goodman et al, 1968; Zienkiewicz et al, 1970; Noorishad et al, 1971; Ghaboussi et al, 1973; Desai and Nagaraj, 1986), it is a very thin element, characterized by a thickness b and length L. Nodal displacements are obtained at all six nodes (Figure 2) while nodal pore pressures and temperatures are obtained only at the corner nodes i, k, l and n. The mechanical behaviour of the element is dictated by its shear and normal stiffnesses $D_{x'y'}$ and $D_{y'y'}$ respectively and its hydraulic and thermal behaviour are governed by the transverse and longitudinal permeabilities $k_{y'y'}$ and $k_{x'x'}$, and the transverse and longitudinal thermal conductivities $\kappa_{y'y'}$ and $\kappa_{x'x'}$, respectively .

3 Code verification

Verification of a computer code is the process by which one assures that the code consistently (and adequately) solves the governing equations. Verification could include such activities as comparison of the code results with analytical solutions or code-to-code comparison (i.e. comparison with the results of other codes which solve the same equations). A set of verification problems for both thermal and isothermal consolidation are given by Nguyen and Selvadurai (1994). For purposes of illustration, we only show here the problem of thermal consolidation of an infinite saturated porous medium containing a cylindrical heat source. An analytical solution for this case is given by Booker and Savvidou (1985). The same problem has been examined by Lewis and Schrefler (1987) using the finite element code PLASCON, using an unspecified but presumably consistent system of units. For comparison purposes, we use here input data identical to those used by Lewis and Schrefler (1987):

radius of cylindical source: 0.3
height of cylindical source: 2
heat generation rate per unit volume of the cylinder q= 1768.39
E=6000
$\nu = 0.4$
$K = 0.4\mathrm{x}10^{-5}$
$\kappa = 1.02857$
$\rho C=40$
$\alpha = 0.5$
$\beta = \beta_s = 0.9\mathrm{x}10^{-6}$
$\beta_f = 0.63\mathrm{x}10^{-5}$

The finite element mesh along with the boundary conditions for this problem are shown in figure 3. Axisymmetric conditions are specified. The results for temperature and pore pressure at three different nodes are shown in figure 4. The FRACON results compare very well with the analytical solutions. It can be seen that at all points, the pore pressure increases due to thermal expansion of the pore fluid, and then gradually dissipates as the medium is allowed to consolidate.

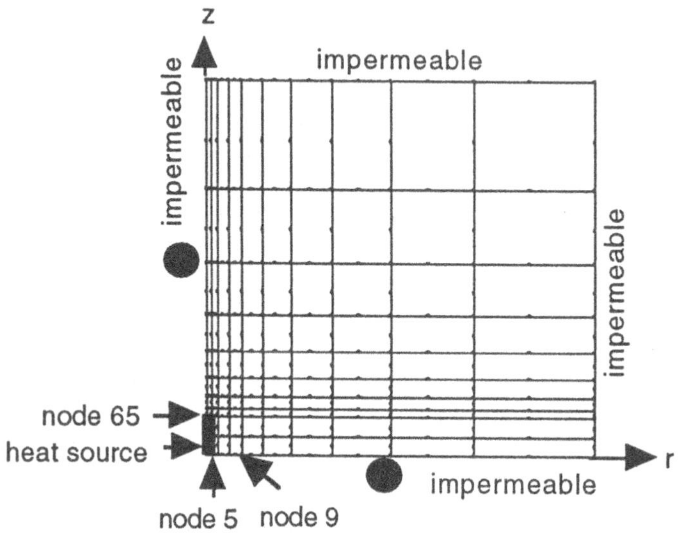

Figure 3 Finite element mesh - Consolidation around a cylindrical heat source

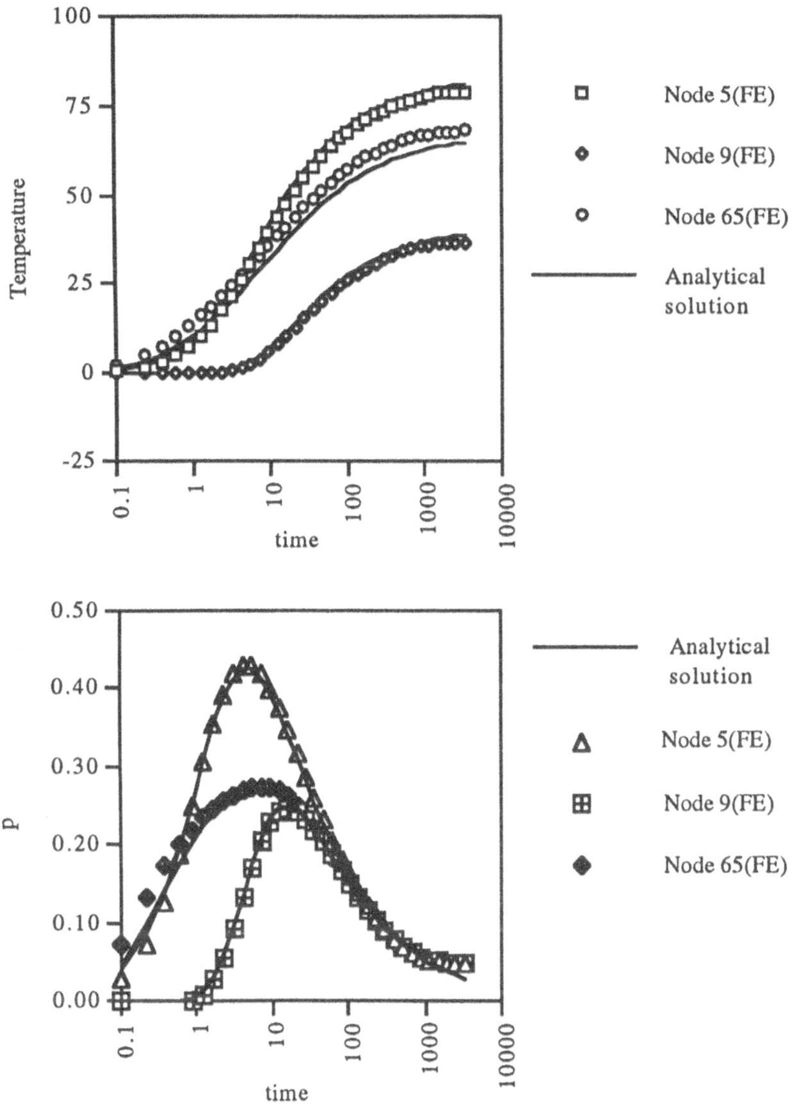

Figure 4 Cylindrical heat source - Temperature and pore pressure evolution

4 Calibration with a Laboratory Experiment on an Intact Synthetic Rock

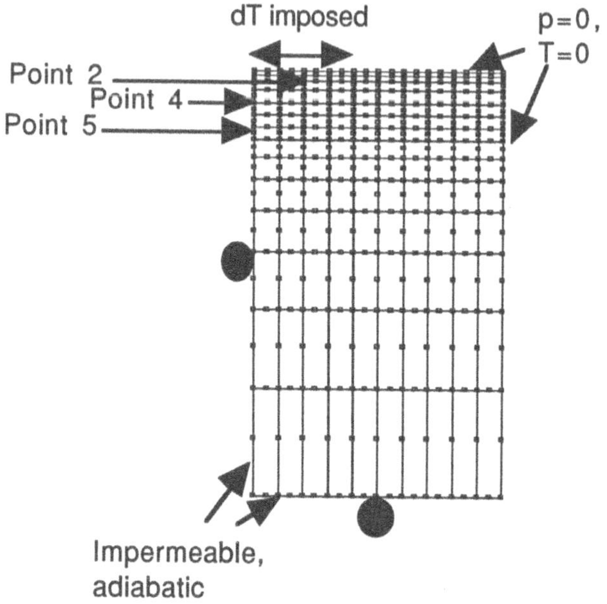

Figure 5 Finite Element mesh for heated cylinder experiment

An experiment was performed at Carleton University, Ottawa, Canada to investigate TMH coupling in intact rock. A cementitious material was used to simulate rock. A cylinder, 50 cm in diameter and 47 cm in height was cast with a special mixture of SIKA grout. Thermocouples and pore pressure transducers were embedded in the cylinder. The cylinder was vacuum saturated after a 28-day curing period. The cylinder was then immersed in water and a heat source was applied at the surface. Temperatures and pore pressures were continuously recorded by a computerized data acquisition system during the experiment.

The properties of the grout material were measured as follows:
Thermal heat conductivity: $\kappa = 0.563$ J/m/s/°C
Heat capacity: $\rho C = 2x10^6$ J/m^3/°C
Poisson ratio: ν=0.24; Young's modulus E = 26.65 GPa.

Coefficient of volumetric thermal expansion of the dry grout: 2×10^{-5}/oC
Hydraulic conductivity : 10^{-14} to 10^{-12} m/s

Further details of the experimental procedures are given by Selvadurai (1994). The laboratory experiment was simulated with the FRACON code, using the above input data. The axisymmetric finite element mesh and the relevant boundary conditions are shown in figure 5. Since the hydraulic conductivity was very low, undrained conditions were assumed in the analysis. The calculated values of temperature agree well with the experimental values (figure 6). In order to match the calculated values of pore pressure with the experimental values, the assumption of slightly imperfect saturation, has to be made. This assumption will result in a modified value of compressibility of the air/water mixture in the pores, as described below.

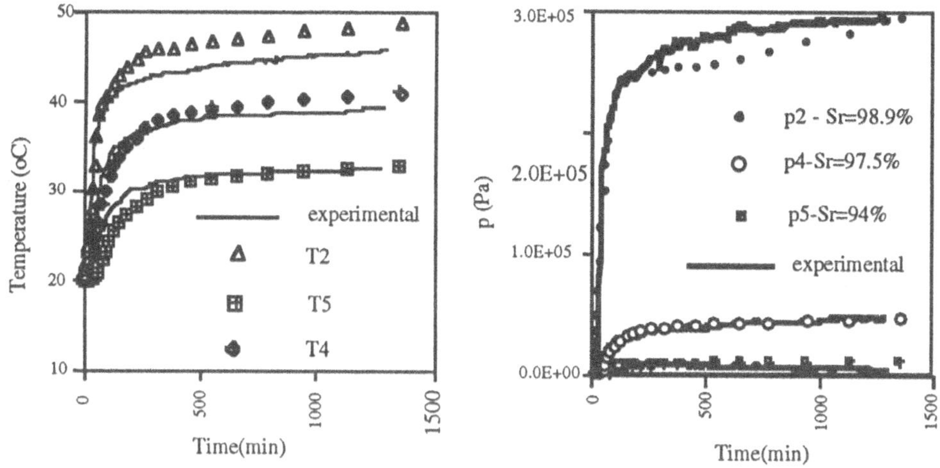

Figure 6 Heated cylinder - temperature and pore pressure evolution

When the degree of saturation Sr is high (80%<Sr<100%) one can neglect capillary and surface tension effects (Vaziri et al, 1992) The porewater/air mixture could then be considered as a homogeneous single phase fluid, albeit with a higher compressibility than pure water. The relationship between the compressibility of the equivalent single phase pore fluid and the degree of saturation could be derived as follows.

Consider a volume V of the porous medium, with a volume V_v of voids. The void space is occupied partially with water and partially with air. By definition:

$V_w = V_v\, Sr$ (4)

$V_a = V_v\,(1-Sr)$ (5)

where V_w and V_a are respectively the volumes of water and air.

When a pressure dp is applied to the water/air mixture, a volumetric compression in the voids dV_v results, such that:

$$\frac{dV_v}{V_v} = \frac{dV_w}{V_w}\,Sr + \frac{dV_a}{V_a}\,(1-Sr) \quad (6)$$

The compressibility of the air/water mixture is, by definition:

$$C_{aw} = \frac{1}{V_v}\frac{dV_v}{dp} \quad (7)$$

From (6):

$C_{aw} = C_w Sr + C_a(1-Sr)$ (8)

where C_{aw}, C_w, C_a are the compressibilities of, respectively, the air/water mixture, water and air.

Let us assume that the ideal gas law is valid for air:

$p_{abs} V_a = mRT$ (9)

where m is the number of moles, R is the universal gas constant, T is the absolute temperature (°K) and p_{abs} is the absolute pressure :

$p_{abs} = p_{atm} + p$ (10)

where p_{atm} is the atmospheric pressure and p is the relative pressure.

Differentiating equation (9) with respect to p, one obtains:

$$C_a = \frac{1}{V_a}\frac{dV_a}{dp} = -\frac{1}{p_{abs}} \quad (11)$$

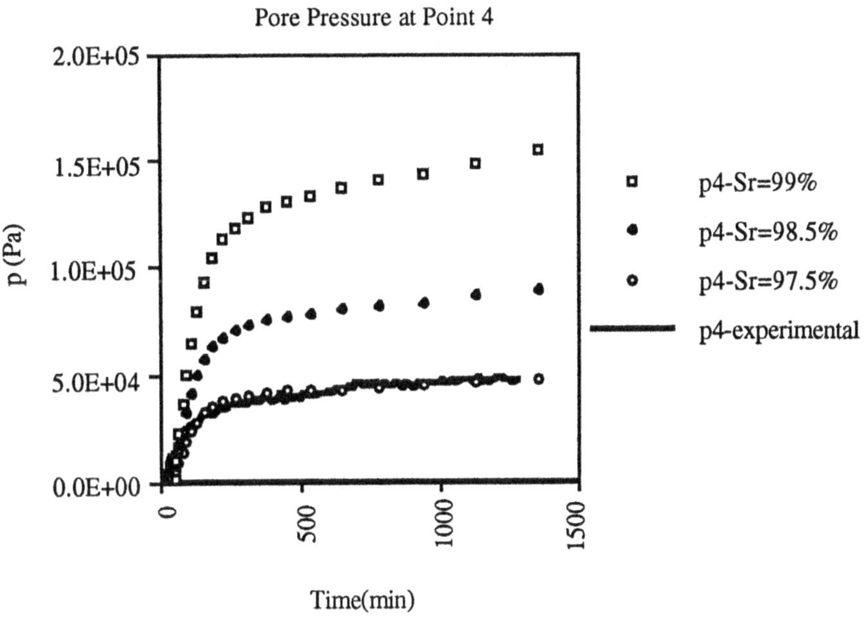

Figure 7 Effect of degree of saturation on pore pressure

Defining compression as positive, and substituting the value of p_{abs} from equation (10) into (11), we obtain:

$$C_a = \frac{1}{p_{atm} + p} \quad (12)$$

Equation (8) then becomes :

$$C_{aw} = C_w Sr + \frac{1-Sr}{p_{atm} + p} \quad (13)$$

In the governing equation (3) , the bulk modulus of the pore fluid would be:

$$K_f = 1/C_{aw} \quad (14)$$

and would depend on the unknown p, thus making the equation nonlinear. A direct iteration method was incorporated in the FRACON code to handle this nonlinear behaviour. We assumed that the degree of saturation of the cylindrical block varies from 94% to 99 % , with a higher degree of saturation near the outer surface. With these modifications to the degree of saturation, a good match is obtained between the calculated and measured pore pressure values as shown in figure 6. Figure 7 shows that the absolute value of the pore pressure is sensitive to Sr, while the time transient is not much affected.

5 Calibration with borehole injection tests in a horizontal fracture

An International project, DECOVALEX (SKI, 1993), was initiated in 1991 , with the objective of developing, verifying and validating computer models for simulating the THM response of fractured hard rocks. Simulation problems are defined to allow the researchers to develop models and compare predictions with a variety of computational solutions, analytical solutions and experimental data. One of the DECOVALEX problems, called TC6, consists of the simulation of borehole injection tests conducted in a single horizontal fracture (Rutqvist, 1993). A vertical borehole, 56 mm in diameter, was drilled in hard crystalline rock, at the Lulea University of Technology, Sweden. The borehole intersected a horizontal fracture at a depth of 356.7 m. A 65 cm section of the borehole, containing the fracture, was isolated by packers. Three types of injection tests were performed by pumping water into the isolated section according to the following methods:

<u>The pulse test</u>: this consists of instantaneously raising the pressure in the isolated section and then closing the valve immediately above it. The pressure in the isolated section will decay to the original pressure as flow into the fracture takes place.

<u>The hydraulic jacking test</u>: this consists of raising the pressure in the section in stepwise manner and measuring the rate of water injection into the section. Each pressure level is maintained for a period of approximately one minute, in order to achieve an apparent steady flow rate, before increasing the pressure to the next level. The test results are presented as pressure versus flow rates (figure 10).

<u>The constant pressure test</u>: this consists of increasing the pressure in the section to certain level and then maintaining it constant during the entire test. The flow rate versus time is monitored (figure 11). In the present case, a constant pressure of 11.5 MPa attained only after approximately 60 seconds of water injection into the isolated section this pressure is then maintained for 10 minutes.

A characteristic nonlinear coupled hydraulic/mechanical behaviour of the rock mass/ fracture system is apparent in its response to the above experiments (Noorishad and Doe, 1982; Rutqvist et al, 1990). This behaviour can be interpreted in the context of Biot's theory of consolidation. The FRACON code was used to simulate these tests and satisfactory results are obtained at the present time for the hydraulic jacking and the constant pressure test.

The following properties of the intact rock were obtained from laboratory tests on drill rock cores:
Young's modulus: $E = 80$ GPa
Poisson Ratio's: $\nu = 0.25$
Density: 2800 kg/m^3

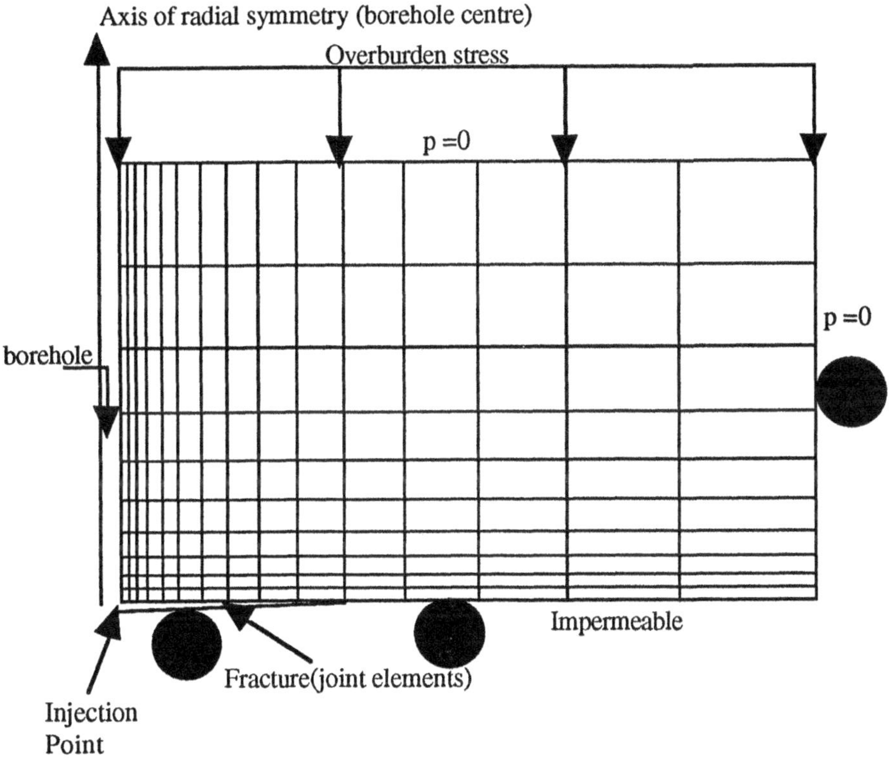

Figure 8 Finite Element Mesh for borehole injection tests

Compression tests on a core containing a fracture were performed and typical curves of normal stress versus normal displacement of the fracture were obtained. A cyclic loading procedure is adopted; the third loading/unloading cycle is generally considered representative of in-situ conditions. The unloading curve from the third cycle is shown in figure 9 (where exceptionally here, compression is considered positive). We only show the unloading curve because during the injection tests, the pore pressure increases and thus the normal effective stress across the fracture decreases. The in-situ total normal stress across the fracture is estimated to be between 8 and 10 MPa. The fracture radius was roughly estimated to be in the 1 m - 5 m range.

The finite element mesh with the assumed boundary conditions is shown in figure 8 . Axisymmetric conditions were assumed. The fracture is represented by joint elements and its radius was assumed to be approximately 5m.

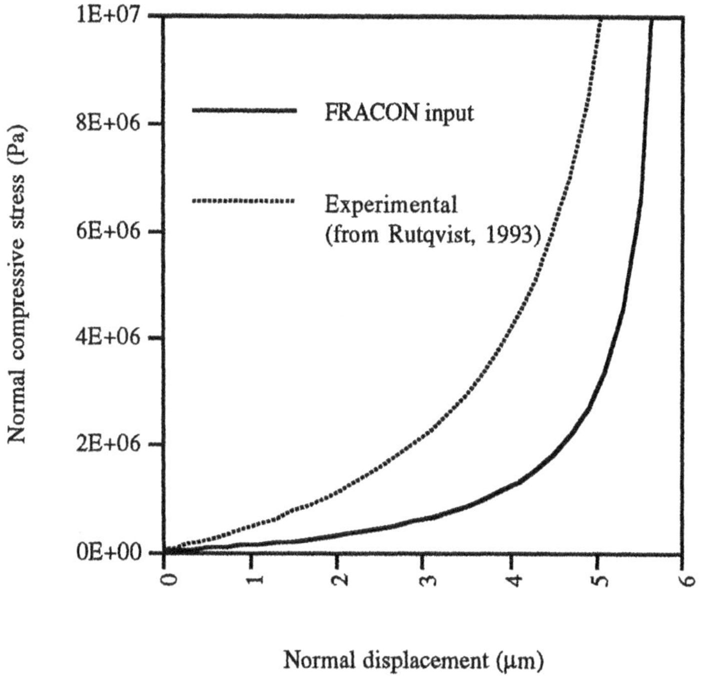

Figure 9 Normal Stress vs Normal Displacement relationship for Fracture

A calibration procedure was performed by varying the properties of the rock mass and the fracture within a representative range. The following input parameters used in the FRACON simulation provided the best fit to date between calculated and experimental results:

Young's modulus of rock mass: E = 47.5 GPa
Poisson Ratio's of rock mass: $\nu = 0.25$
Permeability of rock mass: 10^{-20} m^2

The Young's modulus value of the rock mass used in the analysis is about half of the intact core value. This is considered to be a reasonable assumption due to scale effects and the likely presence of microcracks in the rock mass. The normal stress-normal displacement relationship for the fracture is assumed to follow a hyperbolic function as shown in figure 9. The experimental curve(for unloading during the third cycle of the laboratory experiment) shows a stiffer behaviour of the fracture than the one assumed in the FRACON simulation. Due to scale effects, a less rigid in-situ behaviour as used in the FRACON analysis is considered a reasonable assumption.

The parallel plate model, developed from the application of the Navier-Stokes equation for laminar incompressible flow between two parallel smooth plates, is used to calculate the permeability k of the fracture (see e.g. Benjelloun, 1993):

$k = eh^2/12$ (15)

where eh is the hydraulic aperture of the joint.

Since natural fractures are quite dissimilar to ideal parallel plates, the hydraulic aperture of the fracture is not equal to its mechanical aperture. The relation proposed by Benjelloun (1993) is adopted:

$eh = eh_0 + f\,\Delta em$ (16)

where eh_0 is the initial hydraulic aperture, Δem is the variation in mechanical aperture and f is a correction factor that varies from 0.5 to 1 (as suggested by Benjelloun, 1993).
In the FRACON simulation of the borehole injection tests, we used:

$eh_0 = 18\ \mu m$

f=0.9

The in-situ total normal stress in the FRACON simulation is assumed to be 9.4 MPa.

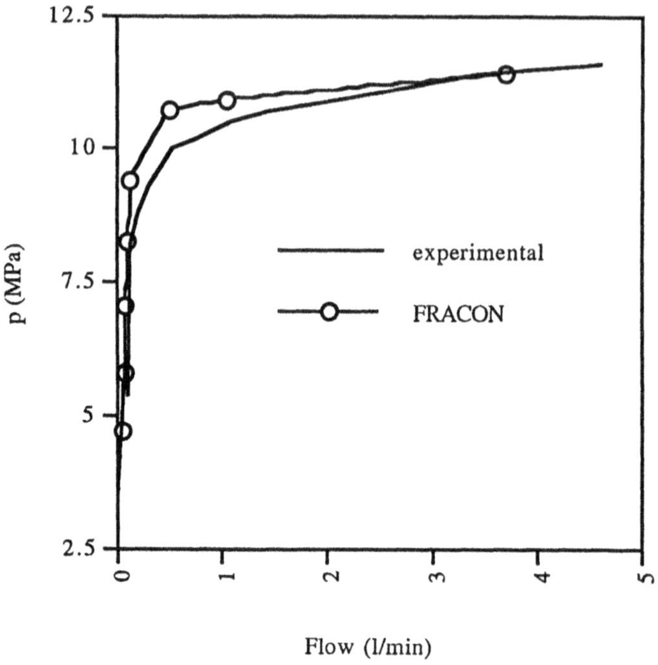

Figure 10 Hydraulic Jacking Test

In figure 10 , the results of the FRACON simulation are compared to the experimental results of the hydraulic jacking test. Satisfactory agreement between the simulated and the experimental data could be observed. Both sets of data show that at injection pressures below the in-situ total normal stress across the fracture, the flow rate is quite low. The resistance to opening of the fracture comes mainly from its normal stiffness, without too much contibution from the surrounding rock mass. For pressures above the in-situ stress, the fracture normal stiffness becomes very low and its aperture increases drastically, resulting in a very rapid increase of its permeability and consequently of the flow rate The rigidity of the system in the latter case is almost entirely due to the surrounding rock mass. It was found during the course of this numerical simulation that the flow rates for this latter case is very sensitive to the Young's modulus of the surrounding rock mass The Young modulus used in the simulation has a value of 47.5 GPa and is lower than the one determined from laboratory tests on intact rock samples (80 Gpa). This is value of the Young's modulus is considered reasonable, since scale effects certainly prevail in the field

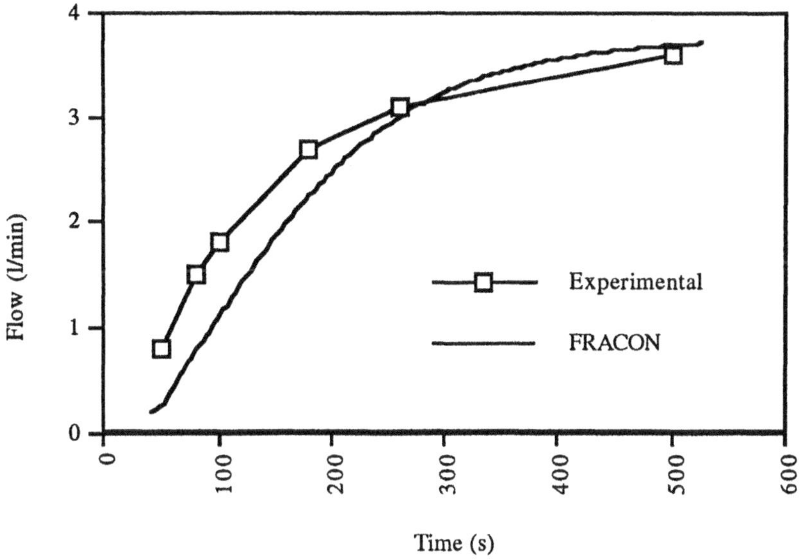

Figure 11 Constant pressure test

In figure 11, the FRACON results are compared with the experimental results for the constant pressure test. Again the simulated results are consistent with the experimental data. For a fracture with constant permeability, analytical solutions (based on Carslaw and Jaeger, 1959) shows that the flow rate sharply increases very early in time, since the hydraulic gradient is very high. This flow then gradually decreases to a steady state-value, as the pore pressure is dissipated radially into the fracture. Both the experimental and simulated results show here a completely different picture from the above behaviour: a gradual increase of the flow rate is observed. This type of behaviour is indicative of the nonlinear nature of the mechanical/hydraulic processes that prevail during the test, because of the high water pressure (11.5 MPa, higher than the in-situ normal stress). The high pressure of the injected water leads to a drastic opening of the fracture, resulting in an increase of its permeability by several orders of magnitude. Consequently, the flow rate increases with time and reach a steady-state value when the aperture distribution in the fracture reaches a stable value, as shown in figure 11.

6 Simulation of Rock Mass Response to a Nuclear Fuel Waste Repository

The FRACON code was used in a preliminary analysis of the performance of a hypothetical repository situated at a depth of 1000 m in a typical pluton of the Canadian Shield. The repository occupies an area of 2000 m by 2000 m. It generates heat at an initial rate of 10.4 W/m^2. Due to the decay in radioactivity of the wastes, this heat production rate decays to 95% after 1 year, 80% after 1000 years and less than 10% after 10000 years. It is assumed that two vertical fracture zones, 20 m thick, exists at 100 m from two opposite edges of the repository. Assuming plane strain conditions, the rock mass response was simulated with the FRACON code. The finite element mesh is shown in Figure 12, along with the boundary conditions. Due to symmetry, only half of the repository was considered. The following thermal-mechanical- hydrological properties of the rock mass are typically applicable to the Canadian Shield (AECL, 1992), and are used in the analysis:
Young's modulus and Poisson's ratio: $E=35$ GPa, $\nu=0.2$,
porosity $n=0.005$, $\kappa=3 W/m/^oC$, $C=900\ J/kg^oC$,
density of solids and fluid: $\rho_s=2700\ kg/m^3$, $\rho_f=1000\ kg/m^3$,
Coefficient of thermal expansion of solid and fluid: $\beta_s=0.24x10^{-4}$, $\beta_f=0.4x10^{-3}$,
Hydraulic conductivity : $K=1x10^{-11}$ m/s

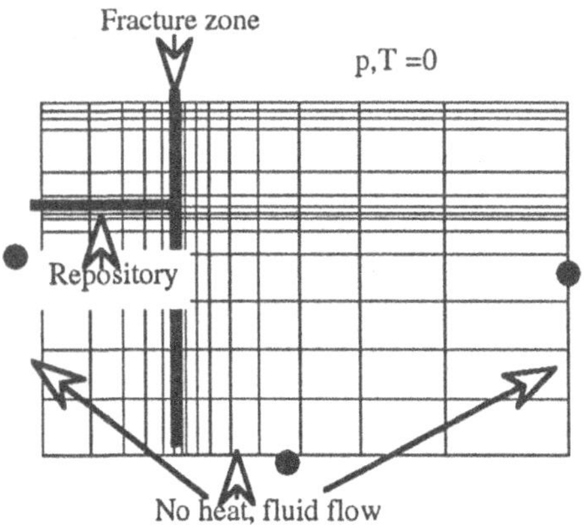

Figure 12 Finite element mesh for waste repository

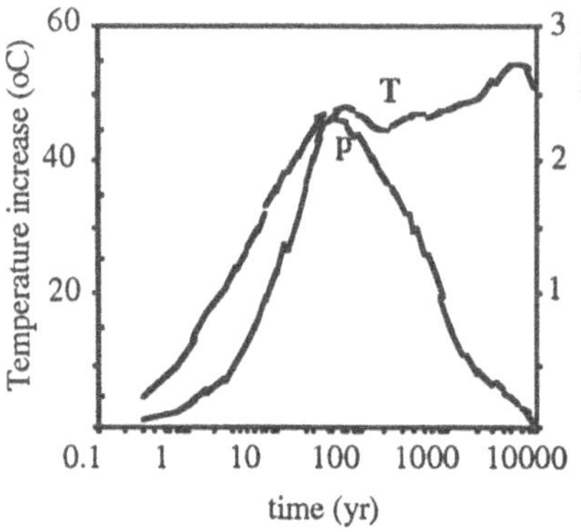

Figure 13 Pore pressure and temperature at centre of repository

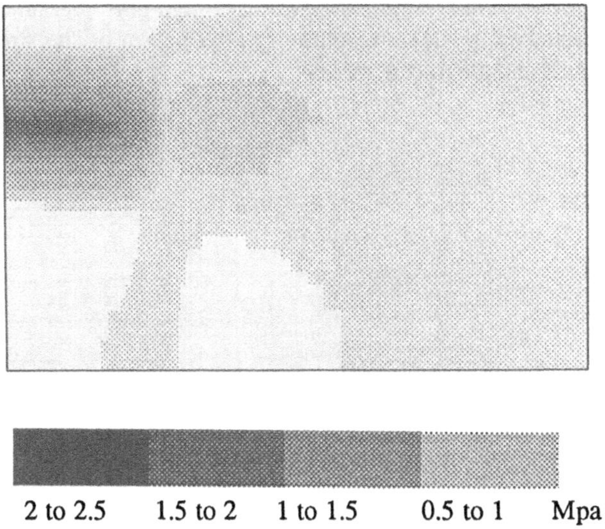

Figure 14 Typical pore pressure contours

The fracture zone is assumed to have axial and transverse hydraulic conductivities which are 3 orders of magnitude higher than those applicable to the competent rock mass. Its shear stiffness is assumed to be: $D_{x'y'}$ = 0.035 GPa (i.e. 3 orders of magnitude lower than the Young's modulus of the competent rock mass).

The temperature and the pore pressure increases at the center of the repository are shown in Figure 13. The temperature increase shows two peaks, 50°C at 70 years and 55°C at 5000 years; this is consistent with results of others (AECL, 1992). The pore pressure increase shows a peak of approximately 2.5 MPa at 40 years.This pore pressure increase is due to the fact that the thermal expansion coefficient of the water is higher than that of the solid matrix. Due to the low permeability of the medium, drainage is slow and the pore water expansion is impeded, resulting in pore pressure increases in early times. With progress of time, drainage of water away from the heat source gradually takes place resulting in pore pressure dissipation. Typical contours of pore pressure increases are shown in Figure 14 (at 40 years). These gradients can attain values of up to 100% (i.e. several orders of magnitude higher than typical regional gradients in the Canadian Shield which are in the order of 0.1%) resulting in increased groundwater velocities diverging from the repository. Thus, the existing groundwater regimes will be significantly modified by the thermal pulse. One can also see that the fracture zone acts as a drainage feature and would constitute a preferential pathway for groundwater exit to the surface area above the repository. Another implication of these high pore pressures is that effective stresses will be reduced, possibly resulting in a reduction of the strength of both the competent rock mass and the fracture zones.

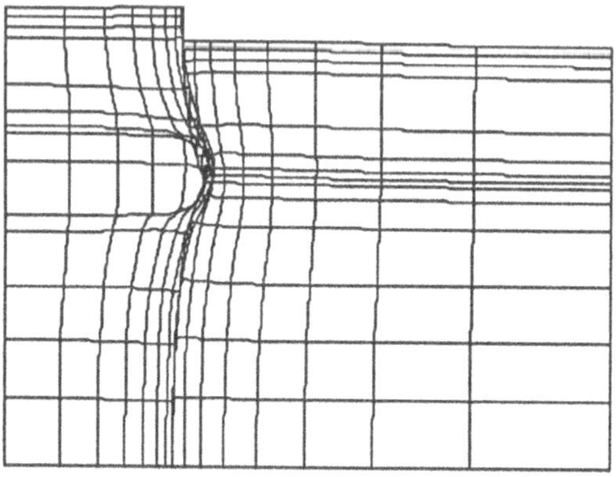

*** Deformation scale is highly exaggerated

Figure 15 Deformed mesh

A typical deformed configuration of the repository region is shown in Figure 15 . Figure 15 shows that shear movements are induced in the fracture zone, an uplift of the ground surface is induced above the repository and thermal expansion of the rock matrix takes place around the repository. These displacements are only of the order of centimetres, but they indicate significant disturbances in the stress regime and the structural integrity of the rock mass in the vicinity of the repository. Efforts will be made by the authors in the near future to further quantify the above hydrological and mechanical perturbations and their implications on the safety of the repository.

7 Conclusions

This paper described the current progress of the authors' research activities on TMH coupling and its implications on the safety of the Canadian Nuclear Fuel waste Disposal concept. As part of this program, the computer code FRACON was developed, tested against analytical solutions, calibrated against a laboratory heater experiment on synthetic rock, calibrated against field hydraulic tests in a fracture, and used in the preliminary assessment of the TMH response of a rock mass around a repository. To gain more confidence in the validity of the conceptual model integrated in the FRACON code, laboratory experiments of TMH processes on intact and fractured granitic blocks are in progress at McGill University for calibration/validation purposes. A detailed analysis to quantify the implications of THM coupling on the safety of a NFW repository will also be performed in the near future.

References

1. Atomic Energy of Canada Ltd, 1992, '*Used Fuel Disposal Centre - A reference Concept*', report TRM-3.

2. Benjelloun, Z. H., 1993, ' Etude Experimentale et Modélisation du Comportement Hydromécanique des Joints Rocheux', Thèse de doctorat, Université Joseph Fourier, Grenoble I.

3. Biot, M. A., 1941, 'General theory of three dimensional consolidation', J. Appl. Phys., 12, 155-164.

4. Booker, J.R. and Savvidou, C. , 1985, 'Consolidation around a point heat source', Int. J. Num. and Analytical Methods in Geomechanics, 9, 173-184.

5. Carslaw, H.S. and Jaeger, J.C. , 1959, ' *Conduction of heat in solids*', Clarendon Press, Oxford.

6. Cook, N.G.W., 1987, ' Coupled processes in geomechanics', in ' *Coupled Processes Associated with Nuclear Waste Repositories*', edited by C.F. Tsang, 39-66, Academic Press Inc.

7. Desai, C.S. and Nagaraj, B.K., 1986, 'Constitutive modelling for interfaces under cyclic loading', in '*Mechanics of Material Interfaces*' (Selvadurai, A.P.S. and G.Z. Voyiadjis, Eds.), Vol.II, Studies in Applied Mechanics, Elsevier, 97-108.

8. Ghaboussi, J., Wilson, E.L. and Isenberg, J., 1973, ' Finite Elements for rock joints and interfaces', J. Soil Mech. Fdn. Div. Proc. ASCE, 99, 833-848.

9. Goodman, R.E.,Taylor, R.L. and Brekke, T.L., 1968, ' A model for the mechanics of jointed rock', J. Soil Mech. Fdn. Div. Proc. ASCE, 94, 637-659.

10. Huyakorn, P.S. and Pinder, G.F., 1983, '*Computational methods in subsurface flow*', Academic Press.

11. Lewis, R.W. and Schrefler, B.A., 1987, ' *The Finite element method in the deformation and consolidation of porous media* ', John Wiley & Sons.

12. Nguyen, T.S., 1992, ' Canadian Concept of Nuclear Fuel waste Disposal - Effectiveness of container, buffer and host rock as contaminant barriers', Proc. of the 1992 Canadian Geotechnical Conference, paper 91.

13. Nguyen, T.S and Selvadurai, A.P.S, 1994, ' The Role of Thermal-Mechanical-Hydrological Coupling on the safety of the Canadian Concept of Nuclear Fuel Waste Disposal' (submitted to the Int. Journ. Rock Mechanics and Mining Sciences)
14. Noorishad J. and Tsang, C.F., 1987, ' Simulation of coupled thermal-hydraulic-mechanical interactions in fluid injection into fractured rocks ', in ' *Coupled Processes Associated with Nuclear Waste Repositories*', edited by C.F. Tsang, 673-678, Academic Press Inc.
15. Noorishad, J., Witherspoon, P.A. and Brekke, T.L., 1971, '*A method for coupled stress and flow analysis of fractured rock masses*', Geotechnical Engineering Publication No. 71-6, University of California, Berkeley.
16. Onishi Y., Shibata, H. and Kobayashi, A., 1987, ' Development of finite element code for the analysis of coupled thermo-hydro-mechanical behaviors of a saturated-unsaturated medium ', in ' *Coupled Processes Associated with Nuclear Waste Repositories*', edited by C.F. Tsang, 679-698, Academic Press Inc.
17. Rutqvist, J., 1993, ' DECOVALEX, Test Case 6, Borehole Injection Test', Doc 93/139, DECOVALEX Secretariat, Engineering Geology, Royal Institute of Technology, Stockholm.
18. Rutqvist, J., Ljunggren, C., Stephansson, O., Noorishad, J. and Tsang, C. F., 1990, ' Theoretical and Field Investigations of Fracture Hydromechanical Response under Fluid Injection', in ' *Rock Joints*', 557-564, edited by N. Barton and O. Stephansson, Balkema.
19. Selvadurai, A.P.S, 1994, ' *Experimental modelling of thermal consolidation around a high level waste repository* ', AECB INFO document (in press), Atomic Energy Control Board, Ottawa, Canada.
20. Selvadurai , A.P.S. and Nguyen, T. S., 1994 , 'Finite element modelling of consolidation of fractured porous media', Proc. of the 1993 Canadian Geotechnical Conference, 79-88.
21. SKI (Swedish Nuclear Power Inspectorate), 1993, '*DECOVALEX - Mathematical Models of coupled T-H-M Processes for Nuclear Waste Repositories, Report of Phase I* ' SKI TR 93:31, Stockholm, Sweden.
22. Smith, I.M. and Griffiths, D.V., 1988, '*Programming the finite element method*' John Wiley & Sons.
23. Vaziri, H., Christian, H.A. and Pitman, T. 1992, ' Generalized Form of Terzaghi's Consolidation Solution', Proc. of the 1992 Canadian Geotechnical Conference, paper 65
24. Witherspoon, P.A., 1987, ' The need for field research in coupled processes ', in *Coupled Processes Associated with Nuclear Waste Repositories*', edited by C.F. Tsang 775-780, Academic Press Inc.
25. Zienkiewicz, O.C., Best, B., Dullage, C. and Stagg, K. 1970, ' Analysis of non linear problems in rock mechanics with particular reference to jointed rock systems' Proc. 2nd Congress Int. Soc. Rock Mech., Belgrade, Yugoslavias, vol. 3, 501-509.

Excavation and Construction Problems Involving Porous Media

J.C. Small and K.L.A. Ng,
School of Civil & Mining Engineering,
The University of Sydney, NSW 2006,
Australia
and
J.P. Hsi,
Coffey & Partners Int. Inc.,
12 Waterloo Road, North Ryde, NSW 2113,
Australia.

Abstract A numerical method is presented which may be used to model excavation in a soil or construction of earthfill embankments in which the soil is assumed to be either an elastic or an elasto-plastic material. The method may be used to compute the appropriate forces to be removed (excavation) or applied (construction) to the finite element mesh, and has the advantage that equilibrium of forces in the mesh is ensured at all times.

1 Introduction

Problems involving the construction of tunnels or excavation of soil beneath the water table are common in Civil Engineering practice. The removal of soil when excavating may lead to plastic failure of soil at the base or the sides of the trench, or may cause large settlements or movements which could damage nearby structures. Water inflows could be large, and lowering of the water table may take place. Potential failure, movements, water inflows and perhaps forces on shoring and bracing need to be predicted for large scale works where the cost of failure could be high.

The problem of adding soil is a similar problem to that of excavation. If wet material is being placed, pore pressures will build up in the material that is placed as well as in the foundation. If these pore pressures become too high, failure of the embankment can occur, and so prediction of pore water pressure build up and dissipation is important, especially for earth fill dams.

Many authors have examined the excavation problem in the past. Early work was carried out by treating the soil as a one phase undrained material (i.e. the excess pore pressures were not treated as separate variables). An early approach proposed by Clough and Mana (1976) and Mana (1976) was widely used for computing the boundary tractions to be removed from the finite element mesh. The method involves computing the nodal forces to be applied to the finite element mesh from the total stresses in the material to be excavated (see Section 2). However this method does not produce theoretically correct

P. S. Selvadurai (ed.), Mechanics of Poroelastic Media, 181–195.
1996 *Kluwer Academic Publishers.*

results (Desai and Sargand (1984)) because it neglects the self weight of the excavated soil in computing the nodal forces.

Other schemes were devised such as that of Christian and Wong (1973) who used direct determination of the nodal forces from extrapolation of stresses in the elements next to the excavation boundary.

The correct method of determining the tractions to be removed from the finite element mesh when simulating excavation in single phase materials was later presented by Ghabousi and Pecknold (1984) and Brown and Booker (1985). The latter method showed how to compute the out of balance forces in a finite element mesh due to removal of some of the elements in such a way the equilibrium of forces in the mesh is always maintained. Such a scheme may also be used for construction where elements are added to the finite element mesh.

Excavation in a two phase material (that is one in which the soil is treated of consisting of pore water and a soil skeleton) has also been of interest to many investigators. Osaimi and Clough (1979), Yong et al. (1989), Banerjee et al. (1988), Borja (1990) and Bolton et al. (1989) have all examined the problem of excavation in a two phase medium. In several of these analyses, the soil is assumed to behave in a non-linear fashion and more advanced constitutive laws are employed.

In most of these analyses the soil around the excavation is assumed to remain saturated and the water table is assumed to remain at its original position. In reality, water will flow into an excavation if the soil is permeable enough and the water table near the excavation will drop. Hsi and Small (1992a,b) have investigated the effects of the water table drop as have Kishnani and Borja (1993).

Papers dealing with the construction of fills consisting of single phase materials, have generally modelled the construction process by "turning on" the self weight of the added elements; a method which is theoretically correct. However, when the soil is treated as a two phase material, the problem becomes more complex. The added soil may not be saturated which means that theories for non-saturated soils must be used. Flow of pore fluid and dissipation of pore pressures will also occur in the soil.

Much research is still needed to overcome the problems involved in modelling the construction of embankments from non-saturated fills and of flow in partially saturated materials. Work done in this area includes that of Li and Desai (1983) who simulated construction of an embankment dam. Deformation during filling was simulated by calculating pore pressures from a seepage analysis and then applying seepage forces to the finite element mesh. Naylor et al. (1988) have also dealt with dam construction and have looked at a dam with a saturated core. Unsaturated analyses of dams have been performed by Kogho and Yamashita (1988) and by Alonso et al. (1988) who consider pore water and air pressures separately in the formulation.

2 Theory

The equations governing the behaviour of a porous medium are well known (Sandhu and Wilson (1969)) and can be written in incremental form

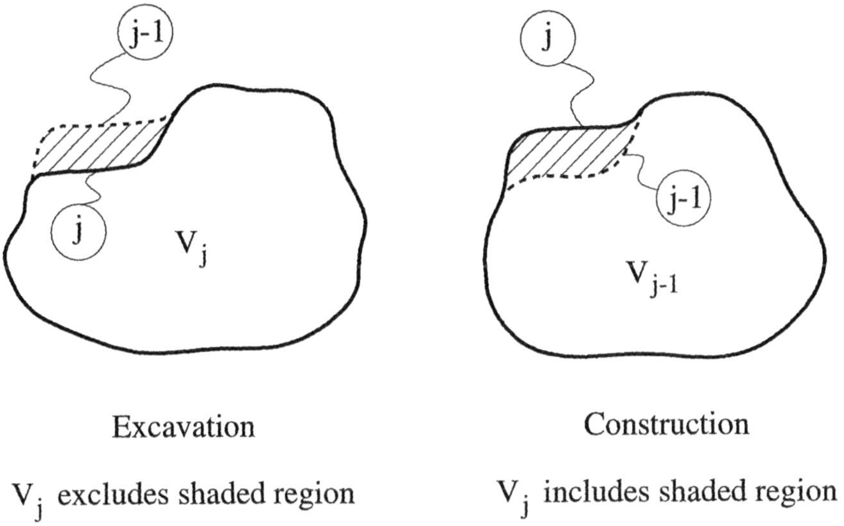

Figure 1: *Volumes considered in excavation or construction problems.*

$$\begin{bmatrix} \boldsymbol{K} & -\gamma_w\,\boldsymbol{L}^T \\ & \\ -\gamma_w\,\boldsymbol{L} & -\gamma_w\,(1-\alpha)\Delta t\,\boldsymbol{\Phi} \end{bmatrix} \begin{bmatrix} \Delta\delta \\ \\ \Delta h \end{bmatrix} = \zeta \tag{1}$$

where $\zeta^T = [\Delta\mathbf{f},\ \gamma_\mathbf{w}\mathbf{t}\boldsymbol{\Phi}\mathbf{h_t}]$.

The equations (1) above have been written in terms of the total head $\boldsymbol{h}$ since the total head is used later for problems where the water table drops. The total head may be written:

$$\boldsymbol{h} = \boldsymbol{h}_w + \boldsymbol{h}_{el} \tag{2}$$

where $\boldsymbol{h}_w$ is the water pressure head and $\boldsymbol{h}_{el}$ is the elevation head.

By using equation (1), the changes in displacements $\Delta\delta$ and in total head Δh over a time interval Δt can be found from the total head h_t at a previous time t. The term α in the equations arises from the time integration scheme used. It has been shown (Booker and Small (1975)), that the process of obtaining the changes in the displacements and total heads from the solutions at a previous time will be unconditionally stable if the value of α is less than or equal to 0.5.

The water pressure head can be made up of two components; the geostatic water pressure head due to the ground water which existed before any excavation or construction, and the changes in water pressure which occur due to any addition or removal of soil. Care must be taken in any definition of 'excess' pore water pressures, as in excavation or construction problems, the water table may drop or rise and so the datum from which the excesses are measured will change. To avoid confusion it is often better to present total water pressure heads h_w.

2.1 Excavation

If we incorporate excavation and drawdown of the water table into the consolidation equations it is necessary to add terms which apply forces to the mesh to simulate the removal of soil. It is also necessary to apply terms which account for the extra flow which occurs across the free surface of the water table as it drops and water drains out of the pores of the soil.

The governing finite element equations for the problem of excavation involving the drawdown of the water table may be written as:

$$\begin{bmatrix} \boldsymbol{K} & -\gamma_w \boldsymbol{L}^T \\ -\gamma_w \boldsymbol{L} & \begin{array}{c} -\gamma_w (1-\alpha)\Delta t\, \boldsymbol{\Phi} \\ -\gamma_w \boldsymbol{G}^{FS} \end{array} \end{bmatrix} \begin{bmatrix} \Delta\boldsymbol{\delta}^{(i)} \\ \Delta\boldsymbol{h}^{(i)} \end{bmatrix} = \zeta^* \tag{3}$$

where

$$\zeta^* = \begin{bmatrix} \Delta\boldsymbol{f} + \Delta\boldsymbol{g} \\ \begin{array}{c} \gamma_w \alpha\Delta t\, \boldsymbol{\Phi}\, \boldsymbol{h}_t + \gamma_w (1-\alpha)\Delta t\, \boldsymbol{\Phi}\, \boldsymbol{h}_{t+\Delta t}^{(i-1)} \\ -\gamma_w \boldsymbol{L}\, \boldsymbol{\delta}_t + \gamma_w \boldsymbol{L}\, \boldsymbol{\delta}_{t+\Delta t}^{(i-1)} \\ +\gamma_w \boldsymbol{G}^{FS} (\boldsymbol{h}_{t+\Delta t}^{(i-1)} - \boldsymbol{h}_t) \end{array} \end{bmatrix} \tag{4}$$

The terms in the above equations (3 and 4) are defined in the Appendix, however certain terms warrant further explanation. The terms containing $\boldsymbol{G}^{FS}$ arise from the need to allow for the effects of water draining out of the pores of the soil as the water table drops. A flow is injected along the free surface as mentioned before, and therefore terms containing the quantity $\boldsymbol{G}^{FS}$ are present to allow for this extra flow.

Within each time step, the position of the free surface needs to be established and so an iterative procedure is used to establish the current position of the free surface. In the above equations, the superscripts (i) or $(i-1)$ refer to the iteration number within each time step i.e. $\boldsymbol{h}^{(i)} = \boldsymbol{h}^{(i-1)} + \Delta\boldsymbol{h}^{(i)}$ etc.

The force vector now has to allow for the out of balance forces which will be created by the removal of soil (simulated by removal of elements from the mesh). This can be done by considering the balance of forces in the soil remaining after excavation of part of the mesh. The self weight of the soil, the applied loads and the pore pressures and effective stresses all contribute to the force balance. The force vector in equation (4) can be evaluated as follows:

$$\Delta \boldsymbol{g} = \int_{V_j} \boldsymbol{B}^T \boldsymbol{\sigma}'_{j-1} \, dV + \gamma_w \boldsymbol{d}^T \boldsymbol{a}(h_{j-1} - h_{el}) \, dV + \int_{V_j} \boldsymbol{N}^T \gamma \, dV + \int_S \boldsymbol{N}^T \boldsymbol{t} \, dS \quad (5)$$

The terms in the above equation are defined in the Appendix, however it may be seen that the out of balance forces $\Delta \boldsymbol{g}$ which must be applied to the mesh, depend on the effective stresses $\boldsymbol{\sigma}'_{j-1}$ and the total heads $\boldsymbol{h}_{j-1}$ existing in the soil at the last increment of excavation. The integrations however are carried out over the present volume V_j. The terms $\Delta \boldsymbol{f}$ and $\Delta \boldsymbol{g}$ are set to zero after the first iteration.

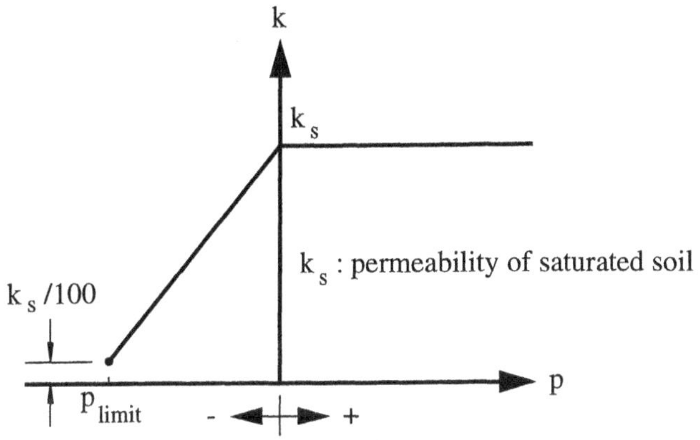

Figure 2: *Pore pressure – permeability relationship.*

This is shown schematically in Figure 1 where the initial volume of soil and final volume of soil remaining after excavation can be seen on the left hand side of the diagram.

The unit weight of the soil γ and any surface tractions $\boldsymbol{t}$ will normally be in equilibrium with the stresses and water pressures existing in the ground, and so the vector $\Delta \boldsymbol{g}$ should be zero at the start of an analysis where only geostatic stresses exist.

Once elements are removed or added, the out-of-balance forces $\Delta \boldsymbol{g}$ will be correctly computed from equation (5). By carrying out the integration over the remaining volume V_j we are ensuring equilibrium of forces in the remaining part of the finite element mesh.

Once the water has drawn down, the soil above the free surface will be partially saturated and negative pore pressures (suctions) will exist. The unsaturated flow problem can be dealt with by using a permeability-pore pressure relationship such as the one shown in Figure 2. The permeabilities in the soil above the free surface are reduced as the calculated total pore water pressure becomes negative. This allows better convergence when finding the position of the free surface using the iterative procedure described above. Once the pore pressures reach a limiting value of suction they are held at that value and no further reduction is allowed to take place. This process models the type of behaviour observed experimentally for many soils.

2.2 Construction

When adding elements (construction), the equations of consolidation equation (3) can be used, however it is necessary to make the right hand side ζ equal to

$$\zeta = \begin{bmatrix} \Delta \boldsymbol{f} + \Delta \boldsymbol{g} \\ \\ \gamma_w \, \Delta t \, \boldsymbol{\Phi} \, \boldsymbol{h}_t \end{bmatrix} \tag{6}$$

where the vector $\Delta \boldsymbol{g}$ is computed for the problem as given in equation 5. In most cases, there are no changes in the external tractions and so $\Delta \boldsymbol{f}$ is zero.

It is often desirable (for problems involving plasticity) to apply the loads as computed from equation (5) in several steps. This in effect means that the self weight of the soil is applied slowly, and this may be necessary to ensure convergence.

This is equivalent to increasing the gravity on the added elements and so care needs to be exercised in evaluating the elevation heads. As the elevation head also depends on the acceleration due to gravity, we must also reflect the increase in gravity by increasing the elevation head for the elements which are added. In equation (5), the increment in unit weight γ and the current elevation head $\boldsymbol{h}_{el}$ are used in the added elements only. In the rest of the mesh, the appropriate full values of these quantities are used.

3 Examples

The following examples illustrate the use of the foregoing theory in solving excavation and construction problems.

3.1 Excavation

Excavation of soil beneath the water table usually involves the drawdown of the water table, since water will flow into the excavation if the soil is permeable enough or a long enough period of time has elapsed. The drawdown of the water table can be taken into account as outlined by Hsi and Small (1992 a,b) and the method appears to be capable of giving reasonably good predictions of measured performance.

As an example, the excavation shown in Figure 3 is to be made in an elasto-plastic material that deforms according to the Modified Cam Clay model. All of the material properties used are shown in Figure 3 and the definitions of these symbols is given in the Appendix. The figure also shows the excavation sequence and when the bracing struts (labelled S1, S2, S3, S4) are installed.

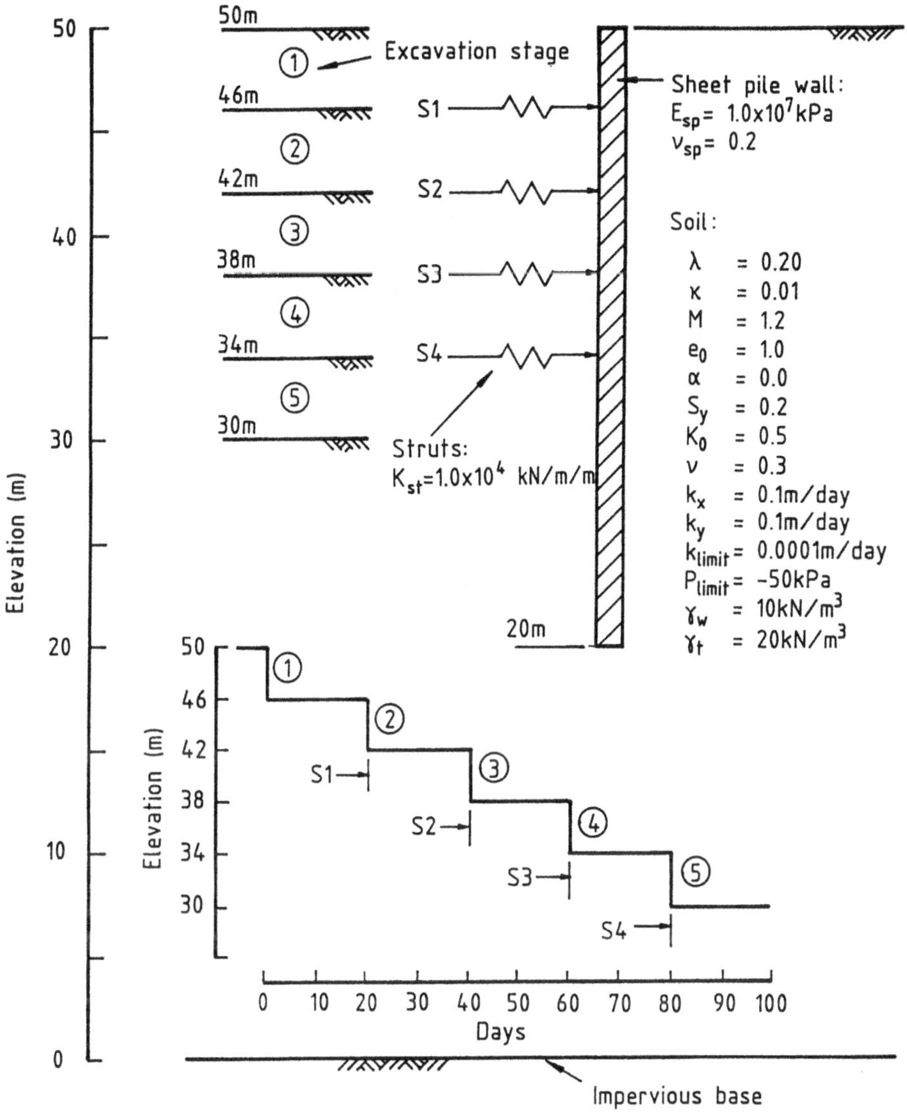

Figure 3: *Excavation problem considered.*

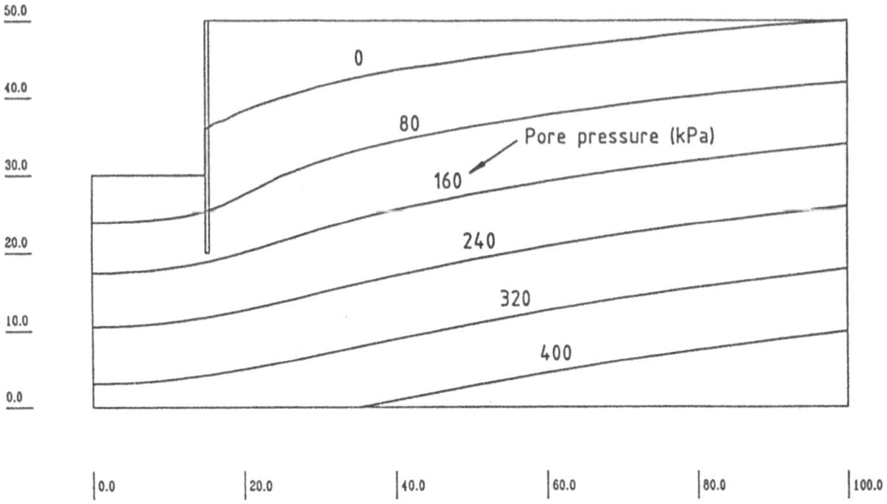

Figure 4: *Pore pressure distribution surrounding excavation.*

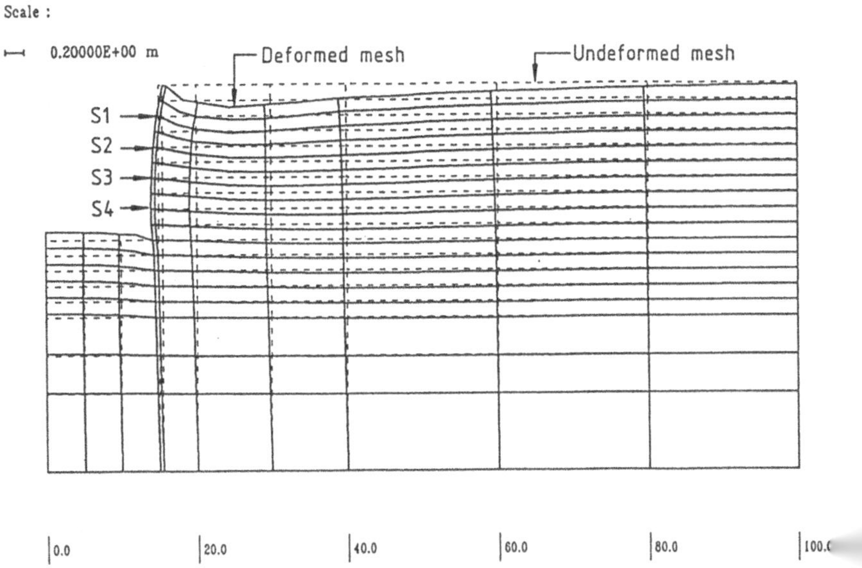

Figure 5: *Predicted movements for soil excavation problem.*

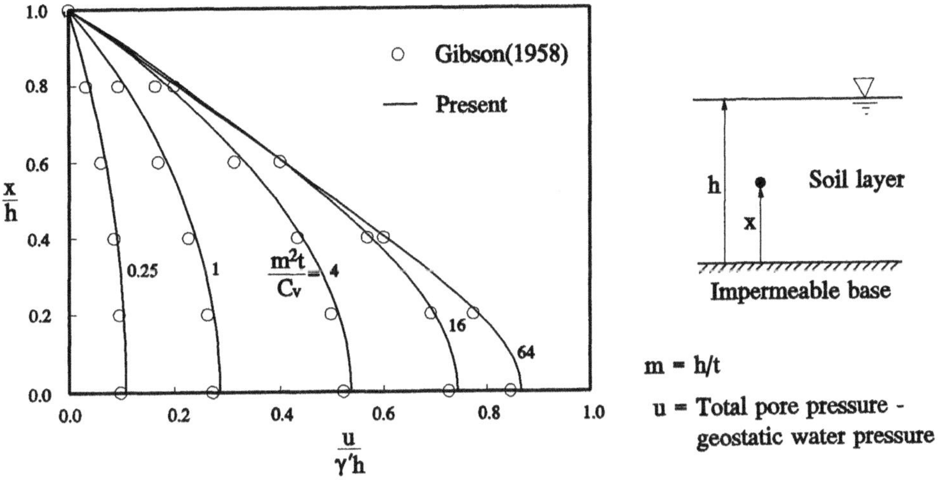

Figure 6: *Excess pore pressures in accreting soil layer.*

The computed pore pressure distribution is shown in Figure 4 (for total water pressures), and predicted movements are shown in Figure 5. One interesting aspect of such analyses is the effect of the hydraulic boundary conditions. In the example presented here, the total head was specified to be constant along the left hand boundary. This corresponds to the water table remaining constant at the distance of the boundary from the excavation. If the boundary were moved, a different solution would be obtained as the water table drop is affected by this condition.

Experience with field cases has shown that the water table will not drop beyond a certain distance from an excavation as there is usually some recharge of the groundwater system. If this point can be found, then reasonable estimates can be made by specifying the total head to be constant along that boundary. A similar conclusion has been reached by Kishnani and Borja (1993).

3.2 Construction Problems

A fundamental construction problem is that of an accreting saturated soil layer, and this has been examined in the past by Gibson (1958). The problem, which is one dimensional in nature, was analysed using the theory presented in this paper and the results for the pore pressures are shown in Figure 6. It can be seen from the figure, that close agreement with Gibson's theoretical solutions were obtained.

Another example is that of the construction of an embankment on soft clay. In this example, the clay beneath the embankment behaves as an elastic–perfectly plastic material with a Mohr-Coulomb failure criterion. The embankment material is treated as granular so that there is no pore pressure build up within the embankment itself.

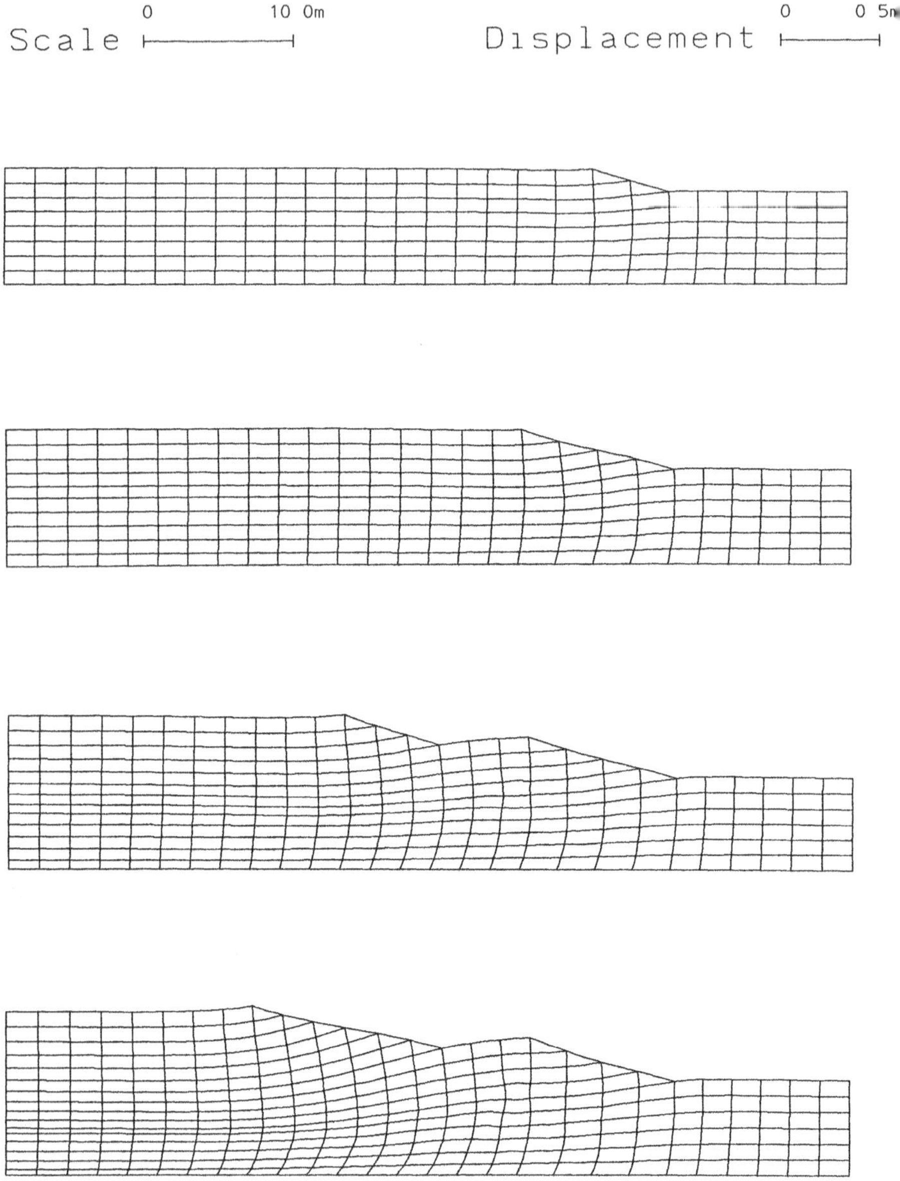

Figure 7: *Deformation of embankment and soil.*

Figure 7 shows the deformation of the embankment and subsoil which occurs as the height of the embankment increases (i.e. elements are added to the mesh). The construction is simulated by generating the complete finite element mesh initially, and then assigning zero stiffness to those elements which have not yet been added to the mesh. Nodes which are not yet included in the mesh are restrained from moving so that the stiffness matrix does not have zero terms on the diagonal corresponding to the zero modulus of the missing elements.

In order to be able to plot the mesh deforming in this way it is necessary to allow the elements which have not been added to the mesh, to "follow" the deformed mesh downwards. This can be simply done by moving the nodes in the missing elements down by the same amount as the upper surface of the embankment. The original coordinates of the elements are used in working out the element stiffnesses however, as the analysis is still a small deformation analysis.

A final example shows the construction of an embankment dam with a vertical core. The construction is assumed to take place rapidly, with a core wet of optimum, so that the soil remains saturated. In this case, the clay core is assumed to behave according to the Cam Clay model.

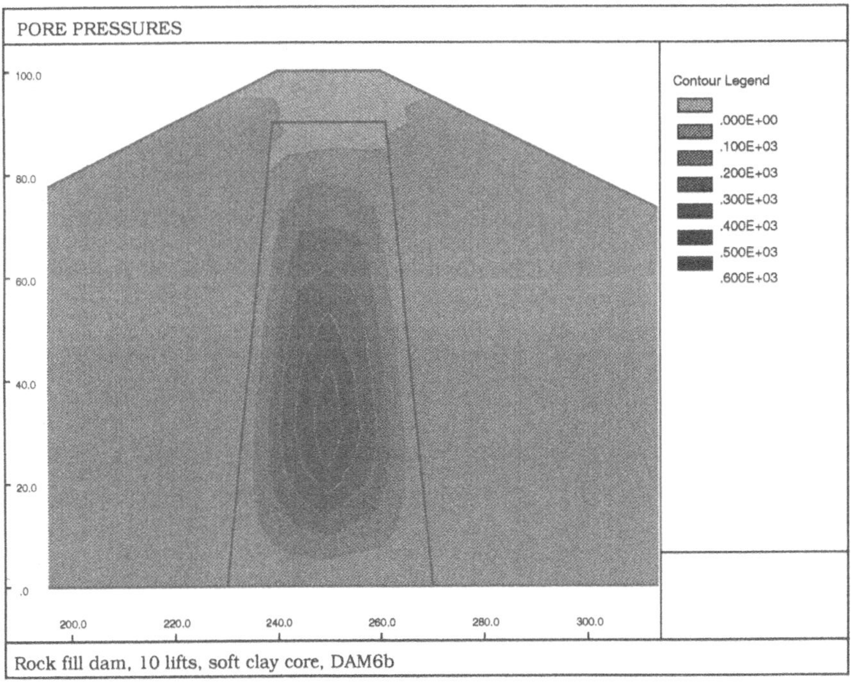

Figure 8: *Pore water pressures in clay core at end of construction.*

Pore water pressure distributions in the clay core at completion of the constructic are shown plotted in Figure 8. This plot shows that high concentrations of pore pressu are locked into the lower part of the core during construction. Prediction of the rate dissipation of these pore pressures can also be made. Loss of of saturation in the core water flows out can be made in a similar way to the excavation problem, where the wat table drops leaving material above the free surface unsaturated.

4 Comments and Conclusions

A method of analysing problems involving excavation or construction in a porous no linear material has been presented. It is shown that the same method can be used f computing the forces to be added (construction) or subtracted (excavation) from the fini element mesh. If these forces are applied, then the remaining finite element mesh will in equilibrium.

Examples are presented which show how the method can be applied to practical pro lems involving porous non-linear materials.

REFERENCES

Alonso, E.E., Batlle, F., Gens, A. and Lloret, A. (1988) Consolidation analysis of partial saturated soils. Proc. Sixth Int. Conf. Num. Meth. in Geomechs., Innsbruck. 2:1303–130

Banerjee, P.K., Kumbhojkar, A.S. and Yousif, N.B. (1988) Finite element analysis of th stability of a vertical cut using an anisotropic soil model. Canadian Geotech. J. 25:119 127.

Bolton, M.D., Britto, A.M., Powrie, W. and White, T.P. (1989) Finite element analys of a centrifuge model of a retaining wall embedded in a heavily overconsolidated cla Computers and Geotechnics. 7:289–318.

Booker, J.R. and Small, J.C. (1975) An investigation of the stability of numerical solutio of Biot's equations of consolidation. Int. J. Solids Structs. 11:907–917.

Borja, R.I. (1990) Analysis of incremental excavation based on Critical State theory. Geotech. Engg. ASCE. 116(6):964–985.

Brown, P.T. and Booker, J.R. (1985) Finite element analysis of excavation. Compute and Geotechnics. 1:207–220.

Christian, J.T. and Wong, I.H. (1973) Errors in simulating excavation in elastic media b finite elements. Soils and Foundations 13:1–10.

Clough, G.W. and Mana, A.I. (1976) Lessons learned in finite element analysis of temp rary excavations. Proceedings 2nd Int. Conf. Num. Meth. Geomech., ASCE. 1:496–510.

Desai, C.S. and Sargand, S. (1984) Hybrid FE procedure for soil-structure interaction. J. Geotech. Engg., ASCE. 110:473–486.

Ghaboussi, J. and Pecknold, D.A. (1984) Incremental finite element analysis of geometrically altered structures. Int. J. Numer. Methods in Engg. 20:2051–2064.

Gibson, R.E. (1958) The progress of consolidation in a clay layer increasing in thickness with time. Géotechnique 8:171–182.

Hsi, J.P. and Small, J.C. (1992a) Simulation of excavation in a poro-elastic material. Int. J. Num. Anal. Meth. Geomechs. 16:25–43.

Hsi, J.P. and Small, J.C. (1992b) Analysis of excavation in an elasto-plastic soil involving drawdown of the water table. Computers and Geotechnics. 13:1–19.

Kishnani, S.S. and Borja, R.I. (1993) Seepage and Soil-structure interaction effects in braced excavations. J. Geotech. Eng. Div. ASCE. 119 GT5:912–928.

Kohgo, Y. and Yamashita, T. (1988) Finite element analysis of fill type dams – Stability during construction by using the effective stress concept. Proc. Sixth Int. Conf. Num. Meth. in Geomechs., Innsbruck. 2:1315–1322.

Li, G.C. and Desai, C.S. (1983) Stress and seepage analysis of earth dams. J. Geotech. Div., ASCE. 109:946–960.

Mana,A.I. (1976) Finite element analysis of deep excavation behaviour. PhD Thesis, Stanford University, Stanford, California.

Naylor, D.J., Knight, D.J. and Ding,D. (1988) Coupled consolidation analysis of the construction and subsequent performance of Monasavu dam. Computers and Geotechnics. 5:95–129.

Osaimi, A.E. and Clough, G.W. (1979) Pore pressure dissipation during excavation. J. Geotech. Engg. Div., ASCE. 105(4):481–498.

Sandhu, R.S. and Wilson, E.L. (1969) Finite element analysis of seepage in elastic media. J. Eng. Mech. Div., ASCE. 95(3):641–652.

Tong, K.Y., Lee, F.H., Parnploy, U. and Lee, S.L. (1989) Elasto-plastic consolidation analysis for strutted excavation in clay. Computers and Geotechnics. 8:311–328.

APPENDIX

Some terms used in the governing finite element equations (Equations 1 and 3) are defined here.

(1) Stiffness matrix $\boldsymbol{K}$:

$$\boldsymbol{K} = \int_V \boldsymbol{B}^T \, \boldsymbol{D} \, \boldsymbol{B} \, dV \tag{A1}$$

(2) Coupling matrix $\boldsymbol{L}$:

$$\boldsymbol{L} = \int_V \boldsymbol{a} \, \boldsymbol{d}^T \, dV \tag{A2}$$

(3) Flow matrix $\boldsymbol{\Phi}$:

$$\boldsymbol{\Phi} = \int_V \boldsymbol{E}^T \, \boldsymbol{k} \, \boldsymbol{E} \, dV \tag{A3}$$

(4) Matrix for the imposed flow $\boldsymbol{G}^{FS}$:

$$\boldsymbol{G}^{FS} = \int_\Gamma \boldsymbol{a} \, \boldsymbol{a}^T \, S_y \, cos\beta \, d\Gamma \tag{A4}$$

where

$\boldsymbol{a}$ is the vector of shape functions,
$\boldsymbol{B}$ is the displacement-strain matrix,
$\boldsymbol{d}$ is the vector relating nodal displacement to volumetric strain,
$\boldsymbol{D}$ is the stress-strain matrix,
$\boldsymbol{E}$ is the matrix relating the nodal total water head to $\boldsymbol{\nabla} h$,
$\boldsymbol{F}$ is the vector of body forces,
$\boldsymbol{G}$ is the matrix for the imposed flow (Eq A4),
$\boldsymbol{h}$ is the vector of total head,
$\boldsymbol{h}_w$ is the vector of water pressure head,
$\boldsymbol{h}_{el}$ is the vector of elevation head,
$\boldsymbol{k}$ is the matrix of permeability coefficients,
$\boldsymbol{K}$ is the stiffness matrix (Eq A1),
$\boldsymbol{L}$ is the coupling matrix (Eq A2),
$\boldsymbol{N}$ is the matrix of element shape functions,
t is time,
$\boldsymbol{t}$ is the vector of surface tractions,
S is the surface loaded by tractions,
S_y is specific yield,
V is the volume of soil,
α is the time integration factor,

β is the angle between the free surface segment and the horizontal direction,
Γ is the free surface contour and
δ is the vector of nodal displacements,
Δ denotes a change in a quantity,
γ_w is the unit weight of water,
γ is the vector of body forces,
σ' is the vector of effective stresses,
$\boldsymbol{\Phi}$ is the flow matrix (Eq A3),
ζ is the vector defined in Eq (1),
ζ^* is the vector defined in Eq (4).

Quantities shown on Figure 3.

e_0 is the initial void ratio,
k_x is the horizontal permeability,
k_y is the vertical permeability,
k_{limit} is the limiting permeability (see Figure 2),
K_0 is the coefficient of earth pressure at rest,
M is the slope of the critical state line (Cam Clay),
P_{limit} is the limiting suction pressure,
S_y is the specific yield of the soil,
γ_w is the unit weight of water,
γ_t is the total unit weight of the soil,
κ is the slope of the swelling line (Cam Clay),
λ is the slope of the normal consolidation line (Cam Clay),
ν is the Poisson's ratio of the soil.

THE EFFECT OF A TIME-DEPENDENT LOAD ON A POROELASTIC SEABED OVER A REGION WITH MOVING BOUNDARIES

G.C.W. SABIN AND W. RAMAN-NAIR
Faculty of Engineering & Applied Science
Memorial University of Newfoundland
St. John's, NF, Canada

Abstract. The displacement effects of a load moving with a constant velocity over the surface of a poroelastic seabed are investigated. Analytic solutions are found for the cases of a point and axisymmetric normal load as well as a shear load. The results immediately beneath the load are shown as a function of depth for various values of a normalized parameter involving the velocity and the consolidation constant. General solutions for the soil displacements u and pore pressure p to Biot's equations of poroelasticity for the quasi-static case are also provided. These solutions are in terms of unknown functions (rather than derivatives of unknown functions).

Key words: poroelastic, moving load, seabed

1. Introduction

The marine environment offers special challenges to engineers. A recurring problem for the oil and gas industry off the Canadian east coast is the threat posed by icebergs to offshore structures, buried pipelines and other installations. The iceberg is a moving load on fluid-saturated porous seabed, and the displacements induced by such a load are of interest to engineers engaged in pipeline design (Been [1]). In order to provide a benchmark for this displacement field the soil is modelled using Biot's theory of poroelasticity [4]. The iceberg can be imagined to be enclosed by an escribed circle and therefore analytic solutions are determined for the soil displacements due to a moving circular load. These solutions should be considered as benchmarks for the sub-scour displacement field. A similar problem for an elastic halfspace has been considered by Eason [2]. Several attempts have been made to provide general solutions to the equations of poroelasticity. The earliest appears to be Biot himself who decoupled the equations by defining new quantities related to displacements and pore pressure. McNamee and Gibson [3] have obtained solutions in terms of displacement functions. Sabin and Raman-Nair [6] have employed displacement functions of the Papkovitch-Neuber solution technique of classical elasticity. A drawback of these approaches is that solutions are obtained for functions whose derivatives are related to the physical quantities of interest: soil displacements u and pore pressure p. This makes it difficult to implement these techniques in a general numerical scheme requiring the specification of physically meaningful boundary conditions. To overcome this difficulty general solutions have been found for $\mathbf{u}$ and p in terms of unknown functions (rather than derivatives of unknown functions) which must be determined from the boundary conditions on the domain of interest.

A. P. S. Selvadurai (ed.), *Mechanics of Poroelastic Media*, 197–213.

The ease of defining boundary conditions makes this approach more attractive for implementation in numerical techniques such as the finite element and boundary element methods.

2. Basic Equations

The general theory of poroelasticity presented by Biot [4] and also by Verruijt [5], takes the following form in the quasi-static case when there are no body forces present (see [6]):

$$\mu\nabla^2\mathbf{u}+(\lambda+\mu)\nabla e=\nabla p \tag{1}$$

$$\sigma'_{ij}=\mu\left(\frac{\partial u_i}{\partial x_j}+\frac{\partial u_j}{\partial x_i}\right)+\lambda e\delta_{ij} \tag{2}$$

$$\frac{k}{\rho_f g}\nabla^2 p=\frac{\partial e}{\partial t} \tag{3}$$

$$e=\nabla\cdot\mathbf{u} \tag{4}$$

where u_i and σ'_{ij} denote the soil displacements, and the effective stresses, respectively, and p the excess pore water pressure. Further, μ and λ represent the Lamé constants for the soil skeleton, and where k is the permeability of soil, ρ_f is the density of water and g is the gravitational constant. The foregoing formulation assumes water is incompressible and the ratio of the bulk modulus of the soil skeleton to the bulk modulus of the soil grains is small [6].

The system of equations (1) to (4) can be manipulated so as to reveal the fundamental nature of the unknown displacements $\mathbf{u}$ and pore water pressure p and their relationship. A general solution can then be obtained in a straightforward manner (see Appendix):

$$\mathbf{u}=\mathbf{x}\phi+\mathbf{\Psi}+\mathbf{u}^2 \tag{5}$$

and

$$p=p_d+2\mu\phi \tag{6}$$

where

$$\nabla^2\phi=0,\quad \nabla^2\mathbf{\Psi}=\mathbf{0},\quad \nabla^2 p_d=\frac{1}{c}\frac{\partial p_d}{\partial t} \tag{7}$$

$$\frac{\partial\mathbf{u}^2}{\partial t}=\frac{c}{\beta}\nabla p_d \tag{8}$$

$$\nabla\cdot(\mathbf{x}\phi+\mathbf{\Psi})=0 \tag{9}$$

The consolidation constant c is defined as

$$c=\frac{k}{\rho_f g}\beta$$

where $\beta=(\lambda+2\mu)$. It should be noted that equations (5-8) give a general solution valid for both two or three dimensional and axisymmetric problems.

3. Solution For A Moving Surface Load

Let x, y, z be the Cartesian coordinate system and for this analysis it is convenient to introduce τ the space-like time coordinate such that

$$\tau = \frac{ct}{a}$$

where a is a parameter of dimension length. As a consequence, time derivatives with respect to t are replaced as follows

$$\frac{\partial}{\partial t} = \frac{c}{a}\frac{\partial}{\partial \tau}$$

The problems to be considered here concern loads moving with constant velocity along the surface of a seabed which is modelled as an isotropic and homogeneous oroelastic half-space of soil. The soil occupies the region $z \geq 0$ with the surface aken as the plane $z = 0$. For the present study, it is assumed that both the soil urface and structures resting on it are permeable and that the surface tractions are nown for all time such that the appropriate boundary conditions are

$$\sigma'_{zz}(x,y,0,\tau) = -p_1(x,y,\tau), \quad \sigma'_{xz}(x,y,0,\tau) = -p_2(x,y,\tau),$$
$$\sigma'_{yz}(x,y,0,\tau) = -p_3(x,y,\tau), \quad p(x,y,0,\tau) = 0 \tag{10}$$

nd the conditions at infinity are

$$\sigma'_{ij}(\mathbf{x},\tau) \to 0 \quad \text{as} \quad \sqrt{x_i x_i} \to \infty \tag{11}$$

To solve these problems a three dimensional integral transform is used which is efined as (see Sneddon [7])

$$\bar{f}(\xi,\eta,z,\omega) = \frac{1}{(2\pi)^{3/2}} \int_{-\infty}^{\infty}\int_{-\infty}^{\infty}\int_{-\infty}^{\infty} f(x,y,z,\tau) e^{i(\xi x + \eta y + \omega\tau)} dx dy d\tau \tag{12}$$

pplying this transform to the equations (7) yields the ordinary differential equations

$$\frac{d^2\bar{\phi}}{dz^2} - (\xi^2 + \eta^2)\bar{\phi} = 0, \quad \frac{d^2\overline{\Psi}}{dz^2} - (\xi^2 + \eta^2)\overline{\Psi} = 0,$$

$$\frac{d^2\overline{p}_d}{dz^2} - (\xi^2 + \eta^2 - i\omega/a)\overline{p_d} = 0$$

using the condition (11) solutions of these equations which remain finite as $z \to \infty$ e given as

$$\bar{\phi} = Ae^{-nz}, \quad \overline{\Psi}_i = B_i e^{-nz}, \quad \overline{p}_d = Ce^{-mz} \tag{13}$$

ere

$$n^2 = \xi^2 + \eta^2 \quad \text{and} \quad m^2 = \xi^2 + \eta^2 - i\omega/a$$

e unknown constants A and B_i are related by the condition (9) , which transforms

$$-i\xi\overline{(x\phi + \Psi_1)} + -i\eta\overline{(y\phi + \Psi_2)} + \frac{\partial}{\partial z}\overline{(z\phi + \Psi_3)} = 0 \tag{14}$$

However it can be shown from the definition (12) and (13) that

$$\overline{(x\phi)} = -i\frac{\partial\overline{\phi}}{\partial\xi} = i\xi zAe^{-nz}/n \quad \text{and} \quad \overline{(y\phi)} = -i\frac{\partial\overline{\phi}}{\partial\eta} = i\eta zAe^{-nz}/n$$

Thus equation (14) becomes

$$-i\xi(i\xi zAe^{-nz}/n + B_1e^{-nz}) - i\eta(i\eta zAe^{-nz}/n + B_2e^{-nz}) + Ae^{-nz} - nzAe^{-nz}$$

$$-nB_3e^{-nz} = 0$$

which simplifies to

$$B_3 = \frac{A - i(\xi B_1 + \eta B_2)}{n}$$

The transformed displacements **u** and pore water pressure p as given by (5-6 can now be written in terms of the unknown constants A, B_1, B_2 and C where th vector $\mathbf{u}^2$ is derived from the condition (8)

$$\overline{u}_x = \left(B_1 + \frac{i\xi zA}{n}\right)e^{-nz} + \frac{a\xi}{\beta\omega}Ce^{-mz}$$

$$\overline{u}_y = \left(B_2 + \frac{i\eta zA}{n}\right)e^{-nz} + \frac{a\eta}{\beta\omega}Ce^{-mz} \qquad (15$$

$$\overline{u}_z = \left(\frac{A - i(\xi B_1 + \eta B_2)}{n} + zA\right)e^{-nz} - \frac{iam}{\beta\omega}Ce^{-mz}$$

$$\overline{p} = 2\mu Ae^{-nz} + Ce^{-mz}$$

By transforming the stresses (2) and boundary conditions (10) and combining the with equations (15), the unknown constants can be determined from the solution the following system of equations

$$2\mu i(\xi B_1 + \eta B_2) + (1 + i\varepsilon n^2 a/\omega)C = -\overline{p}_1$$

$$-\frac{\mu}{n}\left((2\xi^2 + \eta^2)B_1 + \xi\eta B_2\right) - \epsilon m\xi a/\omega C = -\overline{p}_2 \qquad (1$$

$$-\frac{\mu}{n}\left((\xi^2 + 2\eta^2)B_2 + \xi\eta B_1\right) - \epsilon m\eta a/\omega C = -\overline{p}_3$$

$$2\mu A + C = 0$$

where $\varepsilon = 2\mu/\beta = (1 - 2\nu)/(1 - \nu)$, and ν is Poisson's ratio. These equations c be solved to give

$$A = -C/(2\mu)$$

$$B_1 = \frac{(n^2 + \eta^2)\overline{p}_2 - \xi\eta\overline{p}_3}{2\mu n^3} - \frac{m\xi a}{n\beta\omega}C$$

$$B_2 = \frac{(n^2 + \xi^2)\overline{p}_3 - \xi\eta\overline{p}_2}{2\mu n^3} - \frac{m\eta a}{n\beta\omega}C \qquad (1$$

$$C = -\left(1 + \frac{\varepsilon n}{m + n(1-\varepsilon)}\right)\left(\frac{n\overline{p}_1 + i(\xi\overline{p}_2 + \eta\overline{p}_3)}{n}\right)$$

Thus the transformed displacements and pore pressure can be written as

$$\overline{u}_x = \frac{1}{2\mu}\left\{\frac{i\xi}{n}D_1 + \left[\frac{(n^2+\eta^2)\,\overline{p}_2 - \xi\eta\overline{p}_3}{n^3} - \frac{iz\xi}{n}C\right]e^{-nz}\right\}$$

$$\overline{u}_y = \frac{1}{2\mu}\left\{\frac{i\eta}{n}D_1 + \left[\frac{(n^2+\xi^2)\,\overline{p}_3 - \xi\eta\overline{p}_2}{n^3} - \frac{iz\eta}{n}C\right]e^{-nz}\right\}$$

$$\overline{u}_z = \frac{1}{2\mu}\left\{D_2 - \left[\frac{i(\xi\overline{p}_2 + \eta\overline{p}_3)}{n^2} + \frac{(1+zn)}{n}C\right]e^{-nz}\right\} \tag{18}$$

$$\overline{p} = C\left(e^{-mz} - e^{-nz}\right)$$

where

$$D_1 = \varepsilon C\frac{(ne^{-mz} - me^{-nz})}{n^2 - m^2} \tag{19}$$

$$D_2 = \varepsilon mC\frac{(e^{-mz} - e^{-nz})}{n^2 - m^2}$$

The transformed stresses can be found by transforming equations (2) and combining the result with equations (18)

$$\overline{\sigma}'_{xx} = \frac{\xi^2}{n}D_1 + (1-\varepsilon)Ce^{-mz} - \left[\frac{i\left((n^2+\eta^2)\,\overline{p}_2 - \xi\eta\overline{p}_3\right)}{n^3} + \frac{z\xi}{n}C\right]\xi e^{-nz}$$

$$\overline{\sigma}'_{yy} = \frac{\eta^2}{n}D_1 + (1-\varepsilon)Ce^{-mz} - \left[\frac{i\left((n^2+\xi^2)\,\overline{p}_3 - \xi\eta\overline{p}_2\right)}{n^3} + \frac{z\eta}{n}C\right]\eta e^{-nz}$$

$$\overline{\sigma}'_{zz} = -nD_1 + Ce^{-mz} + \left[\frac{i(\xi\overline{p}_2 + \eta\overline{p}_3)}{n} + znC\right]e^{-nz} \tag{20}$$

$$\overline{\sigma}'_{xy} = \frac{\xi\eta}{n}D_1 - \left[\frac{i(\eta^3\overline{p}_2 + \xi^3\overline{p}_3)}{n^3} + \frac{z\xi\eta}{n}C\right]e^{-nz}$$

$$\overline{\sigma}'_{xz} = i\xi D_2 + \left[-\overline{p}_2 + i\xi zC\right]e^{-nz}$$

$$\overline{\sigma}'_{yz} = i\eta D_2 + \left[-\overline{p}_3 + i\eta zC\right]e^{-nz}$$

The solution here is quite general. To find the solution of a particular problem one must give specific functions p_i and perform a transform inversion. It is convenient to carry out the remaining calculations by considering the application of the prescribed stresses separately and combining the results by the principle of linear superposition.

4. Moving Normal Load

If the applied load has only a normal component then $p_2 = p_3 = 0$ and the inverse transform of the displacements and pore pressure given by (18) are

$$2\mu u_x = \frac{1}{(2\pi)^{3/2}} \int_{-\infty}^{\infty}\int_{-\infty}^{\infty}\int_{-\infty}^{\infty} \xi F_1(\omega)\overline{p}_1 e^{-i(\xi x+\eta y+\omega\tau)}\mathrm{d}\xi\mathrm{d}\eta\mathrm{d}\omega$$

$$2\mu u_y = \frac{1}{(2\pi)^{3/2}} \int_{-\infty}^{\infty}\int_{-\infty}^{\infty}\int_{-\infty}^{\infty} \eta F_1(\omega)\overline{p}_1 e^{-i(\xi x+\eta y+\omega\tau)}\mathrm{d}\xi\mathrm{d}\eta\mathrm{d}\omega$$

$$2\mu u_z = \frac{1}{(2\pi)^{3/2}} \int_{-\infty}^{\infty}\int_{-\infty}^{\infty}\int_{-\infty}^{\infty} F_2(\omega)\overline{p}_1 e^{-i(\xi x+\eta y+\omega\tau)}\mathrm{d}\xi\mathrm{d}\eta\mathrm{d}\omega \qquad (21)$$

$$p = \frac{1}{(2\pi)^{3/2}} \int_{-\infty}^{\infty}\int_{-\infty}^{\infty}\int_{-\infty}^{\infty} \overline{p}_1 \left(1 + \frac{\varepsilon n}{m + n(1-\varepsilon)}\right) \cdot \cdot \left(e^{-nz} - e^{-mz}\right) e^{-i(\xi x+\eta y+\omega\tau)}\mathrm{d}\xi\mathrm{d}\eta\mathrm{d}\omega$$

where

$$F_1(\omega) = i\left(1 + \frac{\varepsilon n}{m + n(1-\varepsilon)}\right)\left\{\frac{\varepsilon}{n}\frac{\left(me^{-nz} - ne^{-mz}\right)}{n^2 - m^2} + \frac{z}{n}e^{-nz}\right\}$$

$$F_2(\omega) = \left(1 + \frac{\varepsilon n}{m + n(1-\varepsilon)}\right)\left\{\varepsilon m\frac{\left(e^{-nz} - e^{-mz}\right)}{n^2 - m^2} + \frac{1 + nz}{n}e^{-nz}\right\}$$

4.1. Point Load

If the load is that of a point force of magnitude P moving with velocity V along the x-axis in the positive direction then it can be represented as

$$p_1 = P\delta(y)\delta(x - Vt) = P\delta(y)\delta(x - \alpha\tau), \quad \alpha = aV/c$$

which has the transformation

$$\overline{p}_1 = \frac{P}{\sqrt{2\pi}}\delta(\omega + \alpha\xi) \qquad (22)$$

The equations (21) now have the form

$$u_x = \frac{P}{8\pi^2\mu} \int_{-\infty}^{\infty}\int_{-\infty}^{\infty} \xi F_1(-\alpha\xi) e^{-i(\xi(x-Vt)+\eta y)}\mathrm{d}\xi\mathrm{d}\eta$$

$$u_y = \frac{P}{8\pi^2\mu} \int_{-\infty}^{\infty}\int_{-\infty}^{\infty} \xi F_1(-\alpha\xi) e^{-i(\xi(x-Vt)+\eta y)}\mathrm{d}\xi\mathrm{d}\eta \qquad (23$$

$$u_z = \frac{P}{8\pi^2\mu} \int_{-\infty}^{\infty}\int_{-\infty}^{\infty} F_2(-\alpha\xi) e^{-i(\xi(x-Vt)+\eta y)}\mathrm{d}\xi\mathrm{d}\eta$$

$$p = \frac{P}{4\pi^2} \int_{-\infty}^{\infty}\int_{-\infty}^{\infty} \left(1 + \frac{\varepsilon n}{m + n(1-\varepsilon)}\right)\left(e^{-nz} - e^{-mz}\right) e^{-i(\xi(x-Vt)+\eta y)}\mathrm{d}\xi\mathrm{d}\eta$$

where at this stage

$$m = \sqrt{\xi^2 + \eta^2 + i\xi V/c}$$

The following change of variables is used to evaluate these integrals

$$\begin{aligned} x - Vt = r\cos\theta, \qquad & y = r\sin\theta \\ \xi = \rho\cos\psi, \qquad & \eta = \rho\sin\psi \end{aligned} \tag{24}$$

which gives

$$u_x = \frac{P}{8\pi^2\mu}\int_0^{2\pi}\cos\psi\int_0^{\infty} G_1 e^{-i\rho r\cos(\theta-\psi)}d\rho d\psi$$

$$u_y = \frac{P}{8\pi^2\mu}\int_0^{2\pi}\sin\psi\int_0^{\infty} G_1 e^{-i\rho r\cos(\theta-\psi)}d\rho d\psi$$

$$u_z = \frac{P}{8\pi^2\mu}\int_0^{2\pi}\int_0^{\infty} G_2 e^{-i\rho r\cos(\theta-\psi)}d\rho d\psi \tag{25}$$

$$p = \frac{P}{4\pi^2}\int_0^{2\pi}\int_0^{\infty}\left(1 + \frac{\varepsilon}{m_1 + (1-\varepsilon)}\right)\left(e^{-\rho z} - e^{-m_1\rho z}\right)e^{-i\rho r\cos(\theta-\psi)}\rho d\rho d\psi$$

where

$$G_1 = \left(1 + \frac{\varepsilon}{m_1 + (1-\varepsilon)}\right)\left\{\frac{\varepsilon\rho\left(e^{-\rho m_1 z} - m_1 e^{-\rho z}\right)}{V/c\cos\psi} + i\rho z e^{-\rho z}\right\}$$

$$G_2 = \left(1 + \frac{\varepsilon}{m_1 + (1-\varepsilon)}\right)\left\{\frac{i\varepsilon m_1\rho\left(e^{-\rho z} - e^{-m_1\rho z}\right)}{V/c\cos\psi} + (1+\rho z)e^{-\rho z}\right\}$$

$$m_1 = \sqrt{1 + iV\cos\psi/(c\rho)}$$

The normalized displacements have been calculated as a function of α at various depths beneath the point load i.e., $x - Vt = y = 0 (r = 0)$. These results are shown in figures 1-2 where Poisson's ratio has been chosen as $1/3$. The normalized form of equations (25) for $r = 0$ are given as

$$u_x' = \frac{a u_x \mu}{P} = \frac{1}{2\pi^2}\int_0^{\pi/2}\int_0^1 \mathrm{Re}\left\{H_1(\rho,\psi) + H_1(1/\rho,\psi)/\rho^2\right\}d\rho d\psi$$

$$u_y' = 0 \tag{26}$$

$$u_z' = \frac{a u_z \mu}{P} = \frac{1}{2\pi^2}\int_0^{\pi/2}\int_0^1 \mathrm{Re}\left\{H_2(\rho,\psi) + H_2(1/\rho,\psi)/\rho^2\right\}d\rho d\psi$$

where

$$H_1(\rho,\psi) = \left(1 + \frac{\varepsilon}{m_1' + (1-\varepsilon)}\right)\left\{\frac{\varepsilon\rho\left(e^{-\rho m_1' z'} - m_1' e^{-\rho z'}\right)}{\alpha} + i\rho z'\cos\psi e^{-\rho z'}\right\}$$

$$H_2(\rho,\psi) = \left(1 + \frac{\varepsilon}{m_1' + (1-\varepsilon)}\right)\left\{\frac{\varepsilon\rho m_1'\left(e^{-\rho m_1' z'} - e^{-\rho z'}\right)}{i\alpha\cos\psi} + \left(1 + \rho z'\right)e^{-\rho z'}\right\}$$

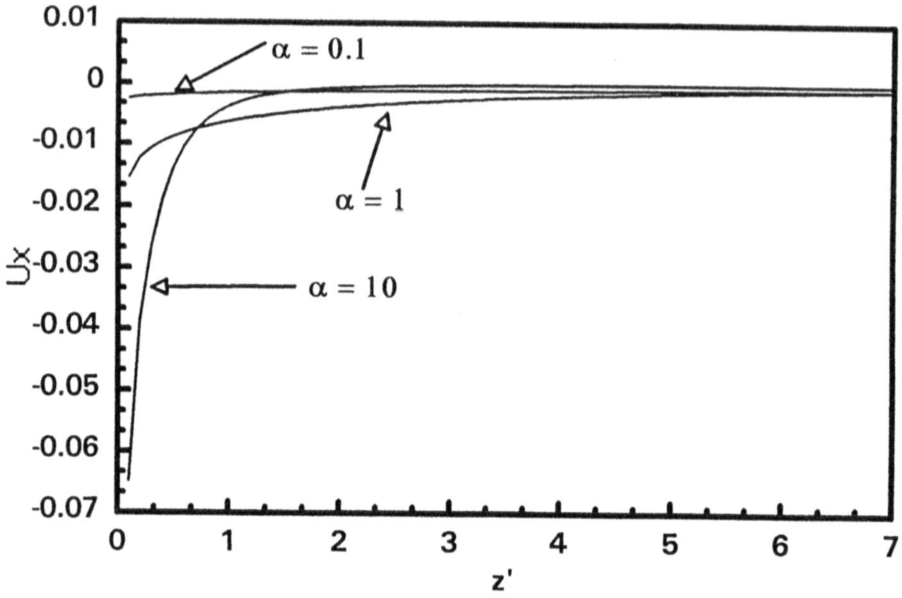

Fig. 1. Normalized displacement u_x' versus depth beneath a moving point load ($\alpha = aV/c$)

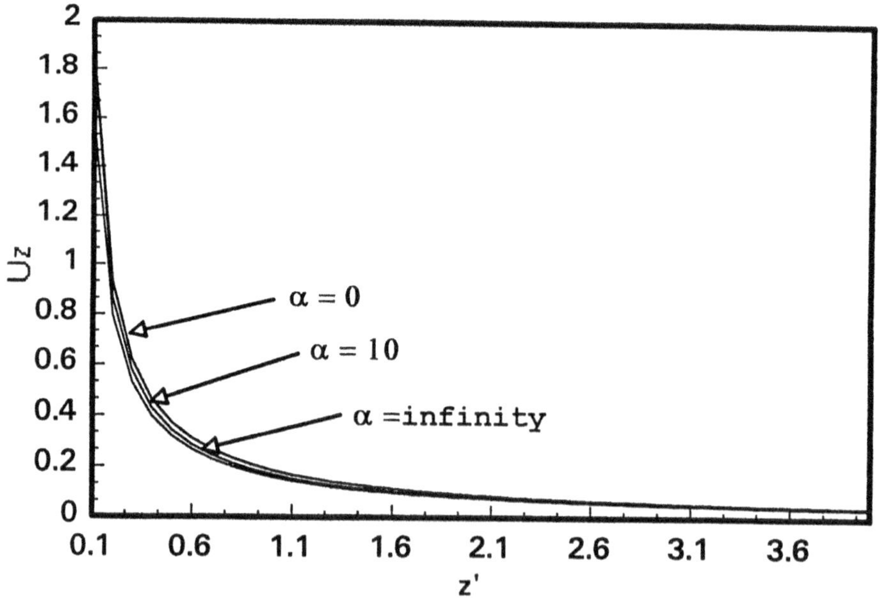

Fig. 2. Normalized displacement u_z' versus depth beneath a moving point load ($\alpha = aV/c$)

$$m_1' = \sqrt{1 + i\alpha \cos\psi/\rho} \text{ and } z' = z/a$$

and Re$\{f\}$ indicates the real part of f. The displacements simplify in the following limiting cases

$$\alpha \to 0: \ u_x' = 0, \ u_y' = 0, \ u_z' = \frac{4-\varepsilon}{4\pi(2-\varepsilon)z'}$$

$$\alpha \to \infty: \ u_x' = 0, \ u_y' = 0, \ u_z' = \frac{1}{2\pi z'} \tag{27}$$

4.2. Axisymmetric Normal Load

This section describes a normal axisymmetric load acting on a disk which is travelling over the surface of the seabed with a velocity V. The appropriate boundary condition for the normal load is

$$p_1 = \begin{Bmatrix} P\left(\sqrt{(x-\alpha\tau)^2 + y^2}\right), & (x-\alpha\tau)^2 + y^2 < a^2 \\ 0, & (x-\alpha\tau)^2 + y^2 > a^2 \end{Bmatrix}$$

where a is the radius of the disk. The integral transform of this function is

$$\overline{p}_1 = \frac{1}{(2\pi)^{3/2}} \int_{-\infty}^{\infty} d\tau \int_{-\infty}^{\infty}\int_{-\infty}^{\infty} P\left(\sqrt{(x-\alpha\tau)^2 + y^2}\right) \cdot$$
$$\cdot H\left(a^2 - (x-\alpha\tau)^2 - y^2\right) e^{i(\xi x + \eta y + \omega\tau)} dx dy$$

where H is the Heaviside function. Let $X = x - \alpha\tau$ and obtain

$$\overline{p}_1 = \frac{1}{(2\pi)^{3/2}} \int_{-\infty}^{\infty} e^{i\tau(\omega + \alpha\xi)} d\tau \int_{-\infty}^{\infty}\int_{-\infty}^{\infty} P\left(\sqrt{X^2 + y^2}\right) \cdot$$
$$\cdot H\left(a^2 - X^2 - y^2\right) e^{i(\xi X + \eta y)} dX dy$$

which by using the result

$$\int_{-\infty}^{\infty} e^{i\tau x} d\tau = 2\pi\delta(x)$$

implifies to

$$\overline{p}_1 = \frac{\delta(\omega + \alpha\xi)}{(2\pi)^{1/2}} \int_{-\infty}^{\infty}\int_{-\infty}^{\infty} P\left(\sqrt{X^2 + y^2}\right) H\left(a^2 - X^2 - y^2\right) e^{i(\xi X + \eta y)} dX dy \tag{28}$$

Iowever the remaining two integrals in (28) represent a double Fourier transform of radially symmetric function which from Sneddon [7] can be related to the Hankel ransform of order zero

$$\frac{1}{2\pi} \int_{-\infty}^{\infty}\int_{-\infty}^{\infty} f\left(\sqrt{x^2 + y^2}\right) e^{i(\xi x + \eta y)} dx dy = \int_0^{\infty} r f(r) J_0(\rho r) dr$$

here $\rho = \sqrt{\xi^2 + \eta^2}$ and $J_0(r)$ stands for the Bessel function of order zero. It ollows that (28) can simplified to

$$\overline{p}_1 = \sqrt{2\pi}\delta(\omega + \alpha\xi) \int_0^{\infty} r P(r) H(a - r) J_0(\rho r) dr,$$

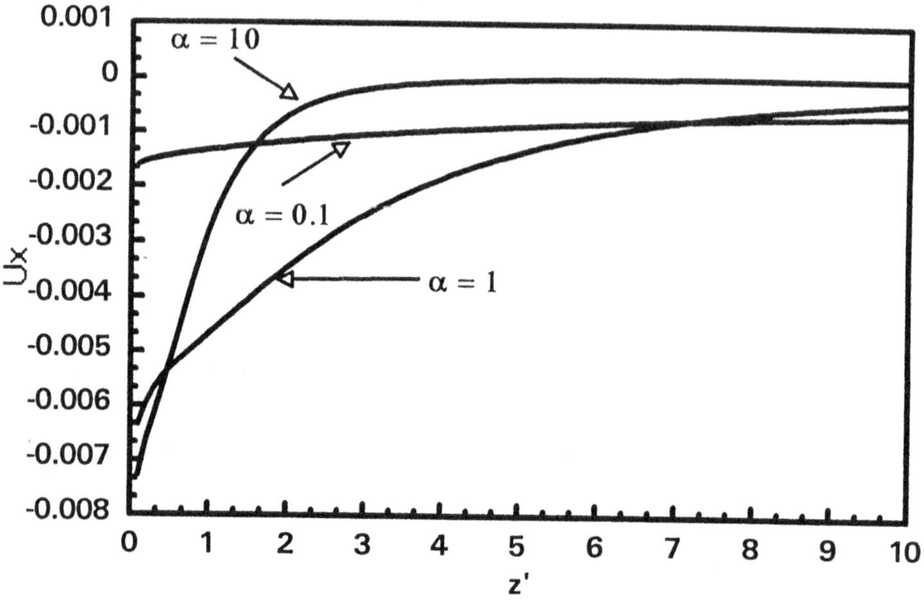

Fig. 3. Normalized displacement u_x' versus depth beneath a moving circular normal lo
($\alpha = aV/c$)

$$\overline{p}_1 = \sqrt{2\pi}\delta\left(\omega + \alpha\xi\right) P_0\left(\rho\right) \qquad (2$$

where

$$P_0\left(\rho\right) = \int_0^a rP\left(r\right) J_0\left(\rho r\right) \mathrm{d}r$$

If (29) is now used in (21) instead of (22) the result comparable to (25), is

$$u_x = \frac{1}{4\pi\mu}\int_0^{2\pi} \cos\psi \int_0^{\infty} G_1 P_0\left(\rho\right) e^{-i\rho r\cos(\theta-\psi)} \mathrm{d}\rho \mathrm{d}\psi$$

$$u_y = \frac{1}{4\pi\mu}\int_0^{2\pi} \sin\psi \int_0^{\infty} G_1 P_0\left(\rho\right) e^{-i\rho r\cos(\theta-\psi)} \mathrm{d}\rho \mathrm{d}\psi \qquad (3$$

$$u_z = \frac{1}{4\pi\mu}\int_0^{2\pi} \int_0^{\infty} G_2 P_0\left(\rho\right) e^{-i\rho r\cos(\theta-\psi)} \mathrm{d}\rho \mathrm{d}\psi$$

$$p = \frac{1}{2\pi}\int_0^{2\pi}\int_0^{\infty} P_0\left(\rho\right)\left(1 + \frac{\varepsilon}{m_1 + (1-\varepsilon)}\right)\left(e^{-\rho z} - e^{-m_1\rho z}\right) e^{-i\rho r\cos(\theta-\psi)} \rho \mathrm{d}\rho \mathrm{d}$$

In the special case that $P\left(r\right) = P_0$, a constant, $P_0\left(\rho\right)$ becomes

$$P_0\left(\rho\right) = \frac{a}{\rho} J_1\left(a\rho\right) P_0 \qquad ($$

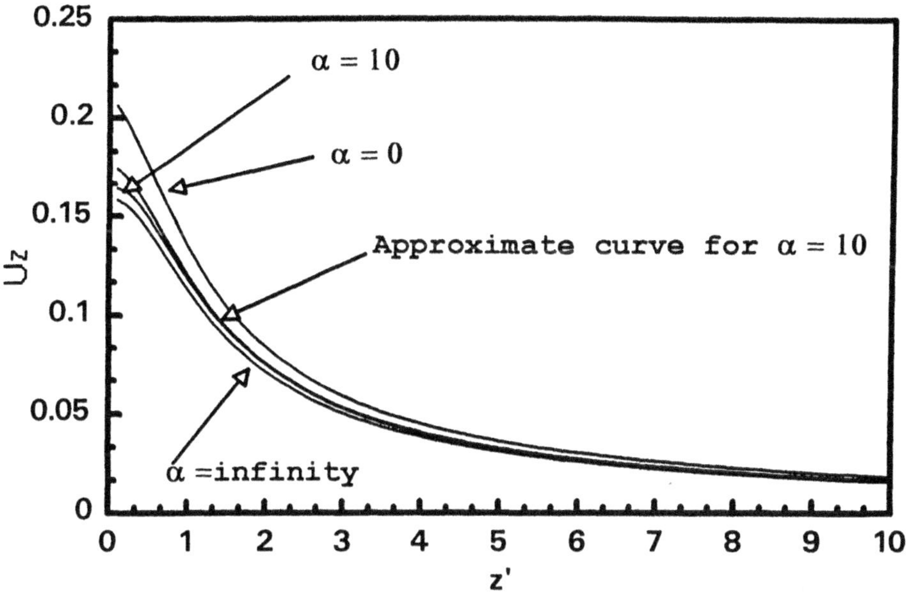

Fig. 4. Normalized displacement u_z' versus depth beneath a moving circular normal load ($\alpha = aV/c$)

Figures 3-5 show typical calculations of the normalized displacements for this case. As before these calculations have been made at points immediately below the centre of the moving disk i.e., $x - Vt = y = 0$ and $\nu = 1/3$. The normalized form of equations (30) for $r = 0$ are given by

$$u_x' = \frac{a u_x \mu}{(\pi a^2 P_0)} = \frac{1}{\pi^2} \int_0^{\pi/2} \int_0^1 \mathrm{Re}\left\{H_1(\rho,\psi) J_1(\rho) + H_1(1/\rho,\psi) J_1(1/\rho)\right\}/\rho \mathrm{d}\rho \mathrm{d}\psi$$

$$u_y' = 0 \tag{32}$$

$$u_z' = \frac{a u_z \mu}{(\pi a^2 P_0)} = \frac{1}{\pi^2} \int_0^{\pi/2} \int_0^1 \mathrm{Re}\left\{H_2(\rho,\psi) J_1(\rho) + H_2(1/\rho,\psi) J_1(1/\rho)\right\}/\rho \mathrm{d}\rho \mathrm{d}\psi$$

where $\pi a^2 P_0$ represents the total load acting on the disk. The displacements simplify in the following limiting cases:

$$\alpha \to 0: \ u_x' = 0, \ u_y' = 0, \ u_z' = u_z'^{\infty} + \frac{\varepsilon}{2\pi(2-\varepsilon)} \frac{1}{\left(z + \sqrt{z^2+1}\right)}$$

$$\alpha \to \infty: \ u_x' = 0, \ u_y' = 0, \ u_z' = \frac{1}{2\pi\sqrt{z^2+1}} = u_z'^{\infty}$$

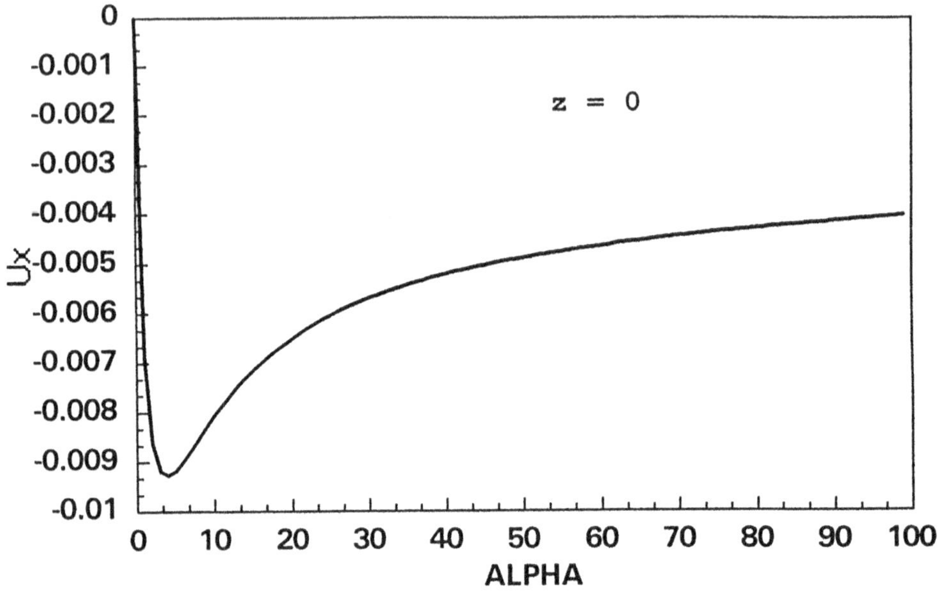

Fig. 5. Normalized displacement $u_z^{'}$ versus $\alpha(= aV/c)$ beneath a moving circular load (for $z = 0$)

After a close examination of (32) the following approximation for $u_z^{'}$ was determined

$$u_z^a = u_z^{'\infty} + \mathrm{Re}\left\{\frac{\varepsilon}{2\pi\left(\sqrt{1+i\alpha}+1-\varepsilon\right)}\right\}\frac{1}{\left(z+\sqrt{z^2+1}\right)} \tag{33}$$

Figure 4 shows that this approximation provides quite accurate results.

5. Moving Shear Load

When only a shear force is applied there are two cases to be considered: shear in the direction of motion (direction of the x-axis) and shear perpendicular to the motion (direction of the y-axis). In this study only the former situation will be considered while the same techniques can be applied to the latter.

5.1. Shear in the Direction of Motion

For this problem, $p_1 = p_3 = 0$ and equations (18) reduce to

$$2\mu u_x = \frac{1}{(2\pi)^{3/2}}\int_{-\infty}^{\infty}\int_{-\infty}^{\infty}\int_{-\infty}^{\infty}\left(i\xi^2 F_1(\omega) + \frac{n^2+\eta^2}{n^2}e^{-nz}\right)\frac{\overline{p}_2}{n}e^{-i(\xi x+\eta y+\omega\tau)}\mathrm{d}\xi\mathrm{d}\eta\mathrm{d}\omega$$

$$2\mu u_y = \frac{1}{(2\pi)^{3/2}}\int_{-\infty}^{\infty}\int_{-\infty}^{\infty}\int_{-\infty}^{\infty}\xi\eta\left(iF_1(\omega) - \frac{e^{-nz}}{n^2}\right)\frac{\overline{p}_2}{n}e^{-i(\xi x+\eta y+\omega\tau)}\mathrm{d}\xi\mathrm{d}\eta\mathrm{d}\omega$$

$$2\mu u_z = \frac{1}{(2\pi)^{3/2}} \int_{-\infty}^{\infty}\int_{-\infty}^{\infty}\int_{-\infty}^{\infty} i\xi \left(F_2(\omega) - \frac{e^{-nz}}{n}\right) \frac{\overline{p}_2}{n} e^{-i(\xi x+\eta y+\omega\tau)} d\xi d\eta d\omega \quad (34)$$

$$p = \frac{1}{(2\pi)^{3/2}} \int_{-\infty}^{\infty}\int_{-\infty}^{\infty}\int_{-\infty}^{\infty} \frac{i\xi}{n}\overline{p_2}\left(1 + \frac{\varepsilon n}{m + n(1-\varepsilon)}\right) \cdot$$

$$\cdot \left(e^{-nz} - e^{-mz}\right) e^{-i(\xi x+\eta y+\omega\tau)} d\xi d\eta d\omega$$

The case considered here is one of a constant shear force acting on a disk which is travelling with a velocity V over the surface of the seabed. The function p_2 is given as

$$p_2 = \begin{Bmatrix} S_0, & (x-\alpha\tau)^2 + y^2 < a^2 \\ 0, & (x-\alpha\tau)^2 + y^2 > a^2 \end{Bmatrix}$$

which has the transformation

$$\overline{p}_2 = \sqrt{2\pi}\delta(\omega + \alpha\xi)\frac{a}{\rho}J_1(a\rho)\, S_0 \quad (35)$$

Following the same procedure as previously outlined the displacements and pore pressure can be written as

$$u_x = \frac{S_0 a}{4\pi^2\mu}\int_0^{2\pi}\int_0^{\infty}\left(i\cos^2\psi G_1 + \left(1+\sin^2\psi\right)e^{-\rho z}\right)\frac{J_1(a\rho)}{\rho}e^{-i\rho r\cos(\theta-\psi)}d\rho d\psi$$

$$u_y = \frac{S_0 a}{4\pi^2\mu}\int_0^{2\pi}\int_0^{\infty}\sin\psi\cos\psi\left(iG_1 - e^{-\rho z}\right)\frac{J_1(a\rho)}{\rho}e^{-i\rho r\cos(\theta-\psi)}d\rho d\psi$$

$$u_z = \frac{S_0 a}{4\pi^2\mu}\int_0^{2\pi}\int_0^{\infty} i\cos\psi\left(G_2 - e^{-\rho z}\right)\frac{J_1(a\rho)}{\rho}e^{-i\rho r\cos(\theta-\psi)}d\rho d\psi \quad (36)$$

$$p = \frac{S_0 a}{2\pi^2}\int_0^{2\pi}\int_0^{\infty} i\cos\psi\left(1 + \frac{\varepsilon}{m_1 + (1-\varepsilon)}\right)J_1(a\rho)\cdot$$

$$\cdot\left(e^{-\rho z} - e^{-m_1\rho z}\right)e^{-i\rho r\cos(\theta-\psi)}d\rho d\psi$$

These quantities have been normalized and evaluated at points immediately below the centre of the moving disk i.e., $x - Vt = y = 0$ and $\nu = 1/3$. The results are shown in figures 6-7. For comparison, the following limiting cases and approximation are also shown.

$$\alpha \to 0: \; u_y' = 0,\; u_z' = 0,\; u_x' = u_x'^{\infty} + \frac{\varepsilon}{4\pi^2(2-\varepsilon)}\frac{1}{\left(z+\sqrt{z^2+1}\right)}$$

$$\alpha \to \infty: \; u_z' = 0,\; u_y' = 0,$$

$$u_x' - \frac{1}{4\pi^2}\left\{\frac{2}{z+\sqrt{z^2+1}} + \frac{1}{\left(z+\sqrt{z^2+1}\right)^2\sqrt{z^2+1}}\right\} = u_x'^{\infty}$$

After a close examination of (36) the following approximation for u_z' was determined

$$u_z^a = u_x'^{\infty} + \mathrm{Re}\left\{\frac{\varepsilon}{4\pi^2\left(\sqrt{1+i\alpha}+1-\varepsilon\right)}\right\}\frac{1}{\left(z+\sqrt{z^2+1}\right)} \quad (37)$$

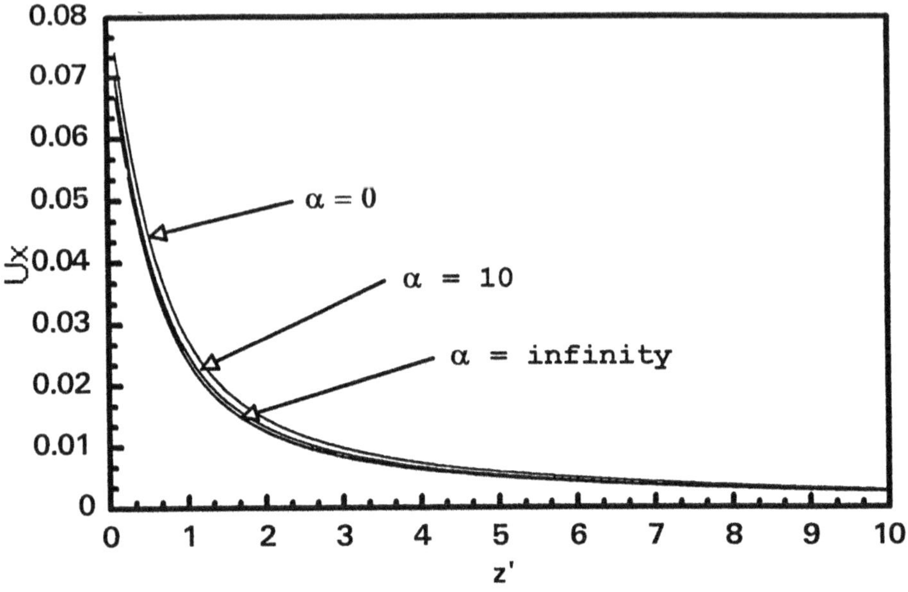

Fig. 6. Normalized displacement $u_x^{'}$ versus depth beneath a moving circular shear load ($\alpha = aV/c$)

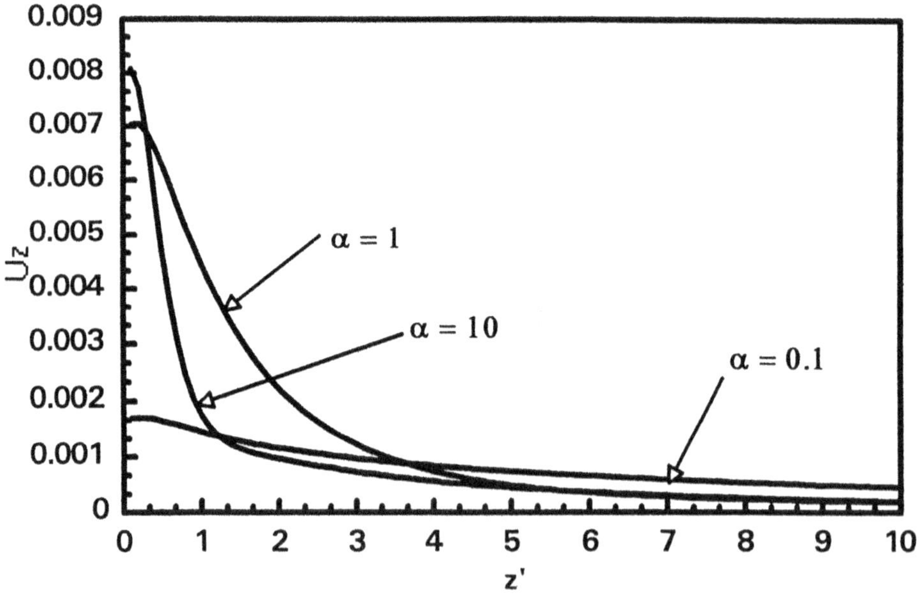

Fig. 7. Normalized displacement $u_z^{'}$ versus depth beneath a moving circular shear load ($\alpha = aV/$

6. Discussion and conclusions

The displacement effects of a load moving with a constant velocity over the surface of a poroelastic seabed have been investigated. Analytic solutions are found for the cases of a point and axisymmetric normal load as well as a shear load. The results immediately beneath the load are shown as a function of depth for various values of a normalized parameter α which is proportional to the velocity and inversely proportional to the consolidation constant. It can be seen from figure 1 that near the surface the horizontal displacement u_x owing to a moving normal point load varies considerably with α. In contrast, the normal displacement u_z for the same problem shown in figure 2, displays little variation between the two extreme cases of $\alpha = 0$ and $\alpha = \inf$ and could be approximated with minimal error by the simple formula

$$u_z^a = \frac{8 - 3\varepsilon}{8\pi(2-\varepsilon)z'}$$

which is just the average of the two limiting cases given by (27). It should be noted that the case $\alpha = 0$ recovers the static elastic solution. It can be observed that the magnitude of the normal displacement is substantially greater than that of the horizontal displacement and could safely be ignored in calculations involving both quantities. Figures 3 and 4 show essentially the same result for the problem of a moving circular normal load. The simple approximating formula for u_z' in this case is given by (33). Figure 5 gives an indication of how the horizontal displacement varies with α on the surface (a similiar behaviour could be expected beneath the surface). In figures 6 and 7 the behaviours of u_x' and u_z' are shown when a constant force is applied to a moving disk in the direction of motion. It can be seen that the variation in u_x' for different α is small. While the differences in u_z' are much greater, the magnitude is much less than u_x'. These results are the reverse from the case of a normal load. A general formula which approximates u_x' is given by (37).

Overall it can be concluded that the largest displacements in each example can be easily approximated by a simple formula which is closely related to the static elastic solution. This observation suggests that for some applications of a moving load over a poroelastic half-space the static elastic displacements would provide an adequate approximation for the corresponding poroelastic ones.

Acknowledgements

This work was supported by the National Sciences and Engineering Research Council of Canada under the operating grant A-4080.

References

Been., K. (1990) Workshop on ice scouring and the design of offshore pipelines. Sponsored by Canadian Oil and Gas Lands Administration (COGLA): 179-191.

Eason, G. (1965) The stresses produced in a semi-infinite solid by a moving surface force. Int. J. Engng. Sci. 2: 581-609.

McNamee, J. and Gibson, R.E. (1960) Displacement functions and linear transforms applied to diffusion through porous elastic media. Q. J. Mech. Appl. Math. 13: 98-111.

4. Biot, M.A. (1962) Mechanics of deformation and acoustic propagation in porous media. J. Appl. Physics 33(4): 1482-1498.
5. Verruijt, A. (1969) Elastic storage of Aquifers. Flow Through Porous Media, ed. R.J.M DeWiest, Academic Press
6. Sabin, G.C.W. and Raman-Nair, W. (1987) The consolidation of a poroelastic seabed beneath an axisymmetric structure. Proc. Sixth International Offshore Mechanics and Arctic Engineering Symposium 1: 403-410.
7. Sneddon, I.N. (1972) The Uses of Integral Transforms. McGraw-Hill, New York.

Appendix

To see the relationship between the pore water pressure and the displacements first take the divergence (∇) of equation (1)

$$\mu\nabla^2(\nabla\cdot\mathbf{u})+(\lambda+\mu)\nabla^2 e=\nabla^2 p$$

and thus, by using (4), find

$$\nabla^2(p-(\lambda+2\mu)e)=0.$$

Hence,

$$p-(\lambda+2\mu)e=2\mu\phi \tag{38}$$

where

$$\nabla^2\phi=0 \tag{39}$$

From equations (3) and (38) it is found that

$$\frac{k}{n_f}\nabla^2 p=\frac{1}{\beta}\left(\frac{\partial p}{\partial t}-2\mu\frac{\partial\phi}{\partial t}\right)$$

Let

$$p=p_d+2\mu\phi$$

and it follows that p_d must satisfy the diffusion equation

$$\left(\nabla^2-\frac{1}{c}\frac{\partial}{\partial t}\right)p_d=0 \tag{40}$$

where for future reference the operator ∇_d^2 is defined as

$$\nabla_d^2=\nabla^2-\frac{1}{c}\frac{\partial}{\partial t} \tag{4}$$

Further it can be seen that

$$e=\frac{p_d}{\beta} \tag{4}$$

and

$$\nabla_d^2 e=0$$

From (38) equation (1) can be rewritten as

$$\mu\nabla^2\mathbf{u}+(\lambda+\mu)\nabla e=\nabla(\beta e+2\mu\phi)$$

which becomes

$$\nabla^2\mathbf{u} = \nabla e + 2\nabla\phi \tag{43}$$

Since $\nabla_d^2 e = 0$ and $\nabla^2\phi = 0$ equation (43) can be reduced to

$$\nabla^4\nabla_d^2\mathbf{u} = \mathbf{0} \tag{44}$$

A general solution for $\mathbf{u}$ that satisfies (44) is given as

$$\mathbf{u} = \mathbf{u}^1 + \mathbf{u}^2 \tag{45}$$

where

$$\nabla^4\mathbf{u}^1 = \mathbf{0} \tag{46}$$

and

$$\nabla_d^2\mathbf{u}^2 = \mathbf{0} \tag{47}$$

By combining equations (45), (42), and (43) one obtains

$$\nabla^2\mathbf{u}^1 + \nabla^2\mathbf{u}^2 = \frac{1}{\beta}\nabla p_d + 2\nabla\phi$$

By employing (39), (40), (46), and (47) this equation naturally decomposes into the following two equations

$$\nabla^2\mathbf{u}^1 = 2\nabla\phi \tag{48}$$

$$\nabla^2\mathbf{u}^2 = \frac{1}{\beta}\nabla p_d \tag{49}$$

A general solution for $\mathbf{u}^1$ which satisfies both (46) and (48) is given as

$$\mathbf{u}^1 = \mathbf{x}\phi + \mathbf{\Psi} \tag{50}$$

$$\nabla^2\mathbf{\Psi} = \mathbf{0}$$

Further by using (47) and (41) equation (49) can be reduced to

$$\frac{\partial\mathbf{u}^2}{\partial t} = \frac{c}{\beta}\nabla p_d \tag{51}$$

From (4), (45) and (42) it follows that

$$\nabla\cdot\mathbf{u}^1 + \nabla\cdot\mathbf{u}^2 = \frac{p_d}{\beta} \tag{52}$$

By taking the divergence of equations (48) and (51) and by using (39) and (40) it can be deduced that

$$\nabla\cdot\mathbf{u}^1 = 0 \tag{53}$$

$$\nabla\cdot\mathbf{u}^2 = \frac{p_d}{\beta}$$

Equation (53) with (50) gives the additional constraint

$$\nabla\cdot(\mathbf{x}\phi + \mathbf{\Psi}) = 0 \tag{54}$$

POROELASTIC RESPONSE RESULTING FROM MAGMA INTRUSION

D. ELSWORTH, B. VOIGHT* and Z. OUYANG
Department of Mineral Engineering
Pennsylvania State University
University Park, PA-16802.

and

A.R. PIGGOTT
National Water Research Institute
Environment Canada
Burlington, Ontario, Canada.

Abstract. The pore pressures that develop around magmatic intrusions in saturated porous media are of interest in both defining the characteristics of the intrusive process and evaluating the potential for destabilization of volcano flanks. Magmatic dike emplacement may be represented as an analog to a moving dislocation in a poroelastic medium, enabling parameters such as intrusion rate, volume, intrusion morphology and location to be determined from the form of the pressure history response. Solutions for moving point and line dislocations within infinite and semi-infinite media are reported as analogs to pencil and sheet dike forms, respectively. Where the mechanical and flow parameters of the system are defined, *a priori*, type curve matching of responses may be used to evaluate both emplacement location and intrusion volume, uniquely. For slow emplacement, the intrusion rate may also be determined. Poroelastic pressures developed around dike intrusions are substantial and may contribute to the triggering of deep seated earthquakes and the development of large-scale instability of volcano flanks. Recognition of the role of emplacement generated pore fluid pressures in the triggering of flank failure addresses the enigma in the failure of shallow inclination flanks of oceanic shield type volcanoes, in the absence of obvious sufficient destabilizing forces.

Key words: Moving dislocation, poroelasticity, geologic processes

1. Introduction

Magmatic emplacement results in a rapid volumetric increase within the saturated medium (Turcotte, 1990; Watanabe, 1983) with secondary volumetric changes possibly resulting from thermal expansion within the saturated porous medium surrounding the intrusion (Delaney, 1982). Together, these processes result in an instantaneous change in pore fluid pressures around the location of emplacement. Although data documenting these processes are rare, two events at Krafla, Iceland have supplied relatively complete transient pore pressure records recorded in a single well. The rapid pressure rise and subsequent slow pressure decay measured in a single well at the Krafla geothermal site may be used to infer the path of magma ascent within the host rock. Indeed, if the processes describing changes in pore pressure that result from intrusion are sufficiently quantified, a surprising array of param-

* Department of Geosciences.

A. P. S. Selvadurai (ed.), *Mechanics of Poroelastic Media*, 215–233.

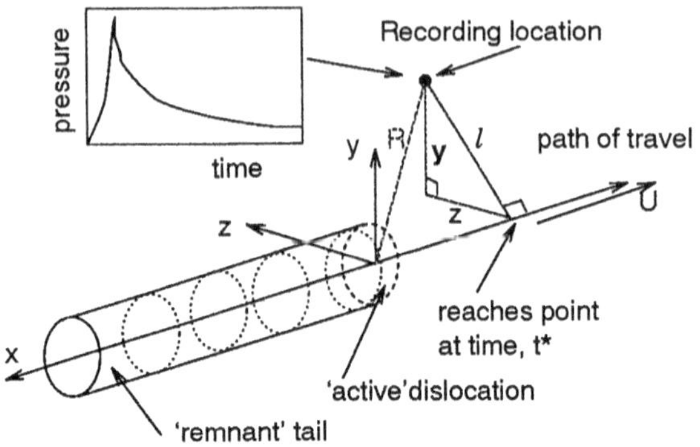

Fig. 1. Local coordinate system for a moving point dislocation (where 'disc' of area, a, is shrunk to a point) relative to a static recording location. Coordinate system is fixed to the tip of the dislocation migrating at rate, U.

eters describing dike morphology may be ascertained. The possibility of inferring dike morphology from pore pressures recorded at a single location in the surrounding medium, is developed in the following.

2. Moving Volumetric Dislocation

Magma intrusion into a saturated porous medium may be represented as a direct analog to a continuous volumetric dislocation that migrates, leaving a remnant 'inflated' tail. This is represented schematically in Figure 1, defining the migration rate, U, relative to the developing dislocation and the static measuring location where excess pore pressures are recorded.

2.1. Infinite Medium

Relationships may be developed to directly define the anticipated pore fluid pressures developed around dislocations moving within infinite and semi-infinite media. These solutions are developed to represent the different morphologies of magma ascent and include propagation of line (pencil shaped) and sheet dike structures. These intrusion morphologies are represented by point and line dislocations, respectively.

2.1.1. *Point Dislocation*

For an infinite medium, the Green's function may be developed directly from the concept of a dilation center that is "inflated" instantaneously and permanently to define the spherically symmetric induced pore pressure field (Cleary, 1977; Rudnicki, 1981). If this static dislocation is integrated in time and space (Elsworth, 1991) then the pore pressure induced as a result of intrusion may be defined in terms of dimensionless quantities as,

$$P_D = \frac{2}{R_D\sqrt{\pi}} \int_{\frac{R_D}{\sqrt{t_D}}}^{\infty} e^{U_D x_D - \eta^2 - (\frac{U_D R_D}{2\eta})^2} d\eta \qquad ($$

for transient behavior where η is a dummy variable defined as $\eta = \frac{R}{2\sqrt{c(t-\tau)}}$ with integration for all time, τ, and dimensionless pressure, P_D is defined as

$$P_D = \frac{4\pi l(p - p_s)}{Ua} \frac{k}{\mu} \tag{2}$$

and $p - p_s$ is the induced pore fluid pressure, a is the cross-sectional area of the dislocation (measured perpendicular to the direction of migration), k is permeability and μ is dynamic viscosity of the saturating fluid. Dimensionless pressure is related to dimensionless intrusion rate, U_D, and dimensionless time, t_D, through

$$U_D = \frac{Ul}{2c} \; ; t_D = \frac{4ct}{l^2} \tag{3}$$

defining the dependence on time, t, hydraulic diffusivity, c, and a diffusive length, l. The geometry of the system is defined in terms of dimensionless cartesian coordinates fixed to the tip of the advancing dislocation as

$$(x_D, y_D, z_D) = \frac{1}{l}(x, y, z) \; ; \; {R_D}^2 = {x_D}^2 + {y_D}^2 + {z_D}^2 \tag{4}$$

where R_D is the non-dimensional radius from the measuring location to the origin of the migrating coordinate system. As the lower integration limit of equation (1) approaches zero, representing long-time conditions, a dynamic steady state is established around the tip of the advancing dislocation with

$$P_D = \frac{1}{R_D} e^{-U_D(R_D - x_D)} \tag{5}$$

for $R_D/\sqrt{t_D} \to 0$. Since both equations (1) and (5) are written with reference to the moving coordinate system, a transformation must be applied to reference a static measuring location. If t^* is defined as the time at which the dislocation is closest to the measuring location, as defined in Figure 1, then the moving coordinate system may be transformed on noting that $x_D = \frac{1}{2}U_D(t_D - t_D^*)$ with $t_D^* = \frac{4ct^*}{l^2}$ enabling dimensionless radius, R_D, to be defined as $R_D^2 = x_D^2 + 1$.

The induced pore pressure distribution of equation (5) is no longer spherically symmetric for significant intrusion rates, as apparent in Figure 2 where a tail of significant 'pore pressures' exist for $U_D > 10^0$. This characteristic is a direct consequence of the advance of the dislocation and directly influences the form of the pressure rise experienced at a static location that is passed by the migrating pressure bulb. The pressure contours are nested within the area around the tip, representing a progressively more spherical form closer to the tip. In this region the most recent increment of spherical dilation has the predominant influence on pressure contour configuration. This change in morphology of the pressure bulb is also recorded where the migrating pressure bulb passes a static measuring location, as illustrated in Figure 3 where pressure build-up is recorded. As the intrusion rate increases above $U_D \geq 10^0$, maximum pore pressure recorded at a static measuring location no longer corresponds to the time of closest passage of the dislocation to that measuring location, i.e. $t_D - t_D^* = 0$.

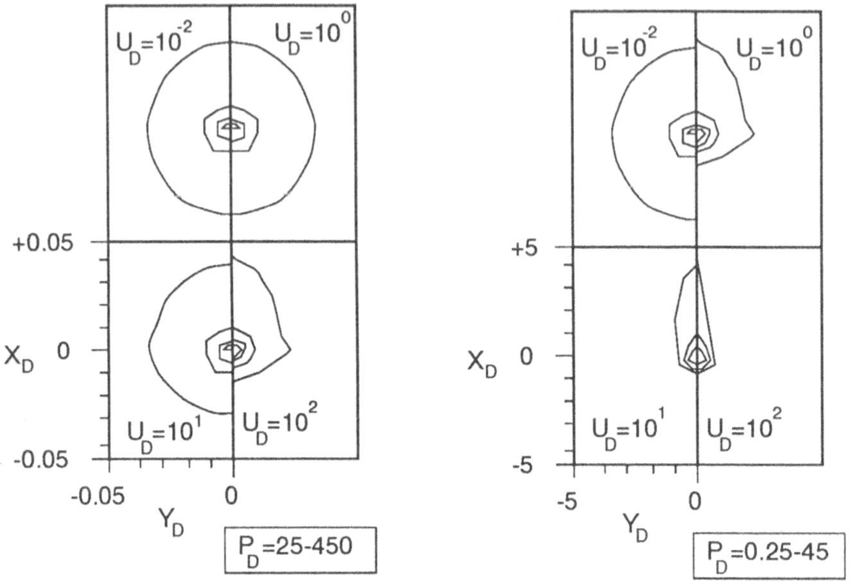

Fig. 2. Form of the steady pore pressure distribution induced around the front of an advancing point dislocation. Left figure is for local view ($x_D < 0.05$) and right figure is for distal view ($x_D < 5.0$). Range of pressure magnitudes, P_D, signify upper and lower bound of contour magnitudes.

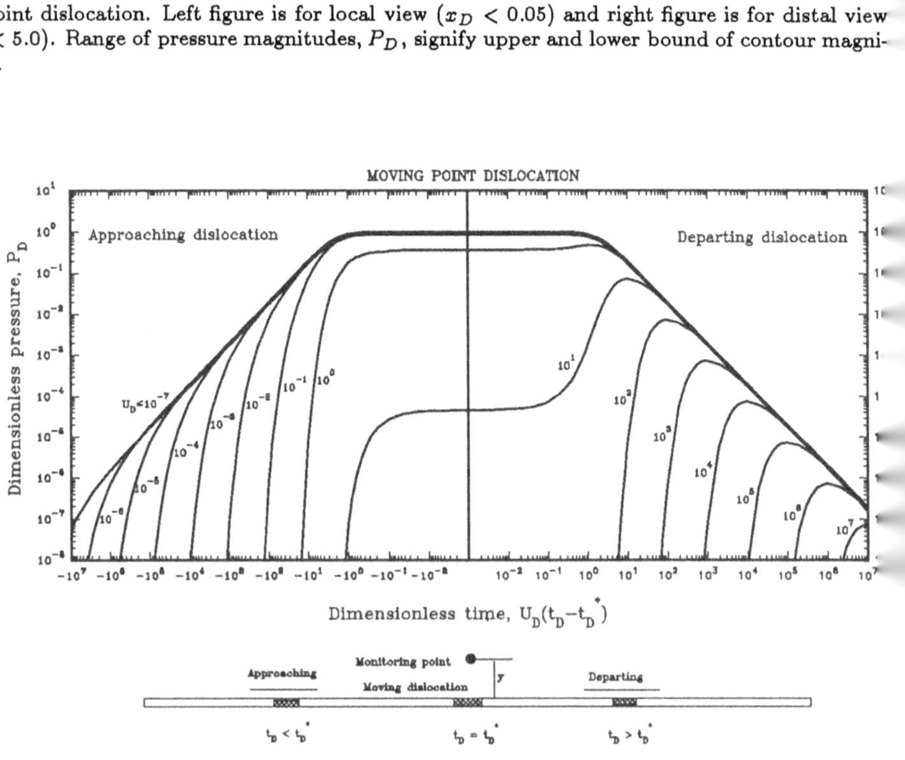

Fig. 3. Form of pore pressure build-up for a static measuring location with an approaching (left) and departing (right) dislocation

2.1.2. *Line Dislocation*

Where the dislocation takes the form of a moving line dilation, the induced pore fluid pressures may be evaluated by integrating equation (5) in the y- direction from $+\infty$ to $-\infty$ to yield

$$P_D = K_0[U_D R_D] e^{U_D x_D} \tag{6}$$

where $K_0[x]$ is the modified Bessel function of the second kind of order zero and all dimensionless parameters are as previously defined except dimensionless pressure. This integration represents the physical situation where a line is 'inflated' instantaneously and permanently to define a cylindrically symmetric increment of induced pore fluid pressure. The line migrates in a direction normal to its length and at rate, U. Dimensionless pore pressure for the line dislocation is now represented as

$$P_D = \frac{2\pi(p - p_s)}{Uw} \frac{k}{\mu} \tag{7}$$

where the dislocation is of width, w.

2.2. SEMI-INFINITE MEDIUM

Where the infinite saturated medium is assumed truncated at a horizontal free surface, behavior may be approximated through application of image dilations. For an impermeable upper boundary the applied image is of the same sense, and for a permeable (constant fluid pressure) boundary the image dislocation is of opposite sense to the *real* dislocation. The "approximation" results from the remnant normal and shear stresses left on the boundary for the impermeable and constant pressure cases, respectively, despite satisfying all the required pressure boundary conditions. Using this approach, the resulting behavior may be represented for point dislocation and line dislocation responses. For the geometry of Figure 4 the dimensionless radii from the static measuring location to the real, R_{RD}, and image, R_{ID}, dislocations may be defined in terms of the geometric parameters, q_D and b_D. Consequently (Ouyang, 1994),

$$R_{ID}^2 = R_{RD}^2 + q_D^2 \tag{8}$$

where for the simplest case of a dislocation moving parallel to the surface,

$$q_D^2 = \frac{4Hh}{l_R^2} \tag{9}$$

where all properties may be evaluated relative to the prescribed geometry of Figure 4. For details incorporating ascent inclined at an arbitrary angle of inclination, θ, the interested reader is referred to Ouyang (1994).

2.2.1. *Point Dislocation*

The magnitude of induced pressure may be straightforwardly evaluated for the steady state from

$$P_D = \frac{1}{R_{RD}} e^{-U_D(R_{RD} - x_D)} \pm \frac{1}{R_{ID}} e^{-U_D(R_{ID} - x_D)} \tag{10}$$

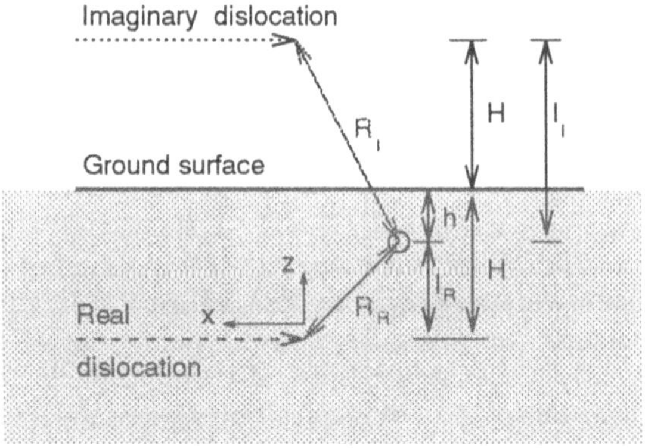

Fig. 4. Geometry for a dislocation moving relative to a free surface. Positions are shown for 'real' and 'imaginary' dislocations. Circle, at depth h, represents position of static measuring location. Minimum separation lengths are denoted as $l_R = H - h$ for the 'real' dislocation and $l_I = H + h$ for the 'imaginary' dislocation. Ground surface represents a constant fluid pressure boundary condition where imaginary dislocation is of opposite sense to the real dislocation and a zero flux condition were of the same sense (i.e. dilational or contractile).

where $R_{RD} = R_D$, as defined in equation (4), and the $\pm$ represents zero flux and constant pressure conditions respectively. Parameter q_D is a measure of recording location 'depth' below the free surface.

As apparent in equation (10), as $q_D \to \infty$, the behavior reverts to that of a dislocation within an infinite medium and equation (5) results. The modification of behavior with changing geometry, conditioned through q_D is described in Figure 5. Apparent from the figure is that the dimensionless time to peak pressure does not vary with geometry, q_D. The influence of the free surface boundary is to reduce the peak pressure magnitude for small q_D, with the magnitude increasing to that of an infinite medium as $q_D \to \infty$.

During the complete propagation process, the total time required for the pressure to rise to its peak magnitude and to subsequently dissipate is proportional to the magnitude of the geometric factor, q_D. Consequently, the pressure response for any desired q_D may be determined by translating the profile laterally in logarithmic space, as apparent in Figure 5.

2.2.2. *Line Dislocation*

Using the same geometric factors as defined in equations (8) and (9) the pressure response around a moving infinite line dislocation may be defined as

$$P_D = [K_0[U_D R_{RD}] \pm K_0[U_D R_{ID}]]e^{U_D x_D} \qquad (11$$

where the dislocation may ascend or descend at any arbitrary inclination, θ, as defined in the full determination of the parameter q_D (Ouyang, 1994).

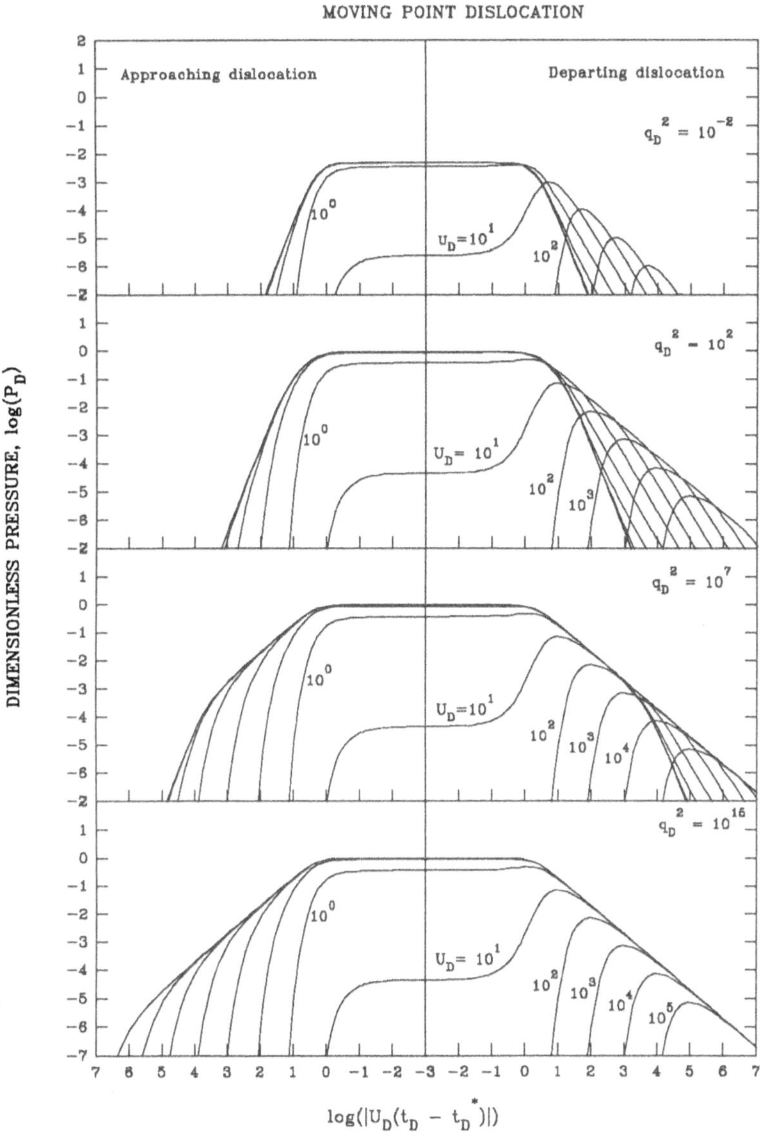

Fig. 5. Form of the induced pressure rise at a static measuring location resulting from a point dislocation migrating within a semi-infinite medium bounded by constant pressure conditions at the surface. t^* is the time when the dislocation passes closest to the measuring location.

3. Type Curve Analyses

Solutions representing behavior in a semi-infinite medium may be straightforwardly obtained through superposition of the results for an infinite medium modified by an influence coefficient, α. This procedure (Ouyang, 1994) directly avoids the necessity of providing multiple charts for type curve data reduction, with each chart representing a different q_D, as in Figure 5. Thus, the results of Figure 5 can be reduced to that indicated in Figure 6. In Figure 6, the type curves are translated in time relative to the time of peak pressure, t^p, since this is the only marker that may be determined from the pressure record alone. Pore pressure rise induced by the real (primary) and secondary image dislocations is evaluated individually and the composite response determined by summation. Two sets of dimensionless parameters, characterized by minimum distances, l_R $(= l)$ and l_I, are defined that are related by conditions of geometric compatibility and pore pressure balance. The influence coefficient, α, and minimum distance, l_I, to the secondary dislocation are respectively defined as

$$\alpha = \frac{l_R}{l_I - l_R}; \quad l_I^2 = l_R^2 + 4Hh\cos^2\theta \tag{12}$$

where the ascent path of the primary dislocation is inclined at an angle θ above the horizontal.

In implementing the type curve match, time of dislocation initiation may be considered generally unavailable, therefore pressure rise must be represented relative to the available time to peak-pressure. This removes the uncertainty of the shape of the recorded data curve. The drawback of this method is that the pre-peak and post-peak pressure data must be simultaneously matched with dual type curves representing the pre- and post-peak pressure response. This artifact of the analysis results from the choice of a zero-time-datum as the time-to-peak pressure, t^p, since zero is undefined in logarithmic space. Well defined pre-peak and post-peak pressure data yield identical matching pairs of $((t_D - t_D^p)_m, (t - t^p)_m)$ and $((P_D)_m, (p - p_s)_m)$ with matched $(U_D)_m$ where the subscript m refers to the "match point" Generally, there exist six unknowns of l_R, a, k/μ, c, U and θ, for the superposition method. Determination of dimensionless parameters from the match points enables the establishment of three equations. To enable solution of these, three unknowns must be defined *a priori* .

4. Test of Theory with Results from the Krafla Geothermal Area

Pore pressure data documenting magmatic intrusions are rare, but data for two separate dike intrusions are available for the geothermal site at Krafla, Iceland as reported by Sigurdsson (1982). The pressure pulses are recorded in a singl observation well, KG-5, completed to 1300 m and cased to a depth of 600 m. Th two magmatic events (sheet intrusions) occurred on September 8, 1977 and July 1C 1978, and are identified by documented peak pressure increases $(p^p - p_s)$ in well KG 5 of 0.76 MPa and 0.34 MPa, respectively. The separation between the recordin well and the surface phenomena associated with the two events was 4300 m for th September event and about 9300 m for the July event as illustrated in Figure 7.

The site geology comprises lower and upper geothermal reservoirs, with interac tion retarded by a confining bed (Stefànsson, 1981). Well KG-5 terminates in th upper single phase reservoir (liquid dominated with a mean water temperature 205°C) while the intrusive events, initiating from a magma chamber below 3 K depth (Ewart et al., 1990), affects also the lower two-phase reservoir (liquid-vap

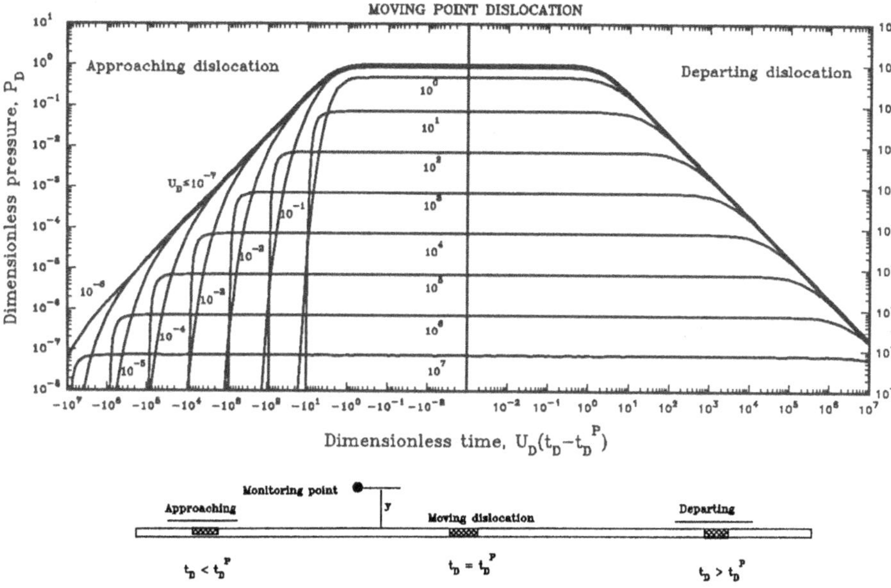

'ig. 6. Type curves used to match data for a moving point dislocation within a semi-infinite ıedium. t^P is the time when the peak pressure is measured, with these curves translated in time om the suite described in Figure 5.

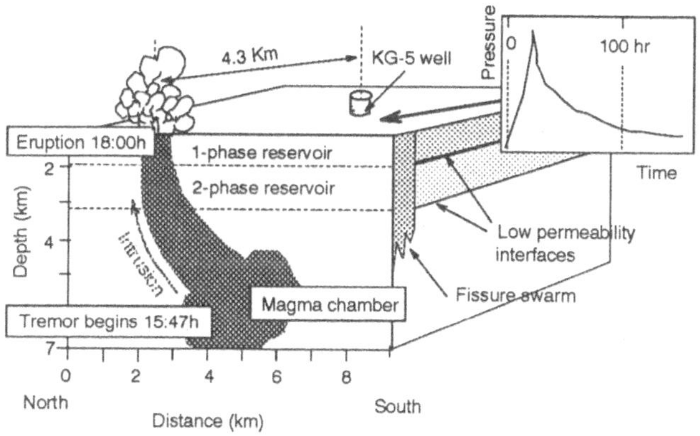

ǥ. 7. Block diagram with front of block aligned along the Krafla fissure swarm showing the ıgma intrusion system, the September 8th, 1977 eruption site and observation well KG-5.

th a mean temperature of $300°C - 350°C$). Pressure transients have not been tected in wells completed only in the lower zone. A schematic diagram indicating e form of the September 1977 event is shown in Figure 7. Ranges of the material rameters are summarized in Elsworth and Voight (1992). Appropriate parameter ıgnitudes are hydraulic diffusivity, c, in the range $7.73 - 15.47 \times 10^2 m^2/s$ and

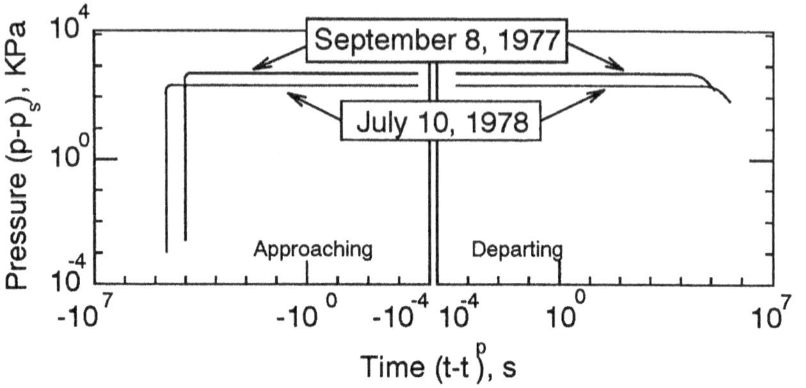

Fig. 8. Transient pore fluid pressure response measured in wellbore KG-5 at Krafla.

permeability, k/μ, in the range $1.09 - 2.18 \times 10^{-11} m^2/Pa.s$, enabling data fits to be attempted for the two events. Pressure rise data for the two events are plotted relative to time to peak pressure $(t - t^p)$ in Figure 8 enabling matching with the type curves of Figure 6.

4.1. Infinite Medium Solution

The point (pencil) and line (infinite sheet) dislocation models are applied to a vertical sheet intrusion of limited length that pressurized the upper reservoir as it broke toward the surface north of the magma storage area on September 8, 1977 (Elsworth and Voight, 1992). The steepness of the pressure rise, apparent in Figure 8, suggests a fit with the point dislocation solution for $U_D > 10^0$. Superimposing the data on the type curves of Figure 6 enables the ambiguous match point for $U_D(t_D - t^p_D) = 10^0$ to be determined and results in the unambiguous magnitude of $U_D(t - t^p) = 2.2 \times 10^4$. An almost perfect fit is achieved. Substituting this magnitude and the magnitude of hydraulic diffusivity from the range defined previously into $l = \sqrt{U_D(t - t^p)4}$ enables the minimum distance to the measuring location, l, to be determined as $l = 4123$ to $5834\ m$. This compares favorably with the distance to the eruptive event reported as $4300\ m$ previously. With this length range determined, the cross sectional area of the dislocation may be evaluated from $a = \frac{\pi l^2(p-p^s)}{e^{-1}c} \frac{k}{\mu}$ to yield a in the range 0.71 to 5.7 m^2. This, however, is difficult to reconcile with a dike 700 long and about 70 cm wide.

As noted previously, no precise determination of the dislocation velocity is possible. A lower bound may however be applied on noting from the form of the type curves that $U_D > 10^0$, therefore the minimum velocity is found from equation (3) $U_{min} = 0.265$ to $0.750\ m/s$. This range compares favorably with $U = 0.4\ m/s$, reckoned from magma ascent along an inclined 4.5 Km path in 193 minutes (data from Brandsdóttir and Einarsson, 1979). These data are reported in Table I.

A similar evaluation may be made for the event of July 10, 1978. The pressure trace is also defined in Figure 8 enabling the parameters defined in Table I to determined from Figure 6. Apparent from both sets of parameters are the surprisi matches with location and velocity, but the poor correspondence with dike volume data. The ability to represent the location data is related to the dependence of these data on the time record alone. The inability to match the dike volume data is due to the overestimation of pressure rise in the infinite medium solution engendered by neglecting the influence of the free surface.

TABLE I
Parameter estimates for intrusive events at Krafla using the model of a migrating point dislocation in an infinite medium ($q_D = \infty$). Bracketed numbers are "best" estimates of field behavior

Parameter	Sept. 8, 1977	July 10, 1978	Units
Length to dislocation, l	4123 – 5834 (4300)	8792 – 12437 (9300)	m
Cross-section area, a	0.71 – 5.72 (500)	1.45 – 11.62	m^2
Minimum velocity, U_{min}	0.27 – 0.75 (0.5)	0.124 – 0.352	m/s

4.2. Semi-Infinite Medium Solution

The cross-sectional area, a_s, (or width, w_s) of the dislocation within a semi-infinite medium is determined from $a_s = \alpha a$ (or $w_s = \alpha w$). Other parameters are determined from the known depth of initiation of magma ascent, $H = 3000m$, and known average depth of the monitoring well, $h = 900m$, together with prescribed hydraulic diffusivity magnitude, c, of 234.5 m^2/s and permeability, k/μ, of 1.09×10^{-11} $m^2/Pa.s$. Assuming $\theta = 80^0$, as the path of the magma ascent, cross-sectional area of the magma is calculated as $519m^2$, a reasonable estimate for "an eruptive dike $700m$ long and $0.7m$ in width". The estimates of propagation velocity and minimum distance to the dislocation are respectively $0.96m/s$ and $4842m$. These estimates compare favorably with the reported data. For the event of July 10, 1978, when $\theta = 69^0 - 80^0$, the estimated cross-sectional area of the dislocation is $1028 - 4330m^2$ for "a dike kilometers long and 1 m in width". Minimum propagation velocity is evaluated as 0.43 m/s. Although independent estimates of minimum advance rate are not available, the estimated minimum velocity appears plausible compared to the results from the event of September 8, 1977. Minimum distance of $l = 9685$ m corresponds well with the reported separation between borehole KG-5 and the center of seismic activity at 9300 m. Time lag of the image dislocation may be ignored since the initiating depth, H, and the depth of monitoring point, h, are small compared to the minimum distances. The above magnitudes are documented in Table II for the intrusive events using the model of a moving point dislocation. For comparison, the data of the two magma intrusive events are reexamined using the model of a moving line dislocation in a semi-infinite medium and the parameters documented in Table III. Solution for the moving point dislocation appears closer to observed results, presumably since this better reflects the geometry of the intrusive event. However, both sets of results for semi-infinite media yield reasonable descriptions of the intrusion, and more specifically enable dislocation volume to be adequately evaluated. This compares with the rather poor estimate of volume available where the infinite medium solution is used.

TABLE II
Parameter estimates for intrusive events at Krafla using the model of a moving point dislocation within a semi-infinite medium. Inclination angle of the ascent path is assumed known and all other parameters are evaluated from the recorded pore pressure response.

Parameter	Sept. 8, 1977	July 10, 1978	Units
Minimum distance, l	4842	9685	m
Cross-sectional area, a	519	1028-4330	m^2
Minimum velocity, U_{min}	.96	.43	m/s
Inclination angle, θ	80	69-80	degrees

TABLE III
Parameter estimates for intrusive events at Krafla using the model of a moving line dislocation within a semi-infinite medium. Inclination angle of the ascent path is assumed known and all other parameters are evaluated from the recorded pore fluid pressure response.

Parameter	Sept. 8, 1977	July 10, 1978	Units
Minimum distance, l	3062	6152	m
Intruded thickness, w	1.23	5.0	$\times 10^{-2}$ m
Minimum velocity, U_{min}	1.53	.76	m/s
Inclined angle, θ	80	80	degrees

5. Destabilization of Volcano Flanks

Observational evidence exists at a number of oceanic-shield type volcanoes indicating massive landslides. These landslide deposits are now recognized as comprising products of the growth and ultimate destruction of shield volcanoes as part of an on-going cyclic process. At Mauna Loa, for instance, debris avalanche deposits more than two hundred kilometers long and five thousand cubic kilometers in volume are apparent in the bathymetric record (Moore et al., 1964, 1989). These slope failures are presumed to accompany all stages of cone building, continue long after dormancy, and develop in the form of both slow moving slumps and fast moving debris avalanches. Despite the clear evidence of these features in the bathymetric record failure remains enigmatic for these shield volcano flanks since estimated frictional resistance on basal failure planes appears too high to enable failure to initiate, or to be sustained (Iverson, 1991). This inconsistency may be explained if all factors affecting shear strength are examined. Most importantly, the factors that affect shear strength may be considered in terms of effective stress analysis whereby both frictional resistance of the basal surface (a material parameter) and reduction of effective stresses through increased pore fluid pressures (an environmental factor) may be considered separately. Consequently, if pore fluid pressures can be increased sufficiently, then failure may initiate and, under appropriate circumstances, may b sustained. One such evolutionary process which has been shown to produce substantial pore fluid pressure increases is that of magmatic dike injection (Voight and Elsworth, 1992). Dike injection may produce pore fluid pressures through the mechanisms of mechanical straining (Stefánsson, 1981; Watanabe, 1983; Elsworth and Voight, 1992) and the complementary process of thermal straining (Björnsson et al., 1977). The anticipated magnitude of the pore fluid pressures developed by each of these processes is quantified in the following with specific regard to the potential to create long run-out flank instability in oceanic-shield type environments.

5.1. Uplift Pressures

The pore fluid pressures, $p - p_s$, induced around a planar dike of width, w, intruded at velocity, U, within a poroelastic medium may be readily evaluated from equation (6) for an infinite medium and equation (11) for a semi-infinite medium. The integrated form of this uplift acts along the base of a delineated failure block, as illustrated in Figure 9 applying a total uplift force, F_{pm}. This uplift will act irrespective of the inclination of the basal failure surface. The applied uplift force may be evaluated as (Elsworth and Voight, 1994),

$$F_{pm} = \int_{-\frac{1}{2}d}^{+\frac{1}{2}d} \int_{0}^{\bar{l}} (p - p_s) dl dx = w_D U_D \gamma_w h_s^3 \int_{-\frac{1}{2}d_D}^{+\frac{1}{2}d_D} \int_{0}^{\bar{l}_D} P_D dl_D dx_D \qquad (13)$$

where induced pore pressures P_D is defined in equations (6) or (11) and the nondimensional width and length of the basal failure surface are d_D and $\bar{l}_D$, respectively. The integration is performed along the $l-$ axis, aligned parallel to the base of the failing block. The nondimensional variables are the same, excepting a new characteristic length is selected, h_s, that represents the height of the flank crest above sea level. This arbitrary diffusive length scale replaces l in the nondimensional coordinates and in U_D as $U_D = \frac{Uh_s}{2c}$. A supplemental variable of dimensionless dike width, w_D is defined as $w_D = \frac{\mu}{k} \frac{wc}{\pi \gamma_w h_s^2}$ where γ_w is the unit weight of water.

Uplift forces may be determined from the pressure bulb forms illustrated in Figure 2, as the migrating dike traverses the rear of the block. Notably, when dimensions of the basal failure plane are sufficiently large to include the entire (significant) extent of the pressure bulb defined in equation (5), and shown in Figure 2, then the mechanical uplift force, F_{pm}, is defined purely in terms of the dimensionless variables of intrusion velocity, U_D, and dike width, w_D. An increase in either of these quantities, given that all other parameters remain constant, results in increased uplift force acting on the free block. Increase in either velocity or width has the net effect of increasing the net volumetric intrusion rate with a corresponding increase in the dilational strain rate in the surrounding porous medium. It is this strain rate, manifest in the surrounding geologic material, that conditions pore pressure generation. Where the basal plane is insufficiently large, then the limits of integration, d_D and $\bar{l}_D$, also condition the uplift force generated by intrusion since the distribution is now truncated at the periphery.

5.2. Flank Failure Geometry

Flank failure may be considered for two specific geometries representing the potential for shallow failure and deep-seated rupture, respectively. These failure modes are

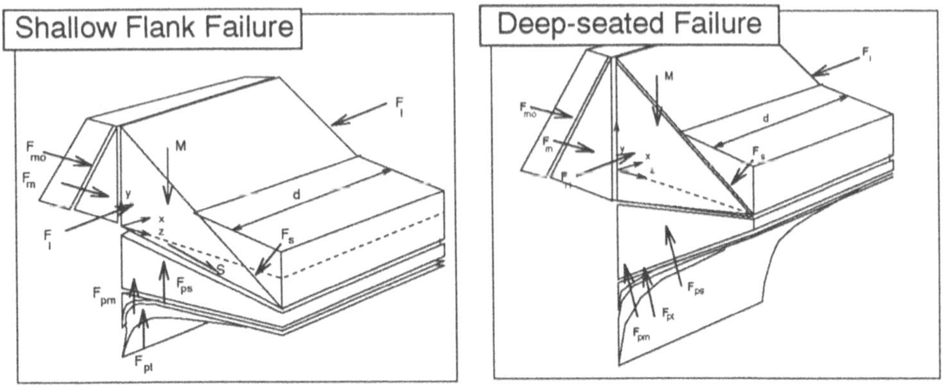

Fig. 9. Generalized geometry for limit equilibrium analysis of shallow and deep-seated flank failure. All quantities defined positive. α is positive for deep seated failure and negative for long runout failure. Forces acting on the system are block weight, M, seawater pressure, F_s, lateral force, F_l, magma force, F_m, magma overpressure force, F_{mo}, static groundwater force, F_{ps}, and induced pore pressure forces that result from mechanical, F_{pm}, and thermal, F_{pt}, strains. Height of block crest above sea level, h_s, and block width, d, define block geometry.

represented in Figure 9. The only difference between the two failure modes is the dip of the potential basal failure surface with respect to the slope of the volcanc flank. For shallow flank failure, the dip of the basal failure surface is congruent with the dip of the flank and provides the potential for long run-out failure of the type observed in debris avalanches. In this the term *shallow* is relative since the rear scarp depth of these failures may be of the order of one kilometer. This mode of failure is in contrast to deep-seated failure where run-out is not possible but where limited rupture close to the rear block scarp may result in the development of limited slump displacements and/or deep earthquakes. These earthquakes may in turn, precipitate seismic destabilization of shallow failures as a result of both induced ground accelerations and potential influence of induced pore-pressure rise or liquefaction of weak materials.

The limit equilibrium behavior of either of these shallow or deep geometries may be described using identical terminology. Delineating a block of width, d, enables a force balance to be performed on a system that includes the effects of block weight M, magma force, F_m, magma overpressure force, F_{mo}, static groundwater force F_{ps}, seawater force, F_s, lateral block force, F_l, and the two important environmental factors representing mechanically induced uplift force, F_{pm}, and thermally induced uplift force, F_{pt}.

Choosing a simple geometry with a flank slope of ten degrees, a crest of the failing block 1,000 meters above sea level, a basal failure plane inclination of five degrees, and a groundwater surface sloping up from sea-level at two degrees enables the stability of a simple geometry to be investigated. The static groundwater seawater, and magma pressures remain relatively invariant under this choice of geometry where reasonable magnitudes of magma ($28 \times 10^{-3} MPa/m$) and rock density ($23 \times 10^{-3} MPa/m$) are selected. Correspondingly, the potential destabilizing influence of dike intrusion through the development of mechanical and thermal po pressures may be evaluated. This evaluation is most conveniently made using a normalized factor of safety, $F_D = F/tan\phi$, to represent behavior for a variety of undefined frictional resistance magnitudes, ϕ. In this configuration, normalized factors of safety in the range 1-2 represent the limiting equilibrium condition for angles of frictional resistance in the range 45-27 degrees, respectively. The response of two different materials to dike intrusion through mechanical effects may be evaluated. The two materials are fractured lava and volcaniclastics, representing reasonable bounds on material behavior.

5.3. Mechanical Pore Pressures

The anticipated stability of a volcano flank comprising both rubbleized volcaniclastic debris and fractured basaltic lava is illustrated in Figure 10. For the volcaniclastic material, an appropriate range of non-dimensional dike widths, w_D, influenced largely by diffusivity, is of the order $10^0 - 10^6$. Apparent in the behavior is the diminishing influence of lateral restraint acting to stabilize the failing wedge as width of the failing block increases. This effect is countered at very large magnitudes of block width where stability again increases as the size of the potential failing block increases beyond the extent of the pressure bulb induced by dike injection. For the larger magnitudes of non-dimensional dike widths anticipated, mechanically induced pore fluid pressures appear sufficient to result in destabilization. These block widths are relatively large representing real block widths in the order of 1 to 100 kilometers ($d_D = 1 - 100$). For the fractured lavas, the diffusivity is such that the appropriate range of non-dimensional dike widths is of the order of 10^{-4} to 10^0 and the corresponding influence on stability is minimal, as illustrated in Figure 10. Correspondingly, the potential destabilizing influence of mechanically induced pore fluid pressures appears closely related to the mechanical and fluid transmissive characteristics of the rock surrounding the intrusion.

5.4. Implications for Flank Stability

From the somewhat idealized geometry and characteristics described previously, the magnitude of mechanically induced pore fluid pressures that may accompany dike intrusion appear sufficient to initiate failure in oceanic-shield type volcanoes. The magnitude of uplift pressures is influenced by the mechanical and fluid transmissive characteristics of the volcano cone material and is therefore controlled by the magnitude or severity of the intrusive event. For the ranges of material characteristics deemed applicable to these environments, the initiation of failure is certainly plausible.

Once initiated, the development of the instability into a debris avalanche requires that the disturbing forces are sustained. Earthquake generated liquefaction in the area of the basal failure plane, pressure generation as result of pore collapse, frictional heating on the failure plane and undrained loading of sediments in the over-run area at the toe of the block may all contribute to maintaining movement of the slide once initiated.

These mechanisms of pore pressure generation offer plausible means of resolving the enigma of failure in low-inclination flanks of shield volcanoes. Although attention in the previous has been restricted to geometries representative of shield volcanoes,

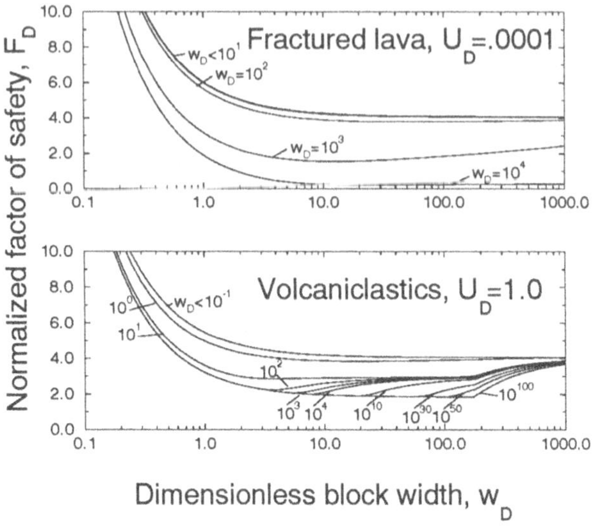

Fig. 10. Variation of normalized factor of safety ($F_D = F/\tan\phi$) for a long runout failure c a volcano flank with mechanically induced pressures resulting from dike intrusion. Basal failu plane is inclined at $\alpha = -5^o$. $h_m/h_s = 1$ with h_m given as the magma column height. $U_D = 10$ is representative of volcaniclastic debris and $U_D = 10^{-4}$, is representative of fractured lava.

similar triggering mechanisms may be shown to act in stratovolcanoes (Voight an Elsworth, 1994). For stratocones, the initially steeper flank inclination implies th a smaller relative pore pressure change is required to bring the flank to limit equ librium, and consequent failure. Correspondingly, the conclusions drawn above fo shield type volcanoes appear more widely applicable to other volcano geometries.

6. Concluding Remarks

The geological process of dike emplacement is associated with significant poroelast effects. Where magma is injected into saturated rock, the strain induced fluid pre sures are of significant magnitude. Indeed, wellbore pressures generated at Kra were of the order of a $70m$ head of water, with the closest passage of the dike b ing $4Km$ remote from the measuring well. Consequently, the magnitude of the generated pore pressures is quite remarkable.

Previously utilized static solutions for dike emplacement are incapable of co rectly representing the essential components of the moving boundary value proble The simple, but mechanistically rigorous, approach defined by the moving dislo tion enables a surprising array of characteristics defining the intrusive process to determined. Ascent velocity is shown to provide an important control on the m phology of the pressure bulb that migrates with the advancing dislocation front. T form of the pressure bulb, in turn, directly influences the transient response record at a static measuring location as the intrusion passes. From the form of the tr ent response both the ascent path and ascent velocity may be determined. In ddition, the magnitude of induced pressures is an index of the volumetric injection ate and enable intrusion volume, or analogous cross-sectional area, to be defined. epresentation as a dislocation traversing an infinite medium had previously yielded nrealistically low estimates of dike width. These low estimates may be discounted here the "true" semi-infinite nature of the system is accommodated. Under these onditions the injected volume estimates appear much more closely constrained and ealistic.

Representation as a *moving* dislocation has important ramifications in the eval- ation of flank instability developed as a result of dike-emplacement induced pore uid pressures. Uplift pressures are dependent on the intrusion rate and the material arameters embodied in the parameters of nondimensional dike width and intrusion elocity. Increasing either of these parameters has the net influence of increasing plift, enabling the pressure increase to be readily evaluated from the representa- ve models reported in the preceding. Estimates of induced fluid pressures appear ufficiently large to initiate, and to potentially sustain failure, and offer a plausible echanistic explanation for the enigmatic failure of shallow angle volcano flanks, uch as in the Hawaiian Islands.

The application of these methods is not limited to dike emplacement, but includes variety of phenomena operating in geologic media. For example, these methods have een applied to define pore pressure behavior that develops around penetrometers, dvanced into the ground as a site investigation technique. Here, the rate of dissi- ation of induced pore fluids from around the tip may be used to directly determine eld permeability and diffusivity magnitudes in homogeneous (Elsworth, 1993) and ayered media (Elsworth, 1992), alike. Similarly, the technique may be used to esti- ate the development of pore fluid pressures around hydraulic fractures developed or hydrocarbon recovery and for application to groundwater remediation (Piggott nd Elsworth, 1994).

cknowledgements

he support of the American Chemical Society under Grant PRF-25922-AC2 and the ational Science Foundation under Grants MSS-9218547, BCS-8904990 and NSF- ainier is gratefully acknowledged.

. References

jörnsson, A., L. Kristjánson, and H. Johnson, Some observations of the Heimaey eep drill hole during the eruption of 1973, *Jökull, 26*, 52-57, 1977.

randsdóttir, B. and E.T. Einarsson, Seismic activity associated with the September 977 deflation of the Krafla central volcano in north-eastern Iceland, *J. Volc. and eothermal Res., 6*, 197-212, 1979.

leary, M. P., Fundamental solutions for a fluid- saturated porous solid, *Int. J. olids Structures, 13*, 785-806, 1977.

elaney, P.T., Rapid intrusion of magma into wet rock: groundwater flow due to pore pressure increases, *J. Geophys. Res., 87(B9),* 7739-7756, 1982.

Elsworth, D., Dislocation analysis of penetration in saturated porous media, *J. E Mech. Div. ASCE, 117(2),* 391-408, 1991.

Elsworth, D. and B. Voight, Dike intrusion as a trigger for large earthquakes an failure of volcano flanks, *J. Geophys. Res.*, submitted for publication, 1994.

Elsworth, D. (1993) Analysis of piezocone data using dislocation based methods. *J Geotechnical Engineering, ASCE, Vol. 119, No. 10,* 1601 - 1623, 1993.

Elsworth, D. and B. Voight, Theory of dike intrusion in a saturated porous solid, *J Geophys. Res., 97(B6),* 9105-9117, 1992.

Elsworth, D. Pore pressure response due to penetration through layered media. *In J. Num. Anal. Meth. in Geomechs., Vol. 16, No. 1,* 45 - 64, 1992.

Ewart, J.A., B. Voight and A. Björnsson, Dynamics of Krafla caldera, north Icelanc 1975-1985. *Magma Transport and Storage, Edited by M.P. Ryan, J. Wiley and Son* 225-276, 1990.

Iverson, R.M., Failure and runout of giant landslides on Hawaiian volcanoes: case of enigmatic mechanics? *Geol. Soc. Amer. Annual Meeting Abstracts, San Dieg* 50, 1991.

Moore, J.G., Giant submarine landslides on the Hawaiian Ridge, *U.S. Geol. Sur Prof. Pap., 501-D,* D95-D98, 1964.

Moore, J.G., D.A. Clague, R.T. Holcomb, P.W. Lipman, W.R. Normark and M.E Torresan, Prodigious submarine landslides on the Hawaiian Ridge, *J. Geophys. Res 94(B12),* 17465-17484, 1989.

Ouyang, Z. An investigation of dislocations propagating in poroelastic media, *Ph.I Thesis, Pennsylvania State University, University Park,* 1994.

Piggott, A.R., and D. Elsworth, Formation fluid displacement induced by hydrauli fracturing, *in Proceedings of the 8th International Conference on Computer Metho and Advances in Geomechanics, A.A. Balkema, Rotterdam,* 1627-1632, 1994.

Rudnicki, J.W., On "Fundamental solutions for a fluid-saturated porous solid" b M.P. Cleary, *Int. J. Solids Structures, 17,* 855-857, 1981.

Sigurdsson, O., Analysis of pressure pulses resulting from volcanic activity in th vicinity of a well, *M.S. Thesis, University of Oklahoma, Norman, Oklahoma,* 75 1982.

Stefánsson, V., The Krafla geothermal field, northern Iceland, in *Geothermal Sy tems; Principles and Case Histories,* edited by L. Rybach and L.J.P. Muffler, p 273-294, John Wiley, New York, 1981.

Turcotte, D.L., On the role of laminar and turbulent flow in buoyancy driven magm fracture, *Magma Transport and Storage, Edited by M.P. Ryan, J. Wiley and Son* 103-111, 1990.

Voight, B. and D. Elsworth, Resolution of mechanics problems for prodigious Hawa ian landslides: magmatic intrusions simultaneously increase driving force and decrease driving resistance by fluid pressure enhancement, *EOS, Transactions, Amer. Geophys, Union, Abstract, 73(43)*, pp. 506, 1992.

Voight, B. and D. Elsworth, Role of magma in volcano collapse mechanics, *Géotechnique*, submitted for publication, 1994.

Watanabe, H., Changes in water level and their implications to the 1977-1978 activity of Usu volcano, Arc Volcanism: Physics and Tectonics, edited by D. Shimozuru and J. Yokoyama, 81-93, 1983.

AXISYMMETRIC INDENTATION OF A MULTILAYERED POROELASTIC SOLID

Z.Q. Yue
Pavements Laboratory, Centre for Surface Transportation Technology, National Research Council, Ottawa, Ontario, Canada K1A 0R6.

A.P.S. Selvadurai
Department of Civil Engineering and Applied Mechanics, McGill University, Montreal, Quebec, Canada H3A 2K6.

1. Introduction

Biot's theory of classical poroelasticity [1,2] has been applied by several researchers to examine the consolidation behaviour of multilayered poroelastic geomaterials [3-6]. A majority of these investigations have focussed on the evaluation of the time-dependent behaviour of a multilayered poroelastic solid which is subjected purely to surface loadings. The present paper in particular extends the study of the axisymmetric indentation of a poroelastic layer given by Selvadurai and Yue [7] to cover the effect of a layered material inhomogeneity. The analysis of the indentation problem related to a multilayered poroelastic solid represents a useful engineering application which models the consolidation behaviour of a relatively rigid foundation resting on the surface of saturated multilayered soils (Fig.1). The rigid foundation is assumed to be in smooth contact with the multilayered poroelastic solid and the entire surface of the poroelastic medium is assumed to be either fully pervious or fully impervious. The paper presents solution representations governing the poroelastic behaviour of a geomaterial layer saturated with a compressible fluid and summarizes the Fredholm integral equations of the second kind governing the mixed boundary-value problems represented in a Laplace transform domain. Numerical results are presented to illustrate the effect of a two-layered poroelastic solid on the consolidation settlement of the rigid foundation.

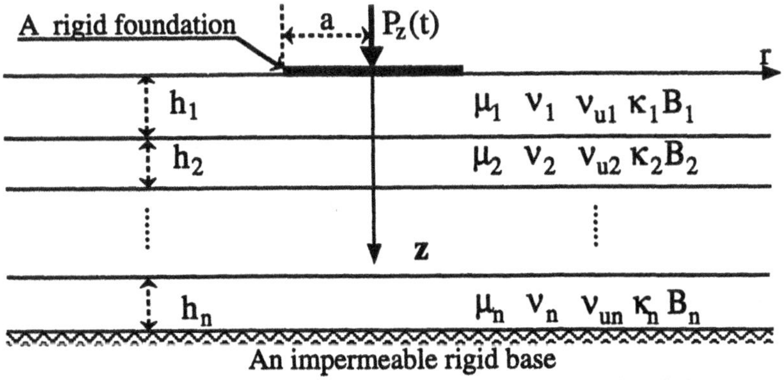

Fig.1 A multilayered poroelastic solid indented by a circular rigid foundation

A. P. S. Selvadurai (ed.), *Mechanics of Poroelastic Media*, 235–241.

2. Governing Equations

Referring to Selvadurai and Yue [7], the following sets of solution representations exist in a cylindrical coordinate system for the field variables in a poroelastic layer saturated with a compressible pore fluid.

$$\begin{aligned}
\mathbf{u} &= \frac{1}{2\pi}\int_0^\infty\int_0^{2\pi}\frac{1}{\rho}\mathbf{\Pi}\mathbf{w}K\rho d\varphi d\rho, & \mathbf{w} &= \frac{\rho}{2\pi}\int_0^\infty\int_0^{2\pi}\mathbf{\Pi}^*\mathbf{u}K^* r d\theta dr \\
\mathbf{T}_z &= \frac{1}{2\pi}\int_0^\infty\int_0^{2\pi}\mathbf{\Pi}\tau K\rho d\varphi d\rho, & \tau &= \frac{1}{2\pi}\int_0^\infty\int_0^{2\pi}\mathbf{\Pi}^*\mathbf{T}_z K^* r d\theta dr \\
\mathbf{v} &= \frac{1}{2\pi}\int_0^\infty\int_0^{2\pi}\rho\mathbf{\Pi}\vartheta K\rho d\varphi d\rho, & \vartheta &= \frac{1}{2\pi\rho}\int_0^\infty\int_0^{2\pi}\mathbf{\Pi}^*\mathbf{v}K^* r d\theta dr \\
p &= \frac{1}{2\pi}\int_0^\infty\int_0^{2\pi}p_w K\rho d\varphi d\rho, & p_w &= \frac{1}{2\pi}\int_0^\infty\int_0^{2\pi}pK^* r d\theta dr
\end{aligned} \tag{1}$$

where $0 \leq z < \infty$, $0 < t < \infty$, and the integrals are interpreted in the sense of a Cauchy principal value. The vector field in (1) consists of the displacement vector $\mathbf{u}(= [u_r, u_\theta, u_z]^T)$, the vertical stress vector $\mathbf{T}_z(= [\sigma_{zr}, \sigma_{z\theta}, \sigma_{zz}]^T)$, and the specific discharge vector $\mathbf{v}(= [v_r, v_\theta, v_z]^T)$. p is the pore fluid pressure. $\mathbf{w}(= [w_1, w_2, w_3]^T)$ $\tau(= [\tau_1, \tau_2, \tau_3]^T)$, $\vartheta(= [\vartheta_1, \vartheta_2, \vartheta_3]^T)$ and p_w are respectively the Fourier transform images of $\mathbf{u}$, $\mathbf{T}_z$, $\mathbf{v}$ and p. The superscript ^{T} stands for the transpose of matrix The Fourier matrix kernel functions K and $\mathbf{\Pi}$ are defined by

$$K = e^{+i\rho r\sin(\theta+\varphi)}, \quad \mathbf{\Pi} = \begin{pmatrix} +i\sin(\theta+\varphi) & +i\cos(\theta+\varphi) & 0 \\ +i\cos(\theta+\varphi) & -i\sin(\theta+\varphi) & 0 \\ 0 & 0 & 1 \end{pmatrix} \tag{2}$$

and their complex conjugates are denoted by K^* and $\mathbf{\Pi}^*$ respectively.

Assuming the initial condition $\zeta_v|_{t=0} = 0$, the governing equations of Biot's theory of poroelasticity for a compressible pore fluid can be re-written as two sets of first-order ordinary differential equations in the Fourier and Laplace transform domains [7]. Solving these ordinary differential equations, an algebraic solution representation of the field variables in the transform domain can be further obtained in terms of the *field variables at the two boundary surfaces of the poroelastic layer*. Furthermore, putting $z = 0$ and $z = h$ in the algebraic solution representation, eight independent boundary algebraic equations can be obtained, which govern the sixteen boundary variables at the two boundary surfaces of the poroelastic layer; *i.e.*,

$$\begin{aligned}
\mathbf{A}\mathbf{U}_v(\rho,\varphi,0,s) + \mathbf{B}\mathbf{U}_v(\rho,\varphi,h,s) = \mathbf{0} \\
\mathbf{Q}\mathbf{U}_u(\rho,\varphi,0,s) + \mathbf{P}\mathbf{U}_u(\rho,\varphi,h,s) = \mathbf{0}
\end{aligned} \tag{3}$$

where $\mathbf{U}_v = [\hat{w}_2, \hat{\tau}_2]^T$; $\mathbf{U}_u = [\hat{w}_1, \hat{w}_3, \hat{\tau}_3, \hat{\tau}_1, \hat{p}_w, \hat{\vartheta}_3]^T$; the circumflex ^ stands for the Laplace transform with respect to t, and s is the Laplace transform parameter and

$$\mathbf{A} = \left[\mathbf{A}_p - \mathbf{A}_q e^{-\rho h}\right]\mathbf{\Lambda}_v, \qquad \mathbf{B} = \left[\mathbf{A}_q - \mathbf{A}_p e^{-\rho h}\right]\mathbf{\Lambda}_v$$
$$\mathbf{Q} = \left[\mathbf{Q}_p + \mathbf{Q}_s - \mathbf{Q}_q e^{-\rho h} - \mathbf{Q}_t e^{-\rho\gamma h} - \rho h \mathbf{Q}_v e^{-\rho h}\right]\mathbf{\Lambda}_u \tag{4}$$
$$\mathbf{P} = \left[\mathbf{Q}_q + \mathbf{Q}_t - \mathbf{Q}_p e^{-\rho h} - \mathbf{Q}_t e^{-\rho\gamma h} + \rho h \mathbf{Q}_u e^{-\rho h}\right]\mathbf{\Lambda}_u$$

where the square coefficient matrices $\mathbf{A}_p$, $\mathbf{A}_q$, $\mathbf{Q}_u$, $\mathbf{Q}_v$, $\mathbf{\Lambda}_v$ and $\mathbf{\Lambda}_u$ have constant elements; $\mathbf{Q}_p$, $\mathbf{Q}_q$, $\mathbf{Q}_s$ and $\mathbf{Q}_t$ are square matrices which are functions of γ $(=\sqrt{\frac{1}{c}\frac{s}{\rho^2}+1}$, where $c = \frac{2\mu B^2(1-\nu)(1+\nu_u)^2\kappa}{9(\nu_u-\nu)(1-\nu_u)}$). The details of these square coefficient matrices are given in the references [7,8].

It is noted that the governing equations (4) are characterized by the *five* independent material parameters which are represented by the drained and undrained Poisson ratios ν and ν_u respectively, the shear modulus μ, Skempton's pore pressure coefficient B, and the constant κ $(=k/\gamma_w$, where k is the coefficient of permeability and γ_w is the unit weight of pore fluid).

3. The Indentation Problem

The mixed boundary value problem considered here is that of a multilayered poroelastic solid subjected to an axisymmetric indentation by a rigid circular foundation with a smooth flat base (see Fig.1). The multilayered poroelastic solid consists of n dissimilar poroelastic layers which adhere to an impermeable rigid bedrock. The interfaces of the n dissimilar poroelastic layers are bonded and the fluid flow across an interface exhibits continuity in the pressure and velocity. Referring to the cylindrical coordinates in Fig.1, it may be noted that the j-th poroelastic layer occupies a layer region of $H_{j-1} \le z \le H_j$ and has the thickness $h_j(= H_j - H_{j-1})$ and has the five material parameters (μ_j, ν_j, ν_{uj}, κ_j and B_j), where $j = 1, 2, 3, \cdots, n$ and $H_0 = 0$. For future reference, the mixed boundary value problem is represented in relation to a generalized asymmetric station of deformation as follows.

The mixed boundary conditions at the surface ($z = H_0 = 0$) of the multilayered poroelastic solid can be written as follows,

$$\begin{aligned} &\sigma_{zz}(r,\theta,0,t) = 0, && a < r < \infty,\ 0 \le \theta < 2\pi \\ &u_z(r,\theta,0,t) = D_z(r,\theta,t), && 0 \le r < a,\quad 0 \le \theta < 2\pi \end{aligned} \tag{5}$$

where $0 < t < \infty$, a is the radius of the rigid circular foundation; $D_z(r,\theta,t)$ is the displacement of the rigid foundation along the z-axis. The shear stresses $(\sigma_{rz}, \sigma_{\theta z})$ at the entire surface ($z = 0$) are zero for all $t \ge 0$. For a drained surface, $p(r,\theta,0,t) = 0$; for an undrained surface, $v_z(r,\theta,0,t) = 0$. At the undrained rigid base ($z = H_n$), $u_r = u_\theta = u_z = 0$, and $v_z = 0$.

The axial load $P_z(t)$ acting on the rigid foundation can be expressed in terms of the contact normal stress $\sigma_{zz}(r,\theta,0,t)$ in the following integral form.

$$P_z(t) = -\int_0^a\int_0^{2\pi} \sigma_{zz}(r,\theta,0,t)\, r\, d\theta\, dr \tag{6}$$

4. Surface Deflection in the Transform Domain

Using the boundary and interface conditons and equations (1-4), it can be shown that $\mathbf{U}_v(z) = \mathbf{0}$. The following algebraic equations in the transform domain govern the unknown variables $\mathbf{U}_u(H_j)$ at the surface and the interfaces ($j = 0, 1, 2, \cdots, n$).

$$\mathbf{M}_u \hat{\mathbf{X}}_u = \mathbf{I}_u \hat{\tau}_3(\rho, \varphi, 0, s) \tag{7}$$

where $\hat{\mathbf{X}}_u = [\mathbf{U}_u(0), \mathbf{U}_u(H_1), \cdots, \mathbf{U}_u(H_n)]^T$, $\mathbf{I}_u = [1, 0, 0, 0, \cdots, 0, 0]^T$. In (7), $\mathbf{M}_u$ is a non-singular square matrix of order $6(n+1) \times 6(n+1)$ and has the following expression.

$$\mathbf{M}_u = \begin{pmatrix} \mathbf{D}_a & \mathbf{0} & \mathbf{0} & \mathbf{0} & \cdots & \mathbf{0} & \mathbf{0} \\ \mathbf{Q}_1 & \mathbf{P}_1 & \mathbf{0} & \mathbf{0} & \cdots & \mathbf{0} & \mathbf{0} \\ \mathbf{0} & \mathbf{Q}_2 & \mathbf{P}_2 & \mathbf{0} & \cdots & \mathbf{0} & \mathbf{0} \\ \vdots & \vdots & \vdots & \vdots & \vdots & \vdots & \vdots \\ \mathbf{0} & \mathbf{0} & \mathbf{0} & \mathbf{0} & \cdots & \mathbf{Q}_n & \mathbf{P}_n \\ \mathbf{0} & \mathbf{0} & \mathbf{0} & \mathbf{0} & \cdots & \mathbf{0} & \mathbf{D}_b \end{pmatrix} \tag{8}$$

where $\mathbf{D}_b = \begin{pmatrix} 1 & 0 & 0 & 0 & 0 & 0 \\ 0 & 1 & 0 & 0 & 0 & 0 \\ 0 & 0 & 0 & 0 & 0 & 1 \end{pmatrix}$; $\mathbf{D}_a = \begin{pmatrix} 0 & 0 & 1 & 0 & 0 & 0 \\ 0 & 0 & 0 & 1 & 0 & 0 \\ 0 & 0 & 0 & 0 & 1 & 0 \end{pmatrix}$ for a fully permeable surface; or $\mathbf{D}_a = \begin{pmatrix} 0 & 0 & 1 & 0 & 0 & 0 \\ 0 & 0 & 0 & 1 & 0 & 0 \\ 0 & 0 & 0 & 0 & 0 & 1 \end{pmatrix}$ for a fully impermeable surface.

Consequently, $\hat{w}_3(\rho, \varphi, 0, s)$ can be expressed in terms of $\hat{\tau}_3(\rho, \varphi, 0, s)$ as follows for the two extreme surface drainage conditions; *i.e.*,

$$\hat{w}_3(\rho, \varphi, 0, s) = \frac{-1}{\mu_1} k_0 \hat{\tau}_3(\rho, \varphi, 0, s) \tag{9}$$

where $\hat{w}_3$ and $\hat{\tau}_3$ are respectively the vertical displacement u_z and normal stress σ_{zz} at $z = 0$ in the transform domain and k_0 can be algebraically evaluated from the linear equations (7).

5. Integral Equations Governing the Indentation Problem

Using the equations (1-9), a two-dimensional (2-D) Fourier transform based system of dual integral equations, in the Laplace transform domain, can be obtained to describe the time-dependent behaviour of the rigid circular foundation with a smooth flat base. For the axisymmetric indentation ($D_z(r, \theta, t) = D_z(t)$), the 2-D integral equations can be reduced to one-dimensional Hankel transform based integral equations by using Fourier series expansions. Isolating the singularity, the sets of Hankel transform based integral equations can be reduced to systems of Fredholm integral equations of the second kind in the Laplace transform domain. Such systems of Fredholm integral equations of the second kind are standard and they are regular integral equations which can be numerically evaluated. Details of the mathematical formulations can be found in [7, 8].

For convenience, the following non-dimensional variables $\hat{\phi}(x, s_1)$, $\hat{\psi}(x, s_1)$, $\hat{X}_z(s_1)$ and $\hat{Y}_z(s_1)$ are introduced: $x = \frac{r}{a}$; $b_j = \frac{h_j}{a}$; $s_1 = \frac{a^2 s}{c_1}$; $\hat{P}_z(s) = P_z \hat{Y}_z(s_1)$; $\hat{\phi}_0(r, s) = \frac{-P_z}{2\pi a}\hat{\phi}(x, s_1)$; $\hat{D}_z(s) = \frac{1-\nu_1}{4\mu_1 a} P_z \hat{X}_z(s_1)$. The systems of Fredholm integral equations of the second-kind, in the Laplace transform domain, can be expressed in the following non-dimensional forms.

$$\hat{\phi}(x, s_1) + \int_0^1 \hat{K}(x, y, s_1)\hat{\phi}(y, s_1)dy = \hat{X}_z(s_1) \tag{10}$$

where $0 \leq x \leq 1$; and the kernel function, $\hat{K}$, can be obtained in the following non-dimensional form,

$$\hat{K}(x, y, s_1) = \frac{2}{\pi}\int_0^\infty \left[\frac{1}{1-\nu_1}k_0 - 1\right]\cos(\rho x)\cos(\rho y)d\rho. \tag{11}$$

It can be shown that the infinite integrals in equations (10) with the independent variables x, y, and s_1 are absolutely convergent provided $Re(s_1) > 0$. The equation (6) reduces to

$$\int_0^1 \hat{\phi}(x, s_1)\, dx = \hat{Y}_z(s_1) \tag{12}$$

In particular, the limiting solutions which pertain to the initial response as $t \to 0^+$ and the final response as $t \to +\infty$ can be obtained by taking the limits of the expressions (10) as $s \to +\infty$ and $s \to 0^+$, respectively. The details of the numerical solution methods applicable to the Fredholm integral equations of the type defined by equations (10) are given by Selvadurai and Yue (1994). In the ensuing, attention will be focused on the presentation of numerical results of interest to geomechanics.

6. Numerical Examples

A two-layered poroelastic solid is selected in the examples. The material properties are assigned as follows: $\mu_1 = \mu$, $\kappa_1 = \kappa$, $\nu_1 = 0.25$, $\nu_{u1} = 0.4$, $B_1 = 0.8$, $h_1/a = 0.5$; $\mu_2 = 4\mu$, $\kappa_2 = 0.25\kappa$, $\nu_2 = 0.3$, $\nu_{u2} = 0.5$, $B_2 = 1.0$, $h_2/a = 0.5$. In particular, $c_1/c_2 = 0.6637$. The rigid foundation is subjected to a constant load of the form $P_z(t) = P_z H(t)$, where $H(t)$ is the Heaviside step function. The degree of consolidation induced displacement of the rigid foundation is defined by

$$U_z(t) = \frac{D_z(t) - D_z(0)}{D_z(\infty) - D_z(0)}. \tag{13}$$

The non-dimensional initial and final settlements $10^2 a\mu D_z(0)/P_z (= 3.3449)$ and $10^2 a\mu D_z(\infty)/P_z (= 5.3245)$ are evaluated by using the limiting equations of (10) as $s \to \infty$ and $s \to 0^+$, respectively. The degree of consolidation induced displacement of the rigid foundation is presented in Fig.2 where the non-dimensional

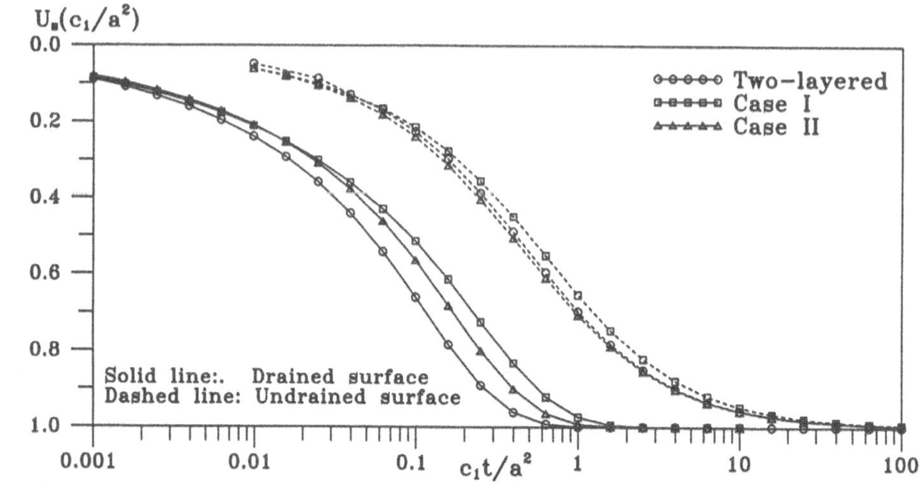

Fig.2 Degree of the consolidation induced settlement

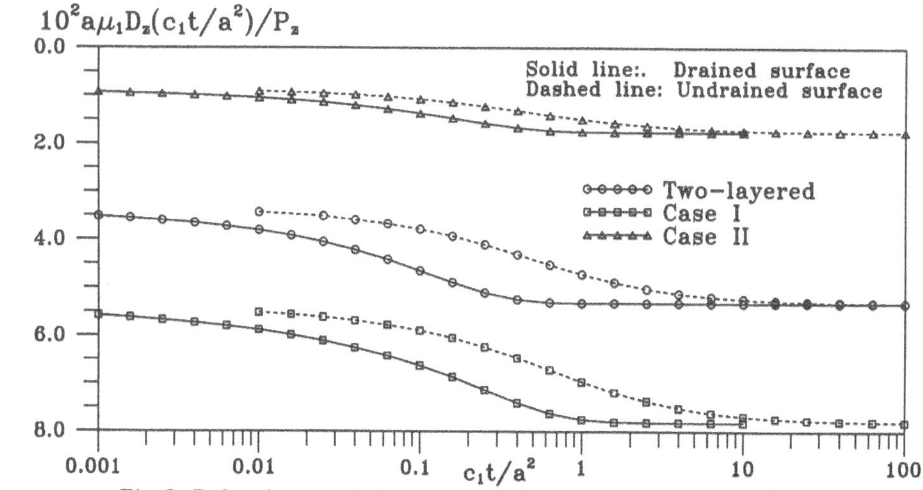

Fig.3 Behaviour of the consolidation induced settlement

time factor c_1t/a^2 is used. Fig.3 illustrates the time-dependent behaviour of th non-dimensional consolidation induced axial displacement of the rigid foundatior

In Fig.2 and Fig.3, in particular, the results of two single-layer poroelasti solids are also presented. The thicknesses of the two single-layer poroelastic solid are the same as that of the above two-layered poroelastic solid (*i.e.*, $(h_1 + h_2)/a =$ 1). The five material parameters are assigned as follows for each of the two single layer poroelastic solids. The Case I has the same material properties as thos of the first layer and the Case II has the same material properties as those c the second layer in the two-layered poroelastic solid. As a result, for the Cas I, $10^2 a\mu D_z(0)/P_z = 5.3690$ and $10^2 a\mu D_z(\infty)/P_z = 7.8197$; and for the Case I

$10^2 a\mu D_z(0)/P_z = 0.86355$ and $10^2 a\mu D_z(\infty)/P_z = 1.7616$.

It can be observed from Fig.2 that under the undrained surface drainage condition, the $U_z(c_1 t/a^2)$ associated with the two-layered poroelastic solid is bounded by those associated with the two single-layer poroelastic solids and is closer to that associated with the Case II. For the drained surface drainage condition, the rate of the consolidation induced settlements of the rigid foundation associated with the two-layered poroelastic solid is faster than those associated with the two single-layer poroelastic solids. The time-dependent behaviour of the rigid foundation associated with the two-layered poroelastic solid is bounded by those associated with the two single-layer poroelastic solids (Fig.3).

7. Concluding Remarks

The mathematical theory for a saturated, porous, elastic solid has been utilized for the evaluation of the axisymmetric settlement of a rigid foundation on a multilayered poroelastic solid. It is shown that the mathematical developments associated with mixed boundary value problems can be reduced to Fredholm integral equations of the second kind in the Laplace transform domain. These equations are amenable to numerical solution and such solutions can be used to develop results of engineering interest. The numerical results indicate that the material nonhomogeneity associated with layering has a significant effect on the consolidation settlement of the rigid foundation.

References

1. Biot, M.A. 1941. General theory of three-dimensional consolidation, *J. Appl. Phys.*, **12**: 155-164.
2. Biot, M.A. 1956. General solution of the equations of elasticity and consolidation for a porous material, *J. Appl. Mech.*, **23**: 91-95.
3. Booker, J.R., and Small, J.C. 1982. Finite layer analysis of consolidation: I; II, *Int. J. Numer. Anal. Methods in Geomech.*, 6, 151-171; 173-194.
4. Booker, J.R., Small, J.C. 1987. A method of computing the consolidation behaviour of layered soils using direct numerical inversion of Laplace transforms, *Int. J. Numer. Anal. Methods in Geomech.*, 11, 363-380.
5. Vardoulakis, I. and Harnpattanapanich, T. 1986. Numerical Laplace-Fourier transform inversion technique for layered-soil consolidation problems: I fundamental solutions and validation, *Int. J. Numer. Anal. Methods in Geomech.*, **10**: 347-365.
6. Harnpattanapanich, T., Vardoulakis, I., 1987. Numerical Laplace-Fourier transform inversion technique for layered-soil consolidation problems: II Gibson soil layer, *Int. J. Numer. Anal. Methods in Geomech.*, 11, 103-112.
7. Selvadurai, A.P.S. and Yue, Z.Q. 1994. On the indentation of a poroelastic layer. *Int. J. Numer. Anal. Methods in Geomech.*, 18, 161-175.
8. Yue, Z.Q. and Selvadurai, A.P.S. 1994. "Eccentric settlement of a rigid foundation on a consolidating thin layer", in *Vertical and Horizontal Deformation of Foundations and Embankments*, eds.: A.T. Yeung and G.Y. Felio, ASCE, 345 East 47th Street, New York. Vol.2, pp. 612-627.

SECTION 4

POROELASTICITY:

APPLICATIONS IN BIOMECHANICS

Porohyperelastic theory and finite element models for soft tissues with application to arterial mechanics.

B.R. Simon, M.V. Kaufmann, M.A. McAfee, A.L. Baldwin*
The University of Arizona,
Department of Aerospace and Mechanical Engineering,
Department of Physiology*,
Tucson, Arizona 85721, USA

Abstract Biological structures composed of soft tissues can be studied using poroelastic models, i.e., the material is viewed as a highly deformable porous solid skeleton that is saturated by mobile tissue fluid. These structures exhibit nonlinear material behavior and undergo finite strains. A porohyperelastic' (PHE) field theory is presented as a natural extension of classical hyperelasticity. Eulerian and Lagrangian forms for the field equaions are given for materials in which both the solid skeleton and the fluid are incompressible. These forms allow proper identification of two fundamental material property functions, the effective strain energy density unction and the hydraulic permeability. Eulerian and Lagrangian mixed inite element models (FEMs) are presented that include the PHE material esponse and finite strains. Applications of PHE theory and FEMs develped using the ABAQUS program are described for the study of large rteries. Representative functional forms for the material properties are iven that include the necessary material constants in the strain energy ensity function and the hydraulic permeability. Both steady-state and ansient cyclic pressurization of a large artery are simulated using ABAQUS FEMs of an axisymmetric segment of a PHE tube constrained in nite plane strain.

ey words: soft tissues, poroelastic models, finite element models, arterial mechanics

Introduction

his paper describes poroelastic models for the study of the mechanics of rge arteries, as well as other biological structures composed of soft tissues.

P. S. Selvadurai (ed.), *Mechanics of Poroelastic Media*, 245–261.

The methods presented can provide a basic knowledge of the mechanics a physiology of these soft tissue structures, as well as initial insights regard the inter-relationships between arterial wall mechanics, transport, a atherogenesis. Large arteries are nonhomogeneous, anisotropic tubes t exhibit a highly nonlinear, history-dependent material response wh undergoing finite deformations. Given such complexities, numerical mod such as the finite element model (FEM) are useful for mechanical a transport analyses of these structures. We have reviewed the uses of FE in arterial mechanics (Simon et al. 1993b) that have been based on b elastic and poroelastic material models. A comprehensive review poroelastic finite element analysis in soft tissue mechanics has been giv by Simon (1992).

We illustrate the approach by considering a typical large artery to b 'porohyperelastic' (PHE) material, i.e., a fluid-saturated porous medium which a fluid flows through the pores of a highly deformable, por hyperelastic solid skeleton. Additional species in the tissue fluid can transported through the arterial wall by diffusion and convection. Su poroelastic transport has been treated in other publications (e.g., see Sin et al. 1994). In this paper we summarize both the Eulerian and Lagrang formulations for the PHE field equations. Details of the development these equations have been presented elsewhere (Simon 1992). T Lagrangian formulation is useful for the development of specialized FE of soft tissue structures. The Eulerian formulation leads naturally to use of the ABAQUS finite element program (Hibbitt, Karlsson and Sorens Inc., Providence, Rhode Island) for finite element modeling. The ABAQ program allows for the consideration of poroelastic and hyperela materials including finite strain. Here we combine these two features carry out finite strain, PHE finite element analyses of soft arterial tissues.

2 Eulerian porohyperelastic field theory

The material is viewed as a continuum in which the total bulk volume, c contains a fluid-saturated porous solid skeleton at the position $x_i = X_i +$ and time t. The displacement of the solid skeleton is u_i. We consider solid skeleton to be an incompressible hyperelastic material saturated by incompressible fluid. The velocities at x_i and t are $\dot{u}_i$ and v_i^f for the so skeleton and the fluid, respectively. The (average) pore fluid velocity, v_i^f defined so that the volume of fluid flowing through a unit area perp dicular to the x_i-axis is nv_i^f. We define a relative fluid velocity as

$$v_i^r = n(v_i^f - \dot{u}_i)$$

The current porosity is $n = dV^f/dV$, assuming the skeleton is saturated by he fluid. For an incompressible solid skeleton, $n = 1 - J^{-1}(1 - n_0)$, where $J = dV/dV_0$ and n_0 = porosity in the reference configuration.

2.1 Eulerian kinematic equations

The strain rates are

$$D_{ij} = \frac{1}{2}\left[\frac{\partial \dot{u}_i}{\partial x_j} + \frac{\partial \dot{u}_j}{\partial x_i}\right], \qquad D^r_{kk} = \frac{\partial v^r_k}{\partial x_k} \tag{2a, b}$$

2.2 Eulerian conservation equations

The total equilibrium and fluid equilibrium (Darcy law) expressions are

$$\frac{\partial \sigma_{ji}}{\partial x_j} = 0, \qquad \frac{\partial \pi^f}{\partial x_i} - k^{-1}_{ij} v^r_j = 0 \tag{3a, b}$$

where k_{ij} is the hydraulic permeability term.

.3 Eulerian constitutive equations

he constitutive law relates total Cauchy stress, σ_{ij}, and fluid stress, π^f = -(pore fluid pressure), to the deformations. We use the 'effective' ress, σ^e_{ij}, defined so that the total Cauchy stress is

$$\sigma_{ij} = \sigma^e_{ij} + \pi^f \delta_{ij}, \qquad \sigma^e_{ij} = J^{-1} \frac{\partial x_i}{\partial X_k} S^e_{km} \frac{\partial x_j}{\partial X_m} \tag{4a, b}$$

here S^e_{ij} is given below. When both the solid skeleton and fluid are compressible, π^f is indeterminate subject to the constraint

$$D_{kk} + D^r_{kk} = 0 \tag{5}$$

iven the material property functions σ^e_{ij} and k_{ij}, Eqs. (2-5) form an initial undary value problem that is to be solved for the field variables u_i and (or π^f, mixed solution) as functions of x_i and t. We also use the draulic permeability defined by ABAQUS as $\hat{k}_{ij}$, which is related to k_{ij} $\hat{k}_{ij} = n^{-1} k_{ij}$

3 Lagrangian porohyperelastic field theory

The Lagrangian formulation is useful for the identification and determination of the material properties and provides the basis for Lagrangia FEMs. The Lagrangian field equations are to be solved for the field variables u_i and $\tilde{v}_i^r$ (defined below) or, using the mixed form, for u_i and π^f a functions of X_i and t. The Lagrangian formulation is based on correspondence rules. We use the classical Lagrangian rule for the total stres which requires equal force in the current and reference configuration Then the first and second Piola-Kirchhoff stresses are T_{ij} and S_{ij} respectively, given by

$$T_{ji} = J \frac{\partial X_j}{\partial x_k} \sigma_{ki} , \qquad S_{ij} = J \frac{\partial X_i}{\partial x_m} \sigma_{mn} \frac{\partial X_j}{\partial x_n} \tag{6a,}$$

Now we use a Lagrangian correspondence rule for the fluid (see Sim 1992), which requires that the relative fluid mass flow is the same in t current and reference configurations, i.e., with $\rho^f = \rho_0^f$ = consta (incompressible fluid)

$$\tilde{v}_i^r = J \frac{\partial X_i}{\partial x_j} v_j^r , \qquad \frac{\partial \tilde{v}_k^r}{\partial X_k} = J \frac{\partial v_k^r}{\partial x_k} \tag{7a}$$

3.1 Lagrangian kinematic equations

Strain measures include the deformation gradient, $F_{ij} = \partial x_i / \partial X_j$, and Green's strain, $E_{ij} = 1/2\, (F_{ki} F_{kj} - \delta_{ij})$. The strain rates are

$$\dot{E}_{ij} = \frac{\partial x_k}{\partial X_i} D_{km} \frac{\partial x_m}{\partial X_j} , \qquad \tilde{D}_{kk}^r = \frac{\partial \tilde{v}_k^r}{\partial X_k} \tag{8a}$$

3.2 Lagrangian conservation equations

The total equilibrium and fluid equilibrium (Darcy law) expressions are

$$\frac{\partial T_{ji}}{\partial X_j} = 0 , \qquad \frac{\partial \pi^f}{\partial X_i} - \tilde{k}_{ij}^{-1} \tilde{v}_j^r = 0 \tag{9a}$$

where the Lagrangian hydraulic permeability is

$$\tilde{k}_{ij} = J \frac{\partial X_i}{\partial x_m} k_{mn} \frac{\partial X_j}{\partial x_n} \tag{10}$$

When k_{ij} is isotropic, $k_{ij} = k\delta_{ij}$, and the corresponding Lagrangian permeability is $\tilde{k}_{ij} = JkH_{ij}$. Here, $H_{ij} = (\partial X_i/\partial x_m)(\partial X_j/\partial x_m)$ = Finger's strain tensor. The deformation measures J and H_{ij} will be available during the solution process. Of course, when k_{ij} is anisotropic, $\tilde{k}_{ij}$ will also be anisotopic.

3.3 Lagrangian constitutive equations

The solid material response is determined using an effective second Piola-Kirchhoff stress, S^e_{ij}. By the definitions of effective stress and the second Piola-Kirchhoff stress

$$S_{ij} = S^e_{ij} + J\pi^f H_{ij}\,, \qquad S^e_{ij} = \frac{\partial W^e}{\partial E_{ij}} \tag{11a,b}$$

where W^e is the 'effective strain energy density function' and π^f is indeterminate subject to the constraint $\dot{E}_{kl}H_{kl} + J^{-1}\tilde{D}^f_{kk} = 0$. The necessary material property functions are

$$W^e = W^e(E_{ij})\,, \qquad \tilde{k}_{ij} = \tilde{k}_{ij}(E_{kl}) \tag{12a,b}$$

Equations (6-12) form an initial boundary value problem to be solved for the field variables u_i and $\tilde{v}^f_i$ (or π^f, mixed solution) as functions of X_i and t.

There is an equivalence (see Simon 1992) of the poroelastic and biphasic models (Mow et al. 1980) used in soft tissue mechanics. The fundamental fields for the biphasic model are the displacement of the solid, $u^s_i = u^i$, and the fluid velocity, v^f_i, with kinematic equations for deformation measures, $D^s_{ij} = D_{ij}$ and $D^f_{kk} = \partial v^f_k/\partial x_k$. The corresponding stress measures are $\hat{\sigma}^s_{ij} = \sigma_{ij} - \hat{\sigma}^f_{ij}$ and $\hat{\sigma}^f_{ij} = n\pi^f\delta_{ij}$, which satisfy equilibrium equations of the form $\partial\hat{\sigma}^\alpha_{ij}/\partial x_j + \hat{\pi}^\alpha_i = 0$ (α = s, f), with $\hat{\pi}^s_i = -\hat{\pi}^f_i = n^2 k^{-1}_{ij}(v^f_j - \dot{u}^s_j) +$ $\partial n/\partial x_i$. Thus, either a biphasic or a poroelastic model can be used for soft tissue mechanics once σ_{ij} and k_{ij} are determined by constitutive laws in which W^e and k_{ij} are related to the appropriate deformation measures.

Porohyperelastic finite element models

FEMs can be developed in a number of ways, including Eulerian and Lagrangian PHE views using mixed and penalty formulations. The mixed

Eulerian and Lagrangian procedures are summarized here. Penalty methods have been described in the literature (e.g., see Prevost 1984, Spilker et al. 1990, Simon 1992) and will not be discussed in this paper. The FEMs are developed from a principle of virtual velocities (PVV).

4.1 Mixed Eulerian formulation

The PVV is $\int_V \delta\dot{u}_i \partial\sigma_{ji}/\partial x_j dV = 0$ and the associated constraint is $\int_V \delta\pi^f (D_{kk} + D^r_{kk})dV = 0$, which become

$$\int_V \delta D_{ij}\sigma^e_{ji} dV + \int_V \delta D_{kk}\pi^f dV = \int_A \delta\dot{u}_i \sigma_{ji}\hat{n}_j dA \quad (13)$$

$$\int_V \delta\pi^f D_{kk} dV - \int_V \frac{\partial(\delta\pi^f)}{\partial x_i} k_{ij} \frac{\partial\pi^f}{\partial x_j} dV = - \int_A \delta\pi^f v^r_i \hat{n}_i dA \quad (14)$$

where $\hat{n}_i$ denotes the unit normal associated with dA. The spatial interpolations for each finite element are

$$u_i = N^u_N u_{Ni} , \qquad \frac{\partial u_i}{\partial x_j} = B^u_{Nj} u_{Ni} \quad (15a,b)$$

$$\pi^f = N^\pi_N \pi_N , \qquad \frac{\partial \pi^f}{\partial x_i} = B^\pi_{Ni} \pi_N \quad (16a,b)$$

where u_{Ni} and π_N are time-dependent nodal quantities with interpolation functions $N^u_N = N^u_N(x_i)$ and $N^\pi_N = N^\pi_N(x_i)$. Also, $B^u_{Nj} = \partial N^u_N/\partial x_j$ and $B^\pi_{Ni} = \partial N^\pi_N/\partial x_i$. Then, elemental approximations to the equilibrium equations and Darcy's law are

$$\mathbf{P}^{int}_e + \mathbf{k}_{u\pi}\, \bar{\boldsymbol{\pi}} = \mathbf{P}^{ext}_u , \qquad \mathbf{c}_{\pi u}\, \dot{\bar{\mathbf{u}}} + \mathbf{k}_{\pi\pi}\, \bar{\boldsymbol{\pi}} = \mathbf{P}^{ext}_\pi \quad (17a,$$

where bold-faced text denotes matrices so that $\bar{\mathbf{u}} = u_{Ni}$, and $\bar{\boldsymbol{\pi}} = \pi_N$, with

$$\mathbf{P}^{int}_e = \int_V B^u_{Nj}\sigma^e_{ij} dV , \qquad \mathbf{P}^{ext}_u = \int_A N^u_N \sigma_{ij}\hat{n}_j dA \quad (18a,$$

$$\mathbf{P}^{ext}_\pi = - \int_A N^\pi_N v^r_i \hat{n}_i dA \quad (1$$

$$\mathbf{k}_{u\pi} = \mathbf{c}^T_{\pi u} = \int_V B^u_{Nk}\delta_{ki}N^\pi_M dV\,, \qquad \mathbf{k}_{\pi\pi} = -\int_V B^\pi_{Ni}k_{ij}B^\pi_{Mj}dV \tag{19a, b}$$

The relative velocity of the fluid can be determined using $v^r_i = k_{ij}B^\pi_{Mj}\pi_M$.

4.2 Mixed Lagrangian formulation

The PVV and the associated constraint are

$$\int_{V_0} \delta\dot{E}_{ij}S^e_{ji}dV_0 + \int_{V_0} \delta\dot{E}_{ij}H_{ij}J\pi^f dV_0 = \int_{A_0} \delta\dot{u}_i T_{ji}\hat{n}_{0j}dA_0 \tag{20a}$$

$$\int_{V_0} \delta\pi^f \dot{E}_{ij}H_{ij}JdV_0 - \int_{V_0} \frac{\partial(\delta\pi^f)}{\partial X_i}\tilde{k}_{ij}\frac{\partial\pi^f}{\partial X_j}dV_0 = -\int_{A_0} \delta\pi^f \tilde{v}^r_i\hat{n}_{0i}dA_0 \tag{20b}$$

where $\hat{n}_{0i}$ denotes the unit normal associated with dA_0. Again, elemental interpolations are

$$u_i = N^u_N u_{Ni}\,, \qquad \frac{\partial u_i}{\partial X_j} = B^u_{Nj}u_{Ni} \tag{21a, b}$$

$$\pi^f = N^\pi_N \pi_N\,, \qquad \frac{\partial \pi^f}{\partial X_i} = B^\pi_{Ni}\pi_N \tag{22a, b}$$

where u_{Ni} and π_N are time-dependent nodal quantities with $N^u_N = N^u_N(X_i)$ and $N^\pi_N = N^\pi_N(X_i)$. Also, $B^u_{Nj} = \partial N^u_N/\partial X_j$ and $B^\pi_{Ni} = \partial N^\pi_N/\partial X_i$. Elemental approximations to the equilibrium equations and Darcy's law are of the form of Eq. (17), with

$$\mathbf{P}^{int}_e = \int_{V_0} B^u_{Nj}\frac{\partial x_i}{\partial X_k}S^e_{ik}\,dV_0\,, \qquad \mathbf{P}^{ext}_u = \int_{A_0} N^u_N T_{ji}\hat{n}_{0j}\,dA_0\,, \tag{23a, b}$$

$$\mathbf{P}^{ext}_\pi = -\int_{A_0} N^\pi_N \tilde{v}^r_i\hat{n}_{0i}\,dA_0 \tag{23c}$$

$$\mathbf{k}_{u\pi} = \mathbf{c}_{\pi u}^{T} = \int_{V_0} B^{u}_{Nk}\delta_{ki}N^{\pi}_{M}\, dV_0\,, \qquad \mathbf{k}_{\pi\pi} = -\int_{V_0} B^{\pi}_{Ni}\tilde{k}_{ij}B^{\pi}_{Mj}\, dV_0 \qquad (24a,b$$

Again, the relative velocity of the fluid can be determined as $\tilde{v}^{r}_{i} = \tilde{k}_{ij}B^{\pi}_{Mj}\pi_{M}$, which provides $v^{r}_{i} = J^{-1}(\partial x_i/\partial X_j)\tilde{v}^{r}_{j}$.

Either the Eulerian or Lagrangian finite element equations can b assembled to obtain the large set of coupled ordinary differential equations

$$\mathbf{C}\dot{\mathbf{r}} + \mathbf{R}^{int} = \mathbf{R}^{ext} \qquad (25$$

which must be integrated subject to initial and boundary conditions. Th ABAQUS program uses a mixed formulation with full backward time inte gration and Newton-Raphson iterations to convergence in each time step.

5 Application of porohyperelastic finite element models to arterial mechanics using ABAQUS

Previously, we reported poroelastic FEMs for arteries (Simon and Gabal 1988, Yuan and Simon 1992, Simon et al. 1993a, b). Our most recent studi of arterial mechanics have been carried out using the ABAQUS prograr which allows the use of hyperelastic materials in conjunction with po pressure elements. Given the equivalence of the biphasic and poroelast formulations, biphasic material properties could also be used in ABAQU for the study of soft tissues.

5.1 Material properties used in finite element analysis

We have described experiments on *in situ* rabbit aortas and the associat data reduction procedures that allow the determination of necessary mater properties for finite element analysis (e.g., see Simon et al. 1993a, McAf et al. 1994). Our experiments allow pressurization of these arteries in t undrained, drained, and steady-state (flow) conditions. The steady-st inflation tests with physiologically buffered saline provide data regardi hydraulic permeability. Two tests are carried out--an inflation of an int vessel followed by inflation of a de-endothelialized vessel. In each test, aorta is inflated to the desired pressure level and tissue fluid flow determined from measurements of the velocity of a bubble moving throu a capillary tube used to cannulate the vessel.

Our initial material models are based on the assumption of *isotropy*. We have previously described (Simon et al. 1993a) the mechanical testing and data reduction methods that provided the necessary isotropic forms for PHE material properties. Here, we view the artery as a homogeneous material (ignoring the intima) when strain energy density is considered and a two-layer tube (intima and combined media-adventitia) when hydraulic permeability is considered. The Fung exponential form for W^e was determined for arterial tissues using the isotropic form

$$W^e = \overline{W}^e(\phi) = \frac{1}{2} C_0(e^{\phi} - 1) \tag{26}$$

$$\phi = C_1'(\bar{I}_1 - 3) + C_2'(\bar{I}_2 - 3) + K'(J - 1)^2 \tag{27}$$

with $K' \neq \infty$ and strain invariants $\bar{I}_1 = J^{-2/3}(3 + 2E_{kk})$ and $\bar{I}_2 = J^{-4/3}[3 + E_{kk} + 2(E_{ii}E_{jj} - E_{ij}E_{ij})]$. Our models use the isotropic hydraulic permeability available in ABAQUS so that

$$\hat{k}_{ij} = \hat{k}\,\delta_{ij} = n^{-1} k \delta_{ij} \tag{28}$$

where $\hat{k}$ is assumed to be a function of voids ratio, $e = n/(1 - n)$. Since $n = 1 - J^{-1}(1 - n_0)$, this is equivalent to assuming that $\hat{k}$ is a function of volume strain, J.

Our experimental data suggest that we can assume the medial and adventitial layers have the same constant permeability, whereas the intima exhibits a nonlinear permeability, i.e.

$$\hat{k}_M = \hat{k}_A = \text{constant} \quad \text{(media-adventitia)} \tag{29a}$$

$$\hat{k}_I = \hat{k}_I(e) \quad \text{(intima)} \tag{29b}$$

Figure 1a shows the strain energy density function and Figure 1b shows the permeability used for this study. The non-dimensional values for some of the material constants used in this paper are $C_0 = 1.0$, $C_1' = 0.5$, $C_2' = 0.35$, $K' = 5.0$, and $\hat{k}_M = 1.0$ The determination of general expressions for W^e and permeability will be described in a future publication.

2 Finite element models

The ABAQUS FEM of a representative aorta is shown in Figure 2. The FEM is a segment of an axisymmetric tube composed of 16 eight-noded quadrilateral axisymmetric pore pressure elements available in the ABAQUS program. The inner two elements represent the relatively thin intimal layer

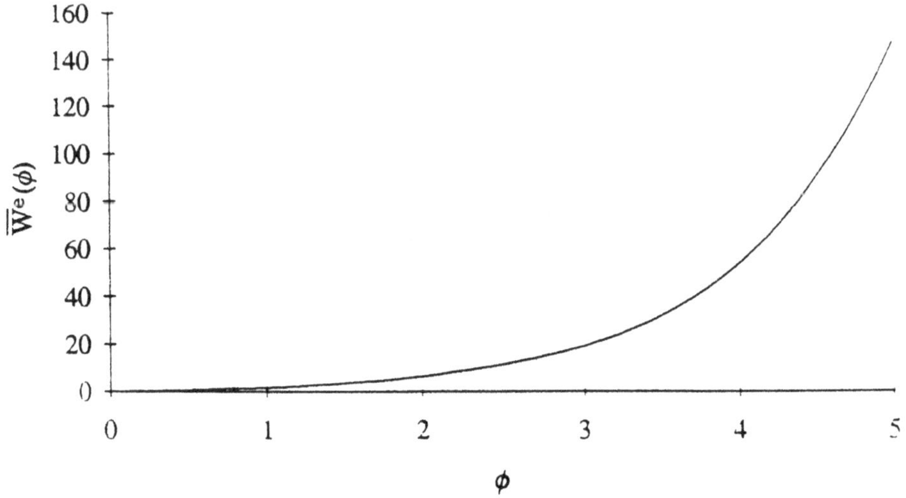

Figure 1a: *Effective strain energy density function,* $\overline{W}^e$, *versus quadra strain function,* ϕ*; isotropic exponential Fung form:* $\overline{W}^e = 1/2\ C_0(e^{\phi} - 1)$ *w* $\phi = C_1'\ (\bar{I}_1 - 3) + C_2'(\bar{I}_2 - 3) + K'(J - 1)^2$.

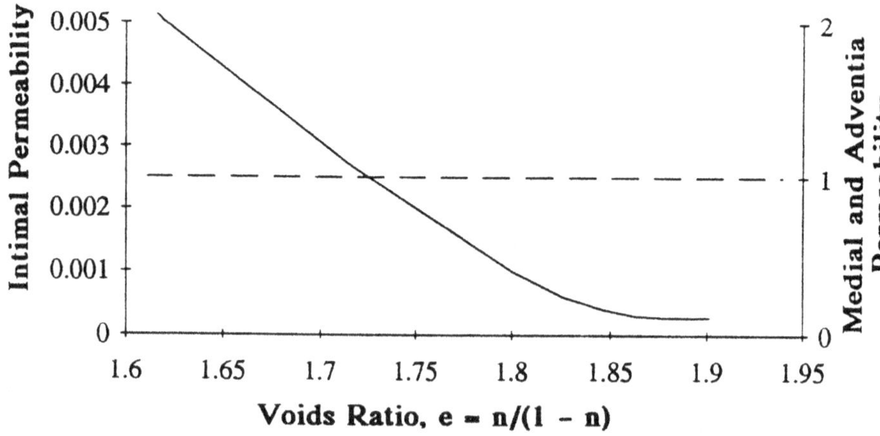

Figure 1b: *Intimal permeability,* $\hat{k}_I$ *(solid line), and medial/adventitial p meability,* $\hat{k}_M = \hat{k}_A$ *(dashed line), versus voids ratio,* $e = n/(1 - n)$ *with* $\hat{k}_M$ $\hat{k}_A = 1.0$ *and* $\hat{k}_I = \hat{k}_I(e)$.

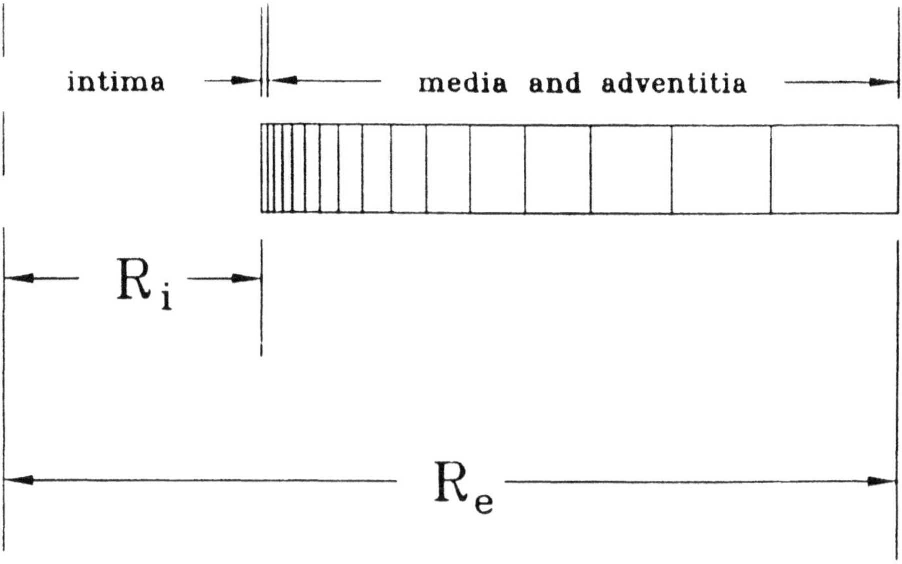

'igure 2: *Axisymmetric plane-strain FEM of the arterial cross section enerated using ABAQUS and a porohyperelastic material model.*

nd the remaining (biased) elements represent the media and adventitia. Iniform axial displacements were prescribed on the top of the model and uppressed on the bottom of the model to impose finite plane-strain condi-ons. The exterior surface (right side) was traction free and the interior urface (left side) was pressurized by prescribing the radial stress (σ_{rr}) and ie pore fluid stress (π^f) to be negative inflation pressure. The PHE iaterial law was introduced by combining the pore pressure element with ie hyperelastic material option in ABAQUS. The fluid and solid skeleton 'ere assumed to be incompressible in the pore pressure element, making the ore fluid pressure an indeterminate quantity. The UHYPER option in BAQUS was used to introduce the exponential Fung form for W^e.

.3 Representative FEM results

ndrained, drained, and flowing (steady-state) conditions. The idrained condition was simulated using a step total pressure and fluid essure at the inside surface. Drained conditions were modeled using a eady-state total pressure and a no-fluid-flow boundary condition at the inside surface. The flowing condition was developed considering the steady-state behavior with both total pressure and fluid pressure equal to the internal pressure in the tube. Figure 3 shows a typical result obtained for the steady-state flowing FEM simulation. Here, we see the nonlinear intimal permeability plotted versus pressure corresponding to the plot in Figure 1b. The free tissue fluid flow through the arterial wall (represented by the bubble velocity in the cannulation tube) is shown as a function of pressure. This flow-pressure trend corresponds to experimental observation of the bubble velocity for normal rabbit aortas, where a characteristic 'dip in flow-pressure response occurs in the physiological pressure range. Figure 4a shows typical step pressure (undrained) and steady-state pressure (flowing condition) versus radius plots obtained from the ABAQUS FEMs.

Cyclic conditions. We have also simulated cyclic (pulsatile) pressures using the simplified (non-dimensional) pressure-time plot shown in Figure 4b. This cyclic loading was imposed on a model in a steady-state flowing condition at representative constant internal pressure (see Figure 4a). Note that the 'up' loading condition was imposed at a higher rate than the 'down' portion of the cyclic pressure. Both normal and abnormally elevated (hypertensive) mean pressure conditions were simulated. Figure 5 shows the pore fluid pressure gradients predicted by the FEM for both of these situations. These plots also illustrate the large deformation associated with changes in mean pressure. On the 'up' portion of the cycle, the tissue fluid pressure gradient is from the external to the internal surface, indicating that

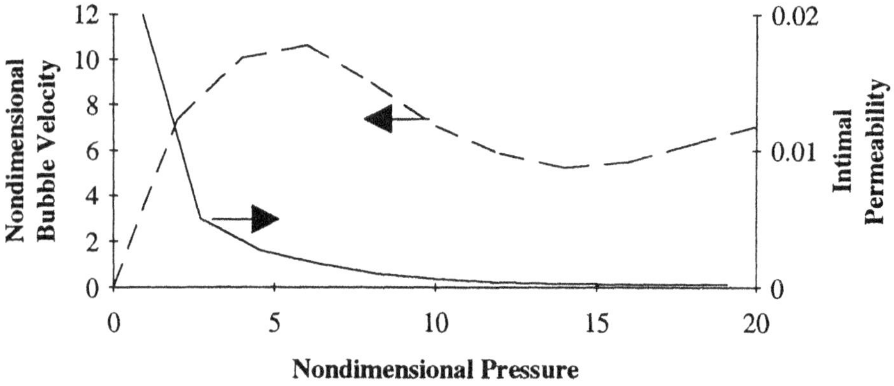

Figure 3: *FEM fluid flux [nondimensional bubble velocity,* $(v_B h_{int})/(\hat{k}_M C_0$ *and intimal permeability,* $\hat{k}_I$*, versus nondimensional pressure,* $(P_i - P_e)/C_0$ *where* v_B *is bubble velocity,* h_{int} *is the original intimal thickness,* $\hat{k}_M$ = *medial permeability, and* C_0 = *elastic constant.*

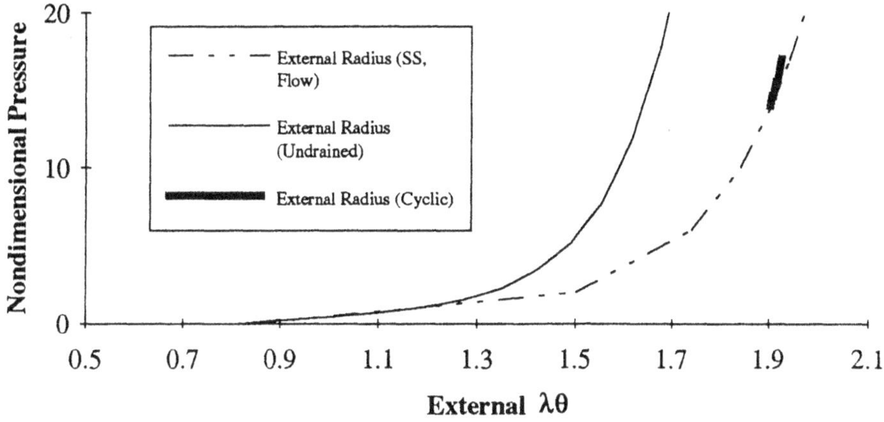

Figure 4a: *FEM pressure versus radius for undrained, steady-state flowing, and cyclic pressurization conditions. Nondimensional pressure =* $(P_i - P_e)/C_0$ *and external* $\lambda_\Theta = r_e/R_e$.

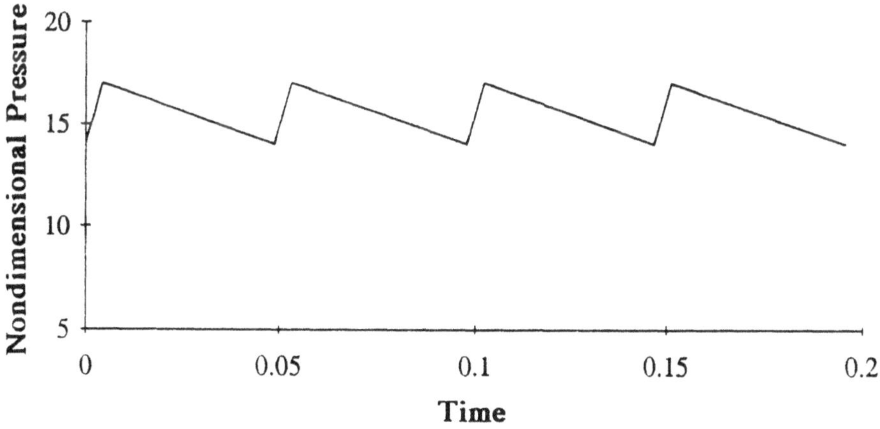

Figure 4b: *FEM pressure versus time for cyclic pressurization conditions. Nondimensional pressure =* $(P_i - P_e)/C_0$.

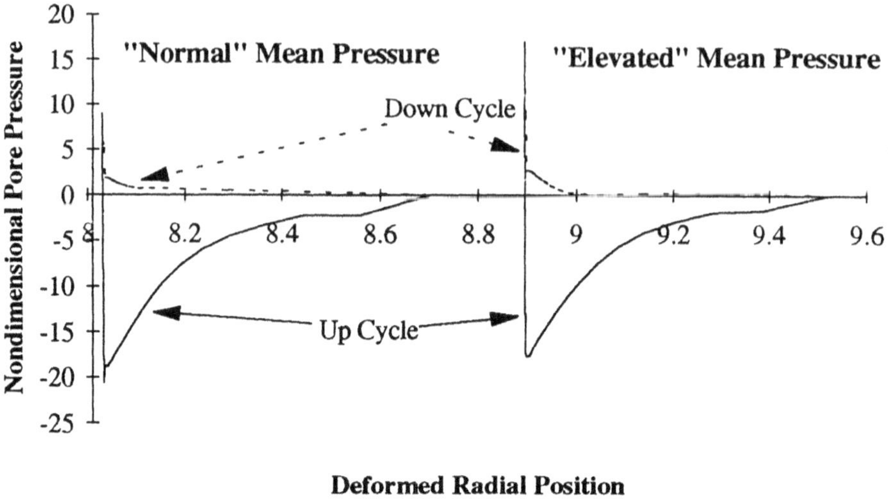

Figure 5: *FEM pore pressure versus location in the arterial wall for cycl pressurization at normal and elevated intraluminal pressure. Nondimension pore pressure =* π^f / C_0.

tissue fluid flows in that direction. At the internal surface, there is an ou ward pressure gradient (and associated relative flow) for all cases. Durir the 'down' portion of the cycle, fluid flows from the internal to the extern surface, following the pressure gradient. Figure 6 is a plot of the curre level of porosity, which gives an indication of the net amount of fluid in t arterial wall (assuming a saturated poroelastic material). These resu demonstrate the barrier function effect of the intima, which has extremely low permeability, thus keeping the the intima 'sealed' under the conditions. It is notable that, because of this flow barrier, the arte imbibes fluid during the 'up' portion of the cycle. Such transport may ha significance in terms of the normal transport (via convection) performan of the arterial wall.

One can also examine the extreme case in which the intima has ve low hydraulic resistance (permeability) associated with a damaged or 'leal endothelium. Figure 7 shows pore pressure gradients for a FEM in whi the intima, media, and adventitia all have the *same* permeability (consta i.e., the intima has been 'removed.' The pore pressure gradients associat with these 'abnormal' conditions are markedly different from the gradient the intact case (see Figure 5, diagram on the right). This prediction of fl field alteration suggests that further analytical and experimental resea should be carried out to discover the influence of the tissue fluid flux normal arterial wall transport and during the periods of time for which intimal permeability is somehow increased (e.g., extreme conditi corresponding to a 'damaged' intima).

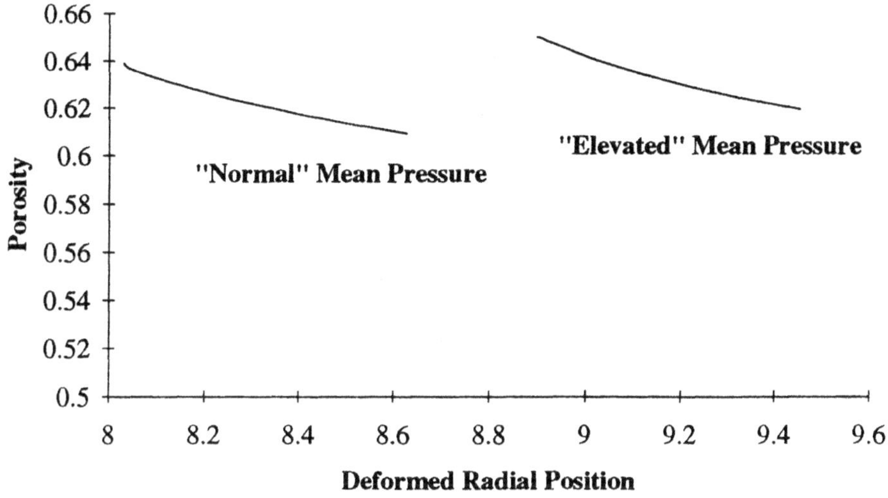

igure 6: *FEM porosity (tissue fluid content),* n = dV/dV_0, *versus location the arterial wall for cyclic pressurization at normal and elevated traluminal pressure.*

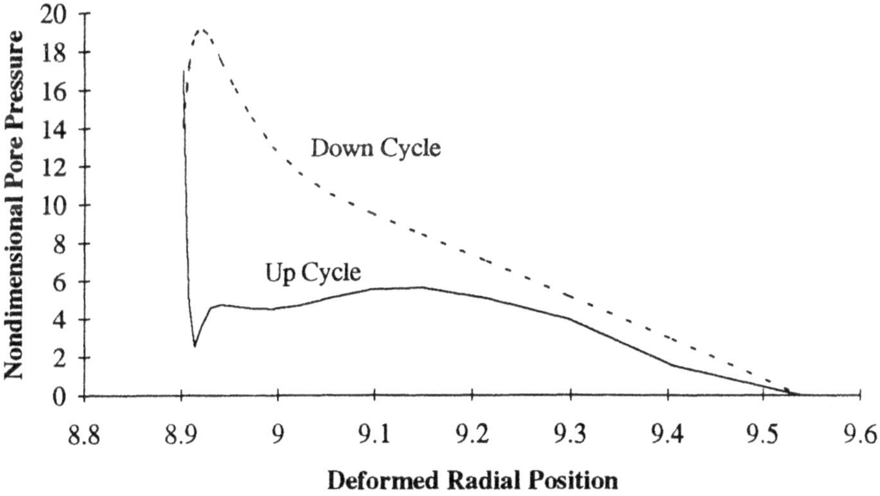

gure 7: *FEM pore pressure versus location in the* de-endothelialized *terial wall for cyclic pressurization at "elevated" intraluminal pressure. ondimensional pore pressure =* π^f/C_0.

6 Conclusions

We have described Eulerian and Lagrangian formulations for the PHE fiel theory and the associated FEMs. The theory is useful in the identificatio and experimental determination of necessary material properties (e.g., W and $\hat{k}$) for PHE modeling of the arterial cross section. We are continuing develop data reduction procedures and additional experimental methods determine appropriate material parameters for rabbit aortas. The PHE fo mulations in this paper are consistent with the ABAQUS finite eleme program, which allowed detailed FEMs to be developed for the arterial wal The numerical modeling approach illustrated for arterial mechanics can applied to analyze various soft tissue structures where a poroelastic materi model is available. Again, these modeling procedures and the ABAQU program can be used to develop FEMs based on available (equivale biphasic material models.

Our research on arterial mechanics is being extended to include furth study of anisotropy, remodeling effects, vascular smooth muscle activity, a factors associated with atherogenesis (hypertension, endothelial damage, etc The future will see the development of poroelastic material models and t associated FEMs that include transport phenomena in soft tissues. We ha described (Simon et al. 1994) a linear poroelastic-transport-swelling (PET that includes the transport of a mobile species in the tissue fluid. equivalence can be demonstrated between the PETS model and the tripha model of Lai et al. (1990). Thus, the PETS theory and FEMs will all analysis of soft tissue structures and transport using either an extend poroelastic view or the equivalent triphasic model of the material. We currently extending this linear model to a nonlinear formulation t includes the finite strain and material nonlinearity observed in la arteries. Supporting experiments are also under development for quantification of the PETS material parameters that are required in FEMs. These nonlinear FEMs (based on PETS theory) will be used simulate transport associated with diffusion and convection in the arte wall.

Acknowledgment: This work was sponsored in part by a grant from National Science Foundation (BCS-9210282). Any opinions, findings, conclusions or recommendations expressed in this paper are those of authors and do not necessarily represent the views of the National Scie Foundation.

References

Lai, M., Hou, J. S., and Mow, V. C. (1990) A triphasic theory for swelling properties of hydrated charged soft biological tiss Biomechanics of Diarthrodial Joints, Vol. 1 (V. C. Mow, A. Ratcliffe, S. L.-Y. Woo, eds.), Springer-Verlag, New York, pp. 283-312.

McAfee, M. A., Kaufmann, M. V., Simon, B. R., and Baldwin, A. L. (1994) Experimental/numerical approach to the determination of material properties in large arteries. 1994 Advances in Bioengineering, ASME, New York (to appear).

Mow, V. C., Kuei, S. C., Lai, W. M., and Armstrong, C. G. (1980) Biphasic creep and stress relaxation of articular cartilage in compression: theory and experiments. J. Biomech. Eng. 102:73-84.

Prevost, J. H. (1984) Nonlinear transient phenomena in soil media. Mechanics of Engineering Materials (C. S. Desai and R. H. Gallagher, eds.), John Wiley & Sons, Ltd., London, pp. 515-533.

Simon, B. R. (1992) Multiphase poroelastic finite element models for soft tissue structures. Appl. Mech. Rev. 45: 191-218 (also, ASME Book No. AMR111).

Simon, B. R. and Gaballa, M. A. (1988) Finite strain, poroelastic finite element models for large arterial cross sections. Computational Methods in Bioengineering (R. L. Spilker and B. R. Simon, eds.), BED-Vol. 9, ASME, New York, pp. 325-334.

Simon, B. R., Kaufmann, M. V., McAfee, M. A., and Baldwin, A. L. (1993a) Determination of material properties for soft tissues using a porohyperelastic constitutive law. 1993 Advances in Bioengineering, BED-Vol. 26, ASME, New York, pp. 7-10.

Simon, B. R., Kaufmann, M. V., McAfee, M. A., and Baldwin, A. L. (1993b) Finite element models for arterial wall mechanics. J. Biomech. Eng. 115:489-496.

Simon, B. R., Liable, J. P., Pflaster, D. S., Yuan, Y., and Krag, M. H. (1994) A poroelastic finite element formulation including transport and swelling in soft tissue structures. J. Biomech. Eng. (accepted subject to revision).

Spilker, R. L., Suh, J. K., Vermilyea, M. E., and Maxian, T. A. (1990) Alternate hybrid, mixed, and penalty finite formulations for the biphasic model of soft hydrated tissues. Biomechanics of Diarthrodial Joints, Vol. 1 (V. C. Mow, A. Ratcliffe, and S. L.-Y. Woo, eds.), Springer-Verlag, New York, pp. 400-435.

Yuan, Y. and Simon, B. R. (1992) Constraint relations for orthotropic porohyperelastic constitutive laws and finite element formulations for soft tissues. 1992 Advances in Bioengineering, BED-Vol. 22, ASME, New York, pp. 203-206.

Prediction of frequency and pore size dependent attenuation of ultrasound in trabecular bone using Biot's theory.

J.L. Williams, M.J. Grimm, F.W. Wehrli*, K.R. Foster**, H-W. Chung*
University of Pennsylvania,
Department of Orthopaedic Surgery,
Department of Radiology*,
Department of Bioengineering**
424 Stemmler Hall, Philadelphia, PA 19104, USA.

Abstract Biot's theory is applied to trabecular bone using the Johnson-Koplik-Dashen (JKD) theory of permeability and tortuosity. This may be the first instance in which the critical physical constants for bone, as determined from experiments, are used with Biot' theory, incorporating the JKD permeability and tortuosity, to calculate velocities and attenuations and compare the results with experimentally determined velocities and attenuations. The velocities are correctly predicted by the theory and there is qualitative agreement with the experimental attenuations. The predicted magnitudes of the attenuations calculated for internal Biot losses due to relative motion of the solid and fluid phases are smaller than the experimentally measured attenuations, which include losses due to porous surface reflection at the front and back of the specimen, scattering due to macroscopic pores, as well as attenuation due to intrinsic losses in the bone and in the marrow phase, which are not directly accounted for by Biot's theory.

1. Introduction

There is considerable interest in the application and interpretation of ultrasound to the heel. It is believed by many that measurements of velocity and attenuation through the heel bone (calcaneus) may detect changes in the bone structure of the porous trabecular bone that occur with age -related or post-menopausal osteoporosis. Several physical studies have shown that the attenuation of ultrasound in trabecular bone is related to the mineral content of the trabeculae (Tavakoli and Evans, 1991) and to the porosity for a fixed mass of bone (Tavakoli and Evans, 1992). Attempts to model the frequency-dependent attenuation using Biot's theory have met with some success as far as predicting the measured fast wave velocity (Williams, 1992) and the attenuation (McKelvie and Palmer, 1991). These models did not use independent measurements of more than a few of many of the parameters used in the theory. We performed independent measurements on human calcaneal bones to acquire most of the parameters needed for the theory. We also measured ultrasound transmission velocity and attenuation over the frequency range of 0.2 to 0.6 MHz.

2. Theory

Biot's theory (Biot, 1956) follows the average displacement of the fluid $\mathbf{U}(\mathbf{r},t)$ and of the elastic porous solid $\mathbf{u}(\mathbf{r},t)$ separately to yield two coupled linear equations of motion, the solutions of which predict a fast and slow longitudinal wave and a shear wave. For the longitudinal motion, the generalized elastic constants P, Q and R and N are as originally defined by Biot.

A. P. S. Selvadurai (ed.), Mechanics of Poroelastic Media, 263–271.

$$P = [\beta(K_s/K_f - 1)K_b + \beta^2 K_s + (1-2\beta)(K_s - K_b)]/(1 - \beta - K_b/K_s + \beta K_s/K_f) + 4N/3 \tag{1}$$

$$Q = (1 - \beta - K_b/K_s)\beta K_s/ (1 - \beta - K_b/K_s + \beta K_s/K_f) \tag{2}$$

$$R = K_s\beta^2/ (1 - \beta - K_b/K_s + \beta K_s/K_f), \tag{3}$$

where K_s is the bulk modulus of the solid bone material, K_f of the fluid, K_b of the skeletal frame (trabecular frame) and N is the shear modulus for the frame and for the solid material. β is thefluid volume fraction.

The density terms used to describe the effects of viscous and inertial drag are complex (after Johnson et al., 1987), but otherwise as defined by Biot,

$$\underline{\rho}_{11} + \underline{\rho}_{22} = (1 - \beta)\rho_s \tag{4}$$

$$\underline{\rho}_{22} + \underline{\rho}_{12} = \beta\, \rho_f \tag{5}$$

$$\underline{\rho}_{12} = -[\underline{\alpha}(\omega) - 1]\, \beta\, \rho_f, \tag{6}$$

where ω is the angular frequency, $\underline{\rho}_{11}$ is the total effective density of the solid, $\underline{\rho}_{12}$ is the additional density of the inertial drag exerted by the fluid on the solid as the solid is accelerated relative to the fluid, ρ_s and ρ_f are the densities of the solid and fluid phases, respectively and $\underline{\alpha}(\omega)$ is the Johnson-Koplik-Dashen dynamic tortuosity,

$$\underline{\alpha}(\omega) = \alpha_\infty + i\eta\, \beta\, (1 - 4i\alpha_\infty{}^2 k_0{}^2 \rho_f\, \omega/\eta\, \beta\, {}^2\Lambda^2)/(\omega\, \rho_f k_0) \tag{7}$$

In this equation η is the viscosity, α_∞ is the experimentally determined tortuosity and k_0 is the experimentally determined d.c. permeability. The Johnson-Koplik-Dashen dynamic permeability is

$$\underline{k}\, (\omega) = i\eta\, \beta/(\underline{\alpha}(\omega)\, \omega\rho_f) \tag{9}$$

The pore size parameter Λ was introduced by Johnson et al. (1986) as an intrinsic measure of dynamically interconnected pore sizes. Johnson et al showed that Λ, α_∞, k_0 and β can be used to calculate the acoustic properties over the entire frequency spectrum. While these four parameters are unrelated and independently measurable, Johnson et al. (1986) showed that for a variety of smooth pore shapes arrangements the parameters may be related approximately as

$$8\, \alpha_\infty k_0/(\beta\, \Lambda^2) = M, \tag{10}$$

where M is of the order of 1 (Johnson et al., 1986; Banavar and Johnson, 1987).

For bulk plane wave solutions, the complex wave speeds for the longitudinal waves are

$$\underline{V}^2 \text{(fast, slow)} = [\Delta \pm \sqrt{\{\Delta^2 - 4(\rho_{11}\rho_{22} - \rho_{12}{}^2)(PR - Q^2)\}}]/\{2(\rho_{11}\rho_{22} - \rho_{12}{}^2), \quad (11)$$

where

$$\Delta = P\,\rho_{22} - 2\,Q\rho_{12} + R\rho_{11} \quad (12)$$

There are two roots to equation (11) giving the fast and slow compressional wave speeds. The real part of the root, q_r, provides the wave speed as ω/ q_r and the imaginary part, q_i, is the attenuation in Np/m.

3. Measurements

The ultrasound attenuation measurements were performed using a Walker Sonix UBA 575 over a frequency range of 0.2 to 0.6 MHz on 1-inch diameter cylinders of trabecular bone (with marrow intact) cored medio-laterally (horizontally) from 22 calcanei (heel bones) of 16 individuals (32-89 years, mean 69 years) at 37°C. The cylinders were trimmed to provide flat and parallel end faces resulting in cylinders of varying lengths. Transmission velocity was measured at room temperature with two matching 1-MHz contact transducers as a time of flight measurement with the marrow intact. Tortuosity, permeability and the static elastic modulus measurements were also obtained at room temperature. The details of these experiments will be reported elsewhere. The elastic modulus of all specimens was measured from the slope of the linear force-deflection plot obtained at a strain rate of 0.001 s^{-1} in a compression test. For ten specimens the tortuosity was determined from the electrical conductivity of the trabecular samples after marrow removal. The ratio of conductivity of saline alone to the conductivity of the sample saturated with saline is the geometrical factor F that relates tortuosity to porosity as (Brown, 1980)

$$\alpha_\infty = F\beta. \quad (13)$$

Permeability was measured using Darcy's law and a gravity-fed permeability chamber with linseed oil as the fluid, thereby providing a d.c measurement of permeability k_0. The specimens were cleaned of all marrow by immersion in an ultrasound bath containing trichloroethylene for a period of four hours prior to the tortuosity and permeability measurements. Measurements of porosity and mean trabecular plate separation (MTPS) were made on the two vertical end faces of each cylinder and on a lengthwise horizontal section.

The elastic modulus (static Young's modulus) was obtained from a low strain rate test. Since the modulus of bone is strain rate sensitive (McElhaney, 1966; Kim and Walsh, 1992), we multiplied the results by a factor of two for the model calculations in the high strain rates expected in the MHz frequency range. Likewise K_s, the bulk modulus of the solid bone, was calculated from a value for solid bone Young's modulus of 25.6 GPa (twice the value extrapolated from the curve fit of modulus versus bone fraction).

Parameters that were not directly measured were K_f, K_s, K_b, N, ν_s, ρ_f, ρ_s, η. The following values were used in the theory:

$K_f = 2.26$ GPa,

$K_s = E_s/3(1 - 2\nu_s)$, with dynamic $E_s = 25.6$ GPa for solid material of trabeculae,

$K_b = E_b/3(1-2\nu_b)$, with E_b taken as 2x curve fit of the static Young's modulus vs porosity,

$N = E_b/2(1+\nu_b)$

$\nu_s = \nu_b = 0.32$

$\rho_f = 930$ kg/m^3

$\rho_s = 1800$ kg/m^3

$\eta = 0.04$ or 0.4 Pa.s (after Bryant et al., 1989).

4. Results and Conclusions

The permeability is a strong function of porosity (Figure 1), while the tortuosity is a weak function of the mean trabecular plate separation (Figure 2).

Figure 1. Permeability vs porosity

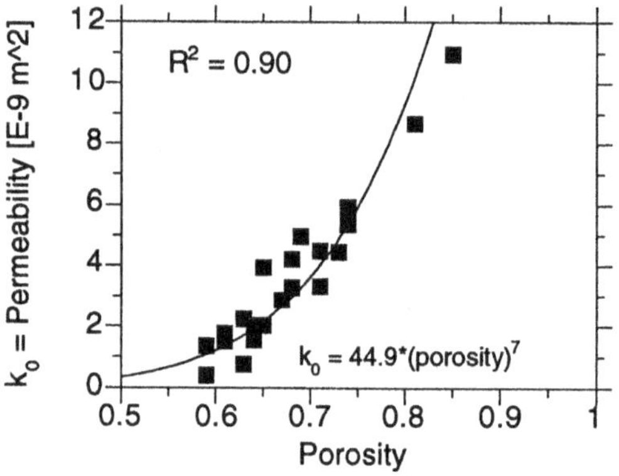

Figure 2. Tortuosity vs mean trabecular plate separation (MTPS)

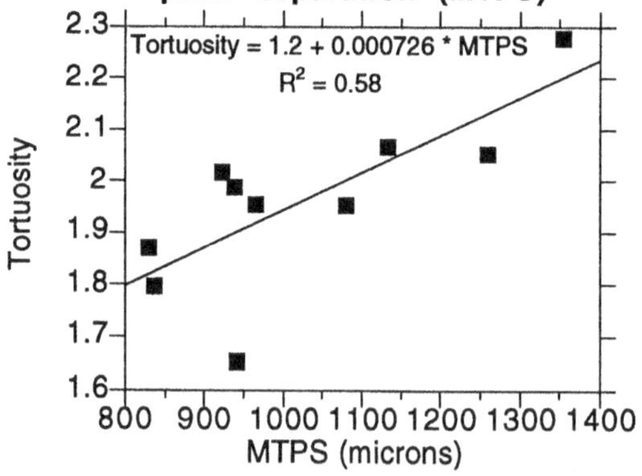

Not shown is the relationship observed between the mean trabecular plate separatio

(MTPS) and porosity (MTPS (in microns) = 2082*porosity^2, R^2 = 0.66) that was used in the model to relate tortuosity indirectly to a specimen of a given porosity. The pore size parameter calculated from (11) correlates well with MTPS (Figure 3). Velocity and attenuation were calculated using values of M ranging from 0.1 to 10. For the calculations reported here M was set equal to 4. The static Young's modulus was curve fit to the volume fraction of bone using nonlinear regression and the relation $E = E_s (1 - \text{por})^n$ (Figure 4).

Figure 3. Pore-size parameter vs MTPS/2

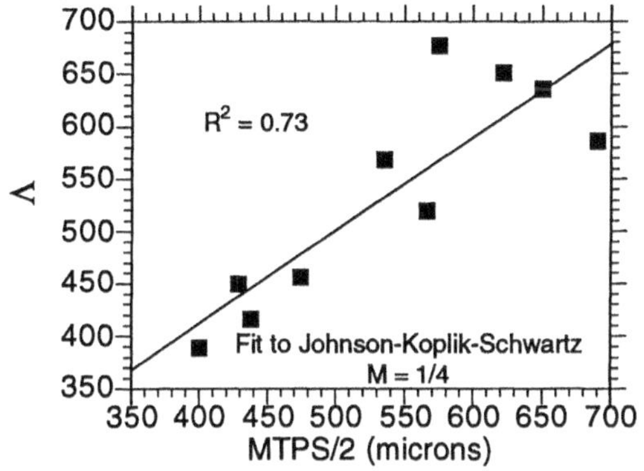

Figure 4. Trabecular Young's modulus

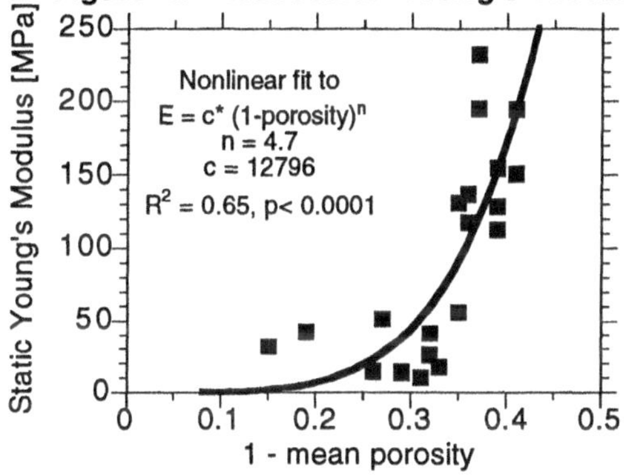

The most interesting aspect of the calculated attenuations (with M = 4) for internal Biot losses can be seen if they are plotted as a function of porosity (Figures 5 and 6). The peak in calculated trabecular bone attenuation occurs at about 65-70% porosity. Most of the experimental data covers a range of porosities from 60% to 90% so only the decreasing portion of the curve would normally be seen. Increasing the marrow viscosity over the reported range at 37°C from 0.04 to 0.4 Pa.s increases the attenuation by a factor of 2 to 3. Increasing the frequency from 0.2 MHz to 0.6 MHz increases the attenuation by a factor of 2.5. It is most likely that this theoretical dependence of attenuation on porosity is partly governed by the dynamic pore size parameter (Λ), since attenuation is a function of the ratio of the viscous skin depth, $\delta = \sqrt{(2\ \eta/\rho_f\ \omega)}$ to the dynamic pore size parameter Λ of Johnson and Plona (1994).

Figure 5. Effect of viscosity at 0.2 MHz

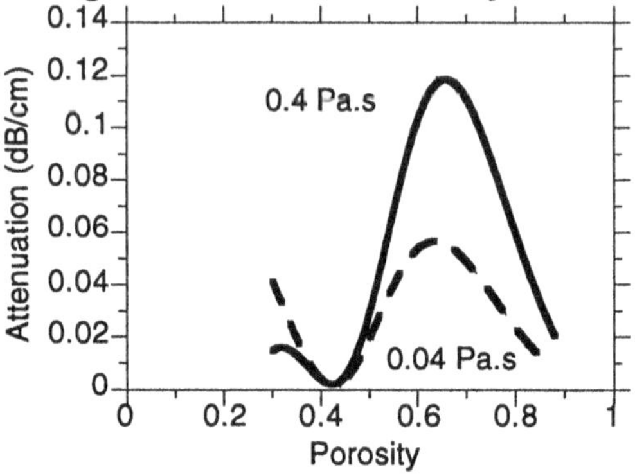

Figure 6. Effect of viscosity at 0.6 MHz

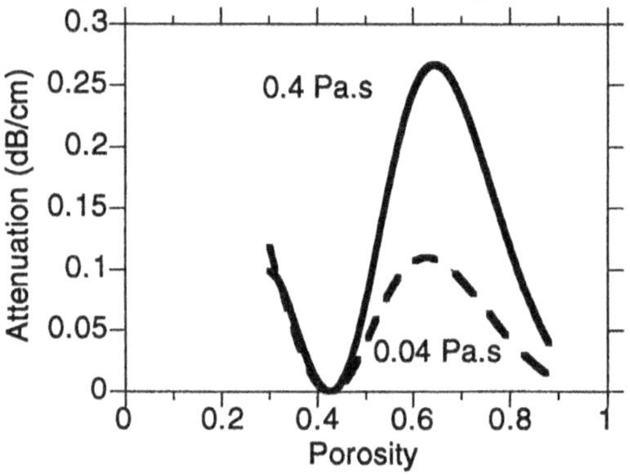

The viscous skin depth is the distance from the pore wall over which the vorticity vanishes. When the viscous depth is small compared with pore size, the fluid follows a quasistatic potential flow. At frequencies where the viscous depth is large compared with the pore size, the fluid follows a quasistatic Poiseuille flow (Johnson and Plona, 1994). The frequency at which the viscous depth is equal to the pore size is the crossover frequency = $\eta/\rho\ k_0\alpha_\infty$ (Johnson and Plona, 1994). For the solution in Figure 5 (M = 4) at 0.2 MHz and 0.4 Pa.s viscosity, for example, the viscous skin depth = 26 microns, whereas the pore size parameter varies from near zero to 330 microns over porosities of 0.3 to 0.88 (Figure 7).

The calculated velocities compare well (Figure 8) with the data from this study and with data from Tavakoli and Evans (1992) even though their specimens came from the bovine femur with properties that may not be modeled well using the relations found for the calcaneus.

Figure 7. Attenuations vs pore size parameter

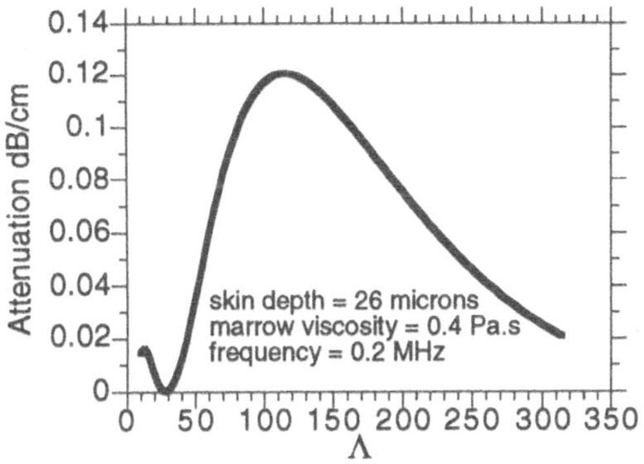

Figure 8. Comparison of experimental and theoretical velocities

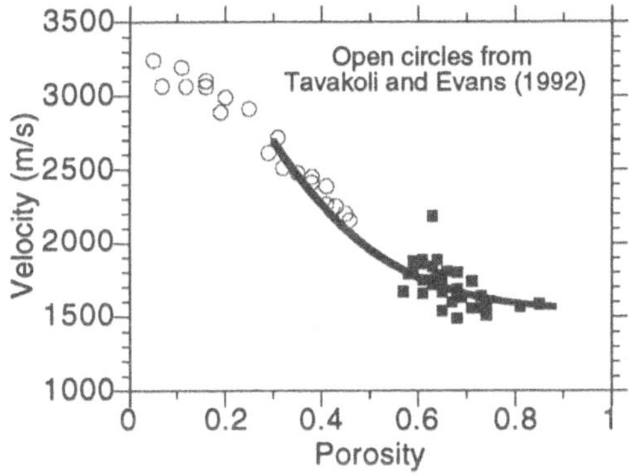

A comparison of calculated attenuations at 0.6 MHz due to internal Biot losses (Figure 6) with our own experimental attenuations at 0.6 MHz (Figure 9) and those of Tavakoli and Evans (1992) reveals a similar qualitative relationship between attenuation and porosity. It is of note that the results of Tavakoli and Evans, showing increasing attenuation with increasing porosity, are difficult to reconcile with the usual finding of the reverse relationship. Biot's theory explains this relationship well, but predicts much smaller attenuations than those measured. Preliminary calculations of reflection losses at both of the flat porous surfaces of these cylindrical samples, using the method of Wu et al. (1990), suggest that losses due to reflections at the porous surfaces of the Biot material dominate over internal Biot losses for the attenuations at 0.2 MHz and that the combination of these losses accounts for 50 to 100% of the measured attenuation at 0.2 MHz. At 0.6 MHz calculated attenuations due to internal Biot losses and porous surface reflection losses account for a smaller part of the measured attenuation (between 15% and 100%, the higher percentage applying to samples of greater porosity). Other attenuation mechanisms, such as those due to marrow pore and trabecular scattering and those due to intrinsic losses in the bone and marrow phases of trabecular bone, appear to be the predominant cause of the measured losses at 0.6 MHz (Grimm, 1995).

Figure 9. Experimental attenuation

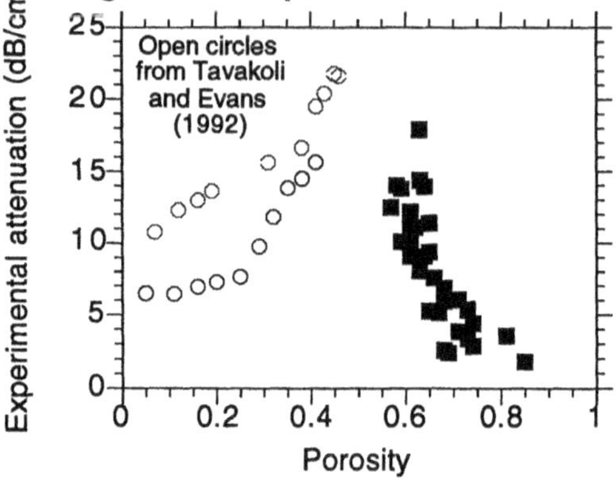

References

Banavar,J.R. and Johnson,D.L. (1987) Characteristic pore sizes and transport in porous media. Physical Review B 35(13): 7283-7286.

Biot, M.A. (1956) Theory of propagation of elastic waves in a fluid-saturated porous solid. II. Higher frequency range. J. Acoust. Soc. Am. 34: 1254-1264.

Brown,R.J.S. (1980) Connection between formation factor for electrical resistivity and fluid-solid coupling factor in Biot's equations for acoustic waves in fluid-filled porous media. Geophysics 45(8): 1269-1275.

Bryant,J.D., David,T., Gaskell,P.H., King,S. and Lond,G. (1989) Rheology of bovine marrow. Proc. Instn Mech. Engrs 203:71-75.

Grimm, M.J. (1995) Ultrasound propagation through the calcaneus: Dependence on "bone quality" and prediction by Biot's theory. Ph.D. Dissertation, University of Pennsylvania, Philadelphia, Pa.

Johnson,D.L., Koplik,J. and Schwartz,L.M. (1986) New pore size parameter characterizing transport in porous media. Physical Review Letters 57(20): 2564-2567.

Johnson,D.L., Koplik,J. and Dashen,R. (1987) Theory of dynamic permeability and tortuosity in fluid-saturated porous media. J. Fluid Mechanics 176: 379-402.

Johnson,D.L. and Plona,T.J. (1994) Probing porous media with first and second sound. II. Acoustic properties of water-saturated porous media. J. Appl. Phys. 76(1): 115-125.

Kim,H.D. and Walsh W.R. (1992) Mechanical and ultrasonic characterization of cortical bone. Biomimetrics 1(4): 293-310.

McElhaney,J.H. (1966) Dynamic response of bone and muscle tissue. J. Appl. Physiol. 21: 1231.

McKelvie,M.L. and Palmer,S.B. (1991) The interaction of ultrasound with cancellous bone. Phys. Med. Biol. 36: 1331-1340.

Tavakoli,M.B. and Evans,J.A. (1991) Dependence of the velocity and attenuation of ultrasound in bone on mineral content. Phys. Med. Biol. 36(11): 1529-1537.

Tavakoli,M.B. and Evans,J.A. (1992) The effect of bone structure on ultrasonic attenuation and velocity. Ultrasonics 30(6): 389-395.

Williams,J.L. (1992) Ultrasonic wave propagation in cancellous and cortical bone: Prediction of some experimental results by Biot's theory. J. Acoust. Soc. Am. 91(2): 1106-1112.

Wu,K.,Xue,Q. and Adler,L. (1990) Reflection and transmission of elastic waves from a fluid-saturated porous solid boundary. J. Acoust. Soc. Am. 87(6):2349-2358.

Load carrying capacity of the pore pressure in a poroelastic beam subject to oscillatory excitation

Dajun Zhang and Stephen C. Cowin,
Department of Mechanical Engineering,
City College of City University of New York,
New York, NY 10031

Abstract Biot theory is employed to analyze a bone beam under combined oscillatory axial compression and bending. It is shown that, for the bone beam considered, the portion of the total axial stress taken by the fluid constituent stress never exceeds the value of porosity. The presence of the pore fluid and pore fluid pressure within a beam of bone doesn't have significant influence on its load-bearing axial total stress. The solid matrix supports most of the axial total stress, which responds almost like a solid elastic beam. The transverse total stress is nonzero but its magnitude is at least one order of magnitude smaller than the axial total stress, and it is closely moderated by the pore fluid pressure.

1 Introduction

The study of fluid-saturated solids (porous media) has been motivated by the need to understand geological materials (Biot, 1941, 1956, Biot and Willis, 1957), and recently, the biological tissues like bone (Salzstein *et al.*, 1987) and cartilage. Living bone tissue is a fluid-saturated solid consisting of a continuous mineralized solid matrix and a continuous bone fluid phase. It has been proposed (Weinbaum, *et al.*, 1993) that the fluid shear stress driven by pore pressure gradient in bone is sensed by the bone cells in contact with the fluid. Further it has been suggested that this sensory activity in the bone is the basis of the mechanotransduction system by which bone tissue senses external mechanical loads and adapts its structure.

An analytical solution of a poroelastic beam under combined oscillatory compression and bending, which has been obtained in a previous study (Zhang and Cowin, 1994), is used here to demonstrate the decomposition of the total stresses into constituent stresses by plotting their time-varying profiles. Restrictions of the results include isotropy, small strains and neglect of inertia terms.

2 Field equation

The Rice-Cleary form of constitutive relations for poroelasticity are, Rice and Cleary (1976),

$$2GE_{ij} = T_{ij} - \frac{\nu}{1+\nu}T_{kk}\delta_{ij} + \frac{3(\nu_u - \nu)}{B(1+\nu)(1+\nu_u)}p\delta_{ij}, \tag{1}$$

A. P. S. Selvadurai (ed.), *Mechanics of Poroelastic Media*, 273–298.

$$\chi - \chi_0 = \frac{3\rho_0(\nu_u - \nu)}{2GB(1 + \nu)(1 + \nu_u)}(T_{kk} + \frac{3}{B}p). \tag{2}$$

There are four material constants in the above constitutive equations: G, ν, B, and ν_u. G and ν are, respectively, the shear modulus and Poisson's ratio when the medium is deformed under the "drained" condition, *i.e.*, when the medium is deformed while maintaining a constant pore fluid pressure. The opposite case is the "undrained" deformation which corresponds to the case when there is no fluid transport between the elements of the medium (*e. g.*, when the time scale of the loading is too small), $\Delta\chi = 0$. The compressibility coefficient B and the undrained Poisson's ratio ν_u are defined in terms of material properties by

$$B = \frac{1/K - 1/K_s}{v_0/K_f + 1/K - (1 + v_0)/K_s} \ , \tag{3}$$

$$\nu_u = \frac{3\nu + B(1 - 2\nu)(1 - K/K_s)}{3 - B(1 - 2\nu)(1 - K/K_s)} \ , \tag{4}$$

where K is the "drained" bulk modulus $K = 2G(1 + \nu)/3(1 - 2\nu)$, K_f is the bulk modulus of the fluid where we have set K_s, K_s', and K_s'' of Rice and Cleary (1976) equal, with no loss of generality to our formulation and solution.

The stress-strain relations (1) can also be written as

$$T_{ij} = 2GE_{ij} + (K - \frac{2G}{3})\delta_{ij}E_{kk} - \zeta p\delta_{ij} \ , \tag{5}$$

where (*c.f.*, *e.g.*, Rudnicki, 1985)

$$\zeta = \frac{3(\nu_u - \nu)}{B(1 - 2\nu)(1 + \nu_u)} = 1 - \frac{K}{K_s} \ . \tag{6}$$

D'Arcy's law is written in the form

$$u_i = -\rho_0\kappa \ \partial p/\partial x_i \ , \tag{7}$$

where u_i is the fluid mass flow rate per unit area in the x_i direction, κ is the permeability and $\kappa = k/\mu$ where μ is the viscosity of the fluid and k is the specific permeability.

The stress/pore fluid pressure conditions of compatibility are given by Rice and Cleary (1976),

$$\nabla^2[(1 + \nu)T_{ij} - \nu T_{kk}\delta_{ij}] + \partial^2 T_{kk}/\partial x_i\partial x_j + \frac{3(\nu_u - \nu)}{B(1 + \nu_u)}[\nabla^2 p\delta_{ij} + \partial^2 p/\partial x_i\partial x_j] = 0. \tag{8}$$

A useful relation between T_{kk} and p is obtained by contracting Eq. (8), thus

$$\nabla^2(T_{kk} + \Lambda p) = 0, \tag{9}$$

where

$$\Lambda = \frac{6(\nu_u - \nu)}{B(1 - \nu)(1 + \nu_u)} = \frac{2(1 - 2\nu)}{(1 - \nu)}\zeta \quad . \tag{10}$$

Another relation between T_{kk} and p can be obtained from the fluid mass conservation equation $\partial u_i/\partial x_i + \partial\zeta/\partial t = 0$, which yields

$$c\nabla^2(T_{kk} + \frac{3}{B}p) = \frac{\partial}{\partial t}(T_{kk} + \frac{3}{B}p) \quad , \tag{11}$$

where

$$c = \kappa[\frac{2G(1 - \nu)}{(1 - 2\nu)}][\frac{B^2(1 + \nu_u)^2(1 - 2\nu)}{9(1 - \nu_u)(\nu_u - \nu)}] \quad . \tag{12}$$

Eqs. (8) and (11) constitute a set of seven coupled equations for the seven scalar unknowns T_{ij} and p. These equations are those used in Zhang and Cowin (1994) to solve the problem of a poroelastic beam under combined oscillatory axial compression and bending.

3 Oscillatory loading of a poroelastic beam

The formulation and an analytical solution for a poroelastic beam (Figure 1), under a combined loading of an oscillatory axial force $N_z = N_0 \sin \omega t$ and an oscillatory bending moment $M_z = M_0 \sin \omega t$, has been given by Zhang and Cowin (1994). Here we only give a brief summary of that work. After the following nondimensionalization (*c.f.*, Figure 1),

$$Y \equiv \frac{y}{d}, \; \tau \equiv \frac{ct}{d^2}, \; T \equiv \frac{\omega d^2}{c}, \; m \equiv \frac{3M_0}{N_0 d}, \; P(\tau, Y) \equiv \frac{2\Psi d}{N_0}p(t, y); \tag{13}$$

the governing equation for the nondimensional pore fluid pressure is found to be

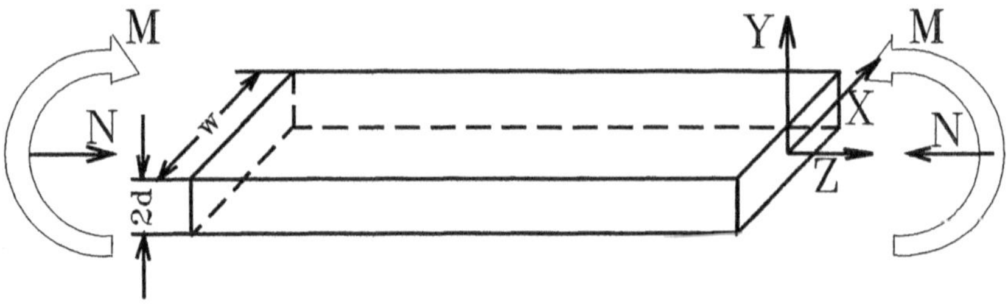

Figure 1: *The beam and the coordinate system employed.*

$$\frac{\partial P}{\partial \tau} - \frac{\partial^2 P}{\partial Y^2} = \frac{\Psi}{\Psi + \Lambda}(1 + mY)T \cos T\tau$$
$$+ \frac{\Lambda}{2(\Psi + \Lambda)}[(\eta + 3Y + 3Y\eta)P(\tau, 1) - (-\eta + 3Y + 3Y\eta)P(\tau, -1)] , \tag{14}$$

where Ψ is defined as

$$\Psi = \frac{3}{B} - \Lambda , \tag{15}$$

and and η is the leakage coefficient. The boundary leakage condition at y = ±d (the upper and lower surface of the beam) is

$$\frac{\partial P}{\partial Y}(\tau, \pm 1) = \mp \eta P(\tau, \pm 1) \ . \tag{16}$$

The steady-state solution of the differential equation (14) subjected to the boundary condition Eq. (16) is

$$P(\tau, Y) = \frac{\Psi}{\Psi + \Lambda}(1 + mY) \sin T\tau$$

$$- \frac{\Psi}{\Psi + \Lambda} Im[\frac{\dfrac{-\eta \cosh \beta Y}{\beta \sinh \beta} + \dfrac{\Lambda \eta}{(\Psi + \Lambda)Ti}}{-\dfrac{\eta \cosh \beta}{\beta \sinh \beta} + \dfrac{\Lambda \eta}{(\Psi + \Lambda)Ti} - 1}$$

$$+m\frac{\dfrac{(1+\eta)\sinh\beta Y}{\sinh\beta-\beta\cosh\beta}+\dfrac{3\Lambda(1+\eta)Y}{(\Psi+\Lambda)Ti}}{\dfrac{(1+\eta)\sinh\beta}{\sinh\beta-\beta\cosh\beta}+\dfrac{3\Lambda(1+\eta)}{(\Psi+\Lambda)Ti}-1}]e^{iT\tau}, \tag{17}$$

where

$$\beta = (1+i)(\frac{T}{2})^{\frac{1}{2}}. \tag{18}$$

The solution (17) is still valid as $\eta \to \infty$, Zhang and Cowin (1994).

The only two nonzero total stress components T_{11} and T_{33} are related to the pore fluid pressure p by

$$T_{11} = \frac{\Lambda}{2}[-p + \frac{wy}{I}\int_{-d}^{d} ypdy + \frac{w}{A}\int_{-d}^{d} pdy], \tag{19}$$

$$T_{33} = -w(\frac{yM_z}{I} + \frac{N_z}{A}) + \frac{\Lambda}{2}[-p + \frac{wy}{I}\int_{-d}^{d} ypdy + \frac{w}{A}\int_{-d}^{d} pdy], \tag{20}$$

where $I = 2wd^3/3$ and $A = 2wd$. w is the width of the beam. After a nondimensionalization for the stresses in the same way the pore fluid pressure is nondimensionalized (*c.f.*, Eq. (13)), the nondimensional total stresses S_{11} and S_{33} are

$$S_{11} = \frac{\Lambda}{2}[-P + \frac{1}{2}\int_{-1}^{1} PdY + \frac{3Y}{2}\int_{-1}^{1} YPdY], \tag{21}$$

$$\begin{aligned} S_{33} &= -(1+mY)\Psi \sin T\tau + \frac{\Lambda}{2}[-P + \frac{1}{2}\int_{-1}^{1} PdY + \frac{3Y}{2}\int_{-1}^{1} YPdY] \\ &= -(1+mY)\Psi \sin T\tau + S_{11}. \end{aligned} \tag{22}$$

The strain components may be obtained from the constitutive relations.

Conceptually the total stresses can be separated into two parts. One (the constituent stress on the solid phase) is the part supported by the solid phase of the porous medium, and the other (the constituent stress on the fluid phase) is supported by the fluid phase. From Eq. (5), we have

$$T_{11} = 2GE_{11} + \frac{2G\nu}{1-2\nu}E_{kk} - \zeta p = \sigma_{11}^{(s)} - \zeta p = \sigma_{11}^{(s)} + \sigma^{(f)}, \tag{23}$$

$$T_{22} = 2GE_{22} + \frac{2G\nu}{1-2\nu}E_{kk} - \zeta p = \sigma_{22}^{(s)} - \zeta p = \sigma_{22}^{(s)} + \sigma^{(f)} = 0, \tag{24}$$

$$T_{33} = 2GE_{33} + \frac{2G\nu}{1-2\nu}E_{kk} - \zeta p = \sigma_{33}^{(s)} - \zeta p = \sigma_{33}^{(s)} + \sigma^{(f)}, \tag{25}$$

where $\sigma_{ij}^{(s)}$ are the constituent stresses on the solid phase which can be interpreted as the portion of the total stress T_{ij} that is supported by the solid part of the medium. In conventional soil mechanics, $\sigma_{ij}^{(s)}$ are known as the effective stresses, Nur and Byerlee (1971). The constituent stress on the fluid phase is hydrostatic, so we use a single expression $\sigma^{(f)}$ to denote it. We use $S_{ij}^{(s)}$ and $S^{(f)}$ to denote the nondimensionalized $\sigma_{ij}^{(s)}$ and $\sigma^{(f)}$.

4 Revision of a definition

Since we have an analytical solution at hand, the total stresses and stresses associated with the fluid and solid phases can be investigated in great detail. In order to evaluate the percentage of the magnitude of the total axial stress that is supported by the fluid pressure, Zhang and Cowin (1994) have defined a pore fluid stress contribution factor Π^e of the following form

$$\Pi^e = \zeta \left(\frac{(P)_{a\tau}}{(S)_{a\tau}} \right)_{aY} , \tag{26}$$

where $S = S_{11} + S_{33}$ and $(\ )_{aY}$, $(\ )_{a\tau}$, are the average operators over the thickness Y and the time period,

$$(\ )_{aY} = \frac{1}{2}\int_{-1}^{1} |\ |\ dY, \quad (\ )_{a\tau} = \frac{T}{2\pi}\int_{0}^{\frac{2\pi}{T}} |\ |\ d\tau. \tag{27}$$

In this paper, we replace S in Eq. (26) by S_{33} because it is shown that S_{33} is the dominant stres component in this problem. Thus Π^e becomes Π_{33}^e on the left side of Eq. (26). It will be show here that this definition has a flaw because it produces misleading numerical artifacts.

The behavior of Π_{33}^e, as the loading composition parameter m and the nondimensiona frequency T change is given in Figures 2 and 3 for two extreme values of the leakag coefficient η, $\eta = 0$ (no leakage) and $\eta = \infty$ (free leakage). Material parameters are chose from those appropriate for a bone beam (*c.f.*, Zhang and Cowin, 1994). Both figures displa

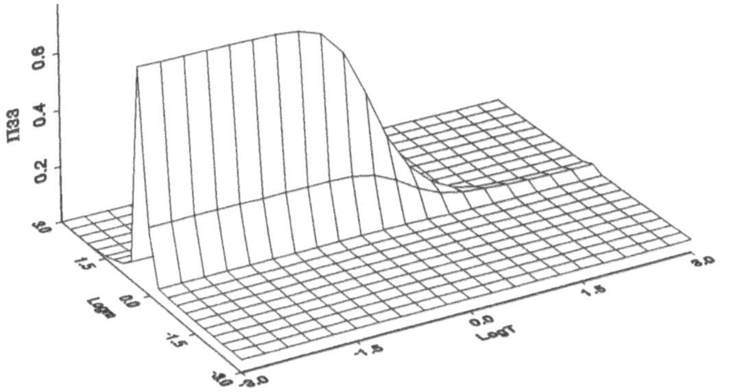

Figure 2: Π_{33}^{e} *versus log m and log T for* η *= 0 (no leakage).*

prominent spikes for certain values of m and T, which, as we suggested (Zhang and Cowin, 1994), might indicate that the pore fluid pressure can share up to 77% (for $\eta = 0$) or 10% (for $\eta = \infty$) of the axial total stress if the external loading is properly tuned, even though the lacunae-canaliculi level porosity of cortical bone is only 5%. We considered this phenomenon to be of physiological significance in Zhang and Cowin (1994).

To verify the existence of the spikes, we plotted the continuous profiles of the total stress S_{33} and the stresses associated with the solid and fluid phases $S_{33}^{(s)}$, $S^{(f)}$ respectively, for the values of m and T at which the spikes in Figures 2 and 3 occur. For the same values of m and T, we also plotted the transverse total stress S_{11} and the constituent stress associated with the solid $S_{11}^{(s)}$. Figures 4a - 4e display the profiles of the nondimensional total stresses and stresses associated with the fluid and solid phases for $\eta = 0$ and m = 2, T = 0.05 as a function of space and time. This combination of m and T is a point associated with the spike in Figure 2. No over-sharing by $S^{(f)}$ (Figure 4c) of the total stress S_{33} (Figure 4a) is observed: $S^{(f)}$ oscillates within a range (± 0.26) much smaller than the oscillating range (around $\pm$ 16) of S_{33}. Further investigation of the profiles of the total and constituent stresses for the spike in Figure 3 also fails to demonstrate any over-sharing of the total stress S_{33} by $S^{(f)}$ (Figures 5a-5e). Now it is clear that the spikes in Figures 2 and 3 are

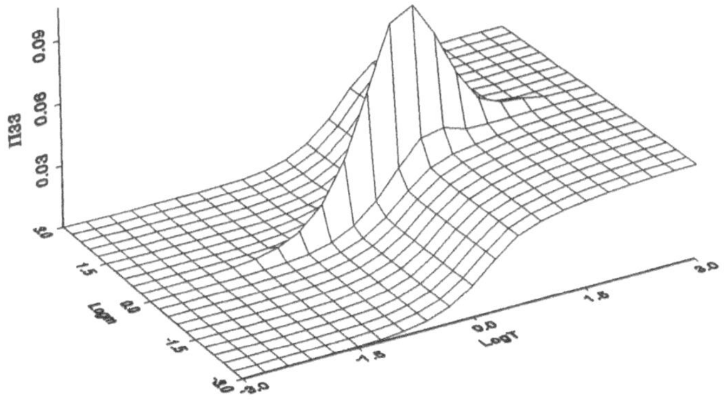

Figure 3: Π_{33}^{e} *versus log m and log T for* η *=* ∞ *(free leakage).*

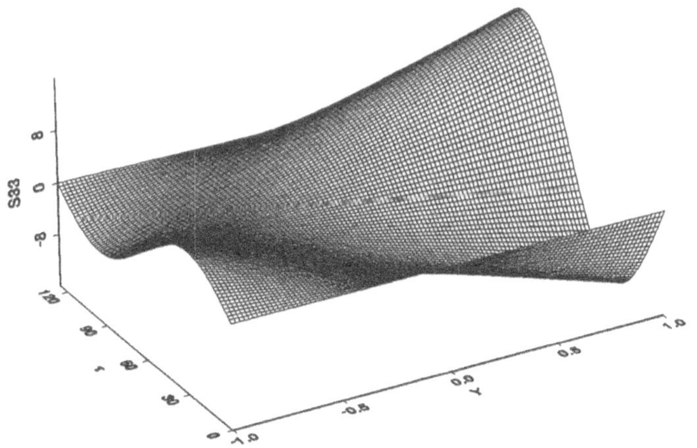

Figure 4a: S_{33} *versus* τ *and* Y *for* $\eta = 0,\ m = 2,\ T = 0.05.$

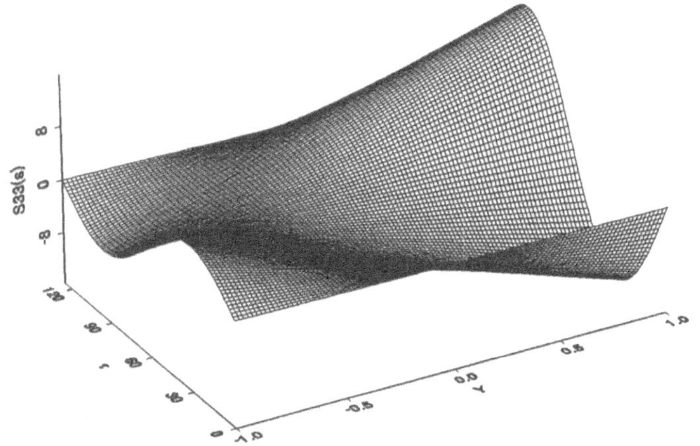

Figure 4b: $S_{33}^{(s)}$ *versus* τ *and* Y *for* $\eta = 0,\ m = 2,\ T = 0.05.$

artifacts due to the way Π_{33}^{e} was defined, Eq. (26). The reason that the spikes occur for some combinations of m and T (m = 2, T < 1 for Figure 2; m = 2, T around 4 for Figure 3) is that, for those loading conditions and certain values of Y, $(S_{33})_{a\tau}$ is nearly zero but $(P)_{a\tau}$ is finite.

In order to get rid of the locally misleading spikes occurring in Π_{33}^{e}, we define Π_{33} as

$$\Pi_{33} = \zeta \frac{\left((P)_{a\tau}\right)_{aY}}{\left((S_{33})_{a\tau}\right)_{aY}}. \quad \textbf{(28)}$$

The behavior of Π_{33} for various values of the leakage coefficient η (η changes from 0 to ∞) is displayed in Figures 6a-6e. Comparing Figures 2 and 3, which employ definition (26), and Figures 6a and 6e, which employ definition (28), we notice that (28) preserves all the physical trend of (26) while clearing away the nonphysical spikes.

5 Results

We demonstrate the solution described above using the material properties appropriate for cortical bone, which is a poroelastic solid due to the presence of free bone fluid within the

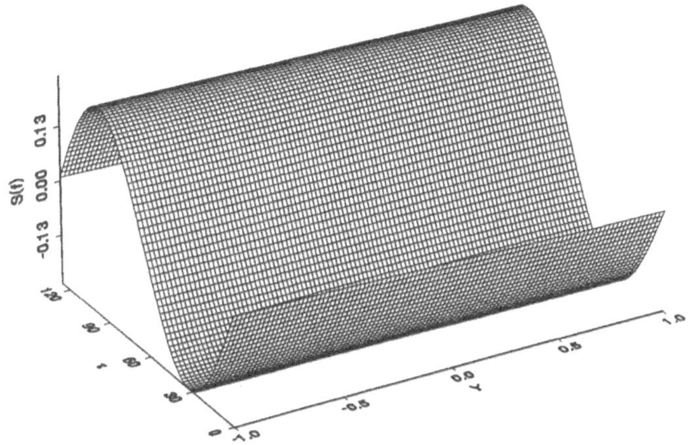

Figure 4c: $S^{(f)}$ *versus* τ *and* Y *for* $\eta = 0$, $m = 2$, $T = 0.05$.

lacunae-canaliculi spaces (pores). Evaluation of the material parameters can be found in Zhang and Cowin (1994), where we have $K_f = 2.3$ GPa, $K = 12$ GPa, $v_0 = 0.05$, $K_s = 16$ GPa, $\nu = 0.25$, $G = 7.2$ GPa, $d = 0.5$ mm. Using Eqs. (3), (4), (6), (10), (12) and (15), we can then obtain $\nu_u = 0.28$, $B = 0.53$, $\Lambda = 0.35$, $\Psi = 5.3$, $\zeta = 0.26$, and $c = \kappa(25.6\text{GPa})$.

Profiles for the nondimensional total stresses S_{ij}, and the constituent stresses $S_{ij}^{(s)}$, $S^{(f)}$ are plotted (Figures 4, 5, 7 - 10) for different loading parameters: m, T and different choices of the boundary leakage coefficient η.

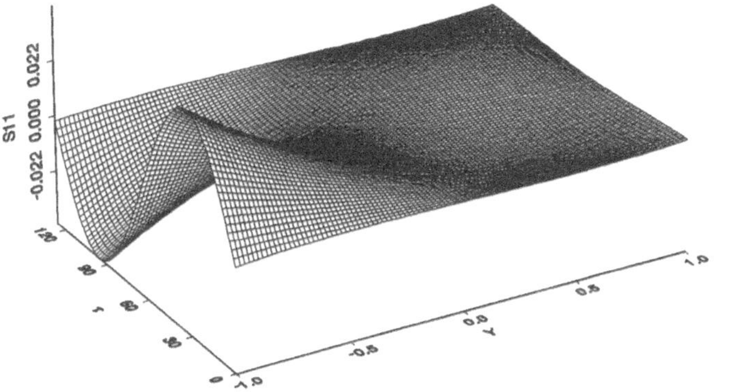

Figure 4d: S_{11} *versus* τ *and* Y *for* $\eta = 0$, $m = 2$, $T = 0.05$.

Behavior of the pore fluid stress contribution factor in the axial total stress is plotted for two different definitions given by (26) for Π_{33}^{e} (Figures 2 and 3) and by (28) for Π_{33} (Figure 6) . We focus our attention on the axial stress S_{33} because it is the dominant stress in our problem.

6 Discussion

Figure 6a illustrates the behavior of the contribution factor of the fluid constituent stress $S^{(f)}$ in the total axial stress S_{33}, Π_{33}, when there is no boundary leakage ($\eta = 0$) at the upper and lower surfaces ($Y = \pm 1$) of the beam. Figure 6a is similar to Figure 2 in that there is a wide-

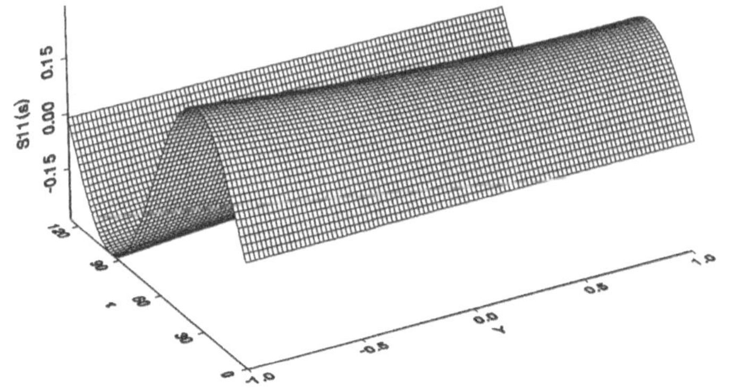

Figure 4e: $S_{11}^{(s)}$ *versus* τ *and* Y *for* $\eta = 0,\ m = 2,\ T = 0.05.$

spread plateau in both figures at $\Pi_{33} = 4.6\%$, which is nearly the value of the porosity (5%). This means that when $\eta = 0$ (no leakage) the portion of the total stress S_{33} shared by the fluid never exceeds the porosity (5%). When $m < 1$ ($\log m < 0$), the nondimensional frequency T has no influence on Π_{33} (Figure 6a): it stays at 4.6%. When $m > 1$ ($\log m > 0$), Π_{33} is a monotonically increasing function of T until the plateau is reached for $T > 10$ ($\log T > 1$). The fact that Π_{33} displays totally different behaviors for the cases $m > 1$ and $m < 1$ shows that the poroelastic beam behaves differently in driving the pore fluid pressure when the external loading is axial-force dominant ($m < 1$) or bending-moment dominant ($m > 1$), if no boundary fluid leakage is allowed ($\eta = 0$). The $m = 1$ line is defined as the threshold line. If the external loading is axial-force dominant ($m < 1$), the pore fluid pressure has very little gradient along the beam thickness, thus little pore fluid shift occurs within the beam. In that case the pore fluid pressure builds up easily. However, if the external loading is bending-moment dominant, the tensile side of the beam has lower pressure than the compressional side. So pore fluid shift within the beam is significant. The pore fluid pressure is more difficult to build up when the nondimensional frequency T is small. When T is large, the pore fluid pressure has no time to relax, thus the pore fluid pressure also builds up even when the external loading is bending-moment dominant ($m \gg 1$). This is why Π_{33} also reaches the

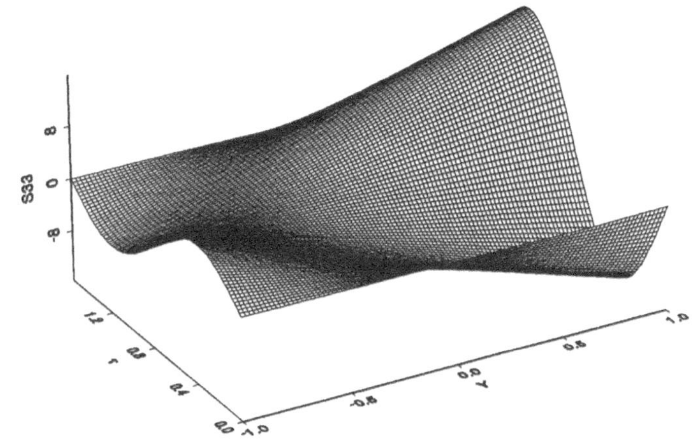

Figure 5a: S_{33} *versus* τ *and* Y *for* $\eta = \infty,\ m = 2,\ T = 4.$

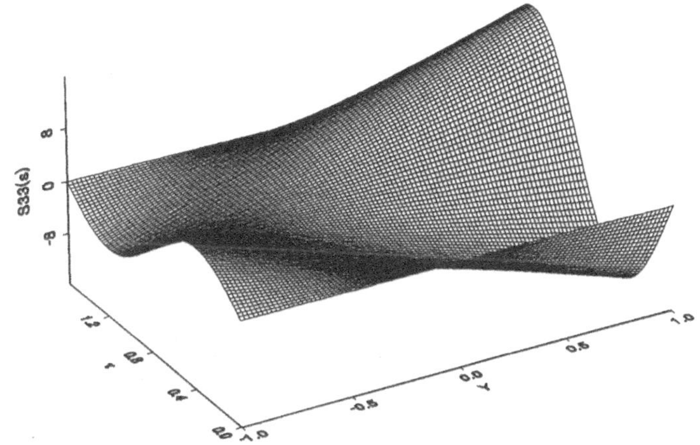

Figure 5b: $S_{33}^{(s)}$ *versus* τ *and* Y *for* $\eta = \infty$, $m = 2$, $T = 4$.

plateau (4.6%) for the m > 1 case (Figure 6a) when T is large enough.

As the upper and lower surfaces of the beam become increasingly leaky (η increases, Figures 6b - 6e), we can see the plateau area in Π_{33} for m < 1 (log m < 0) and log T < 0 shrinks continuously as η increases until eventually it almost joins the plateau edge for the m > 1 (log m > 0) side when $\eta = \infty$. This means that when the surfaces of the beam become increasingly leaky, the pore fluid pressure within the beam becomes more and more difficult to build up so that it takes an increasingly smaller portion of the total axial stress. Since Π_{33} has a maximum value of about 4.6% for the no leakage case (η = 0), this maximum prevails for all values of η, as shown in Figure 6. This means that for the bone beam considered, the portion Π_{33} of the axial total stress taken by the pore fluid constituent stress never exceeds the porosity of the beam (5%). This conclusion corrects the error in Zhang and Cowin (1994) associated with the misleading spikes in Π_{33}^{e}. In that paper we erroneously concluded that the pore fluid pressure can share a much higher portion of the total stresses than the value of the porosity, based on those spikes.

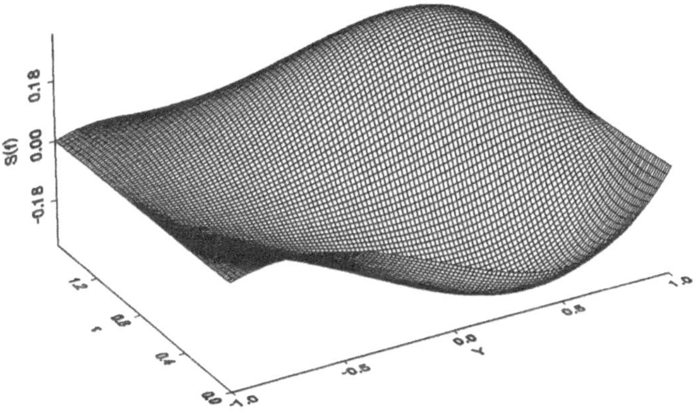

Figure 5c: $S^{(f)}$ *versus* τ *and* Y *for* $\eta = \infty$, $m = 2$, $T = 4$.

The fact that the threshold (Figure 6) of the two distinctly different behaviors of Π_{33} for different values of m happens at m = 1 makes sense physically. For a solid, elastic beam, m = 1 corresponds to the case when one edge of the beam (Y = - 1) experiences zero axial stress

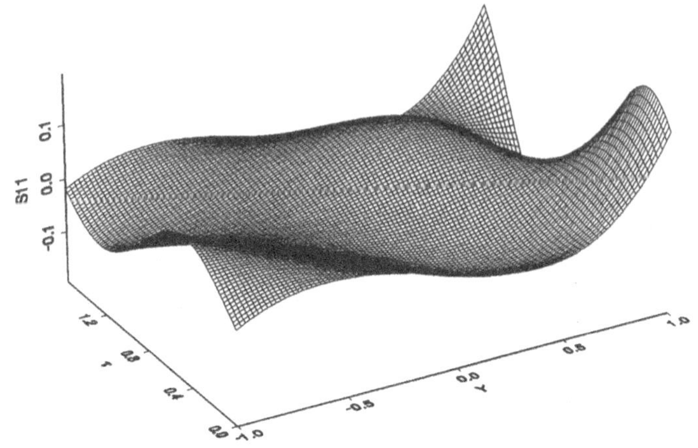

Figure 5d: S_{11} *versus* τ *and* Y *for* $\eta = \infty$, $m = 2$, $T = 4$.

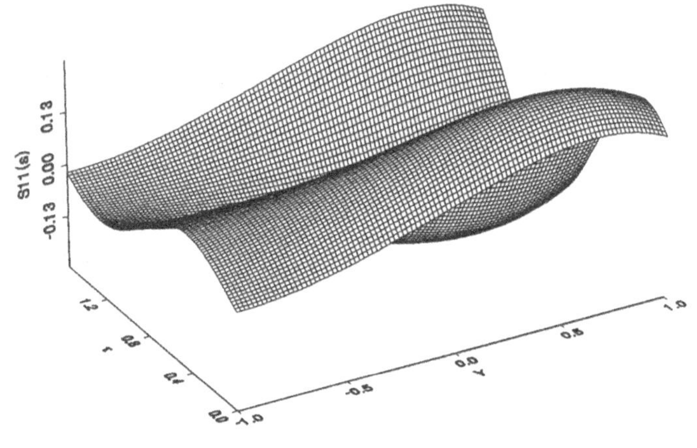

Figure 5e: $S_{11}^{(s)}$ *versus* τ *and* Y *for* $\eta = \infty$, $m = 2$, $T = 4$.

and the axial stress across the thickness of the beam is of a same sign all the time. For our bone beam of low porosity, this situation still applies. If $m > 1$, the axial total stress will change signs across the thickness, and it doesn't if $m < 1$.

The axial total stress S_{33} has a spatial distribution almost like that of a solid, elastic beam: it is linear in space (Figure 4a) and closely follows the external loading in phase. Comparing Figures 4a and 4b, we find the difference between the total axial stress S_{33} and its correspondent solid constituent stress $S_{33}^{(s)}$ is indiscernible to the eye. The reason for this is that the fluid constituent stress $S^{(f)}$ (Figure 4c) has a magnitude almost 2 orders lower than that of S_{33} and $S_{33}^{(s)}$. This resemblance between S_{33} and $S_{33}^{(s)}$ is observed for all the other combinations of parameters η, m, and T. Comparing Figures 4a and 4c, we can see that the fluid constituent stress $S^{(f)}$ (or the negative of the nondimensionalized pore fluid pressure p) doesn't always follow the axial total stress S_{33}: $S^{(f)}$ is almost uniform in its spatial distribution, while S_{33} has a change of sign. This means that at the tensile side of the beam we can still have a positive pore fluid pressure. This occurs because no fluid leakage at the boundary surfaces is allowed ($\eta = 0$) and the loading frequency is very low ($f \approx (T/5)\text{Hz} = 0.01\text{Hz}$, almost static); thus the pore fluid pressure has enough time

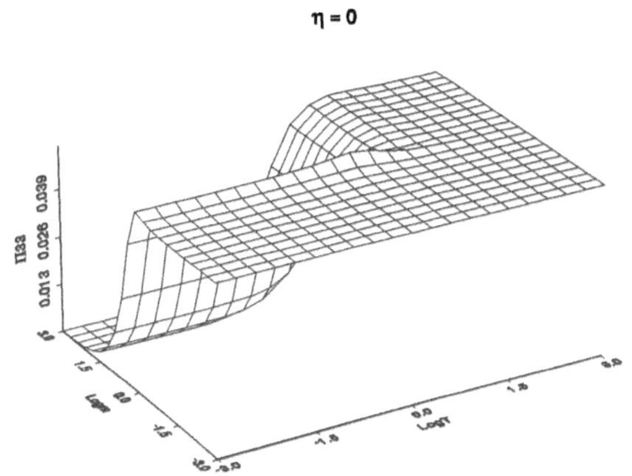

Figure 6a: Π_{33} *versus log m and log T for* $\eta = 0$.

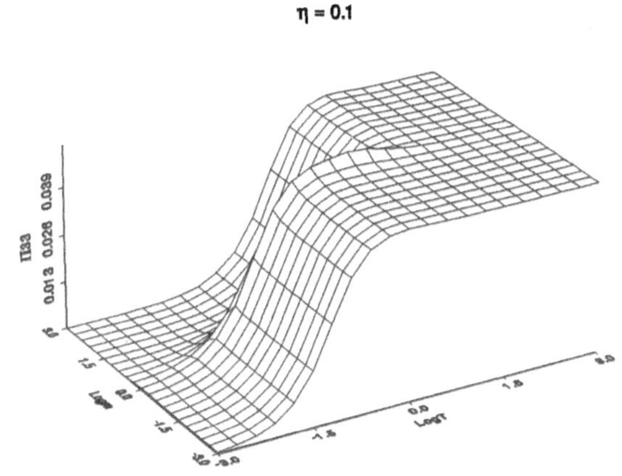

Figure 6b: Π_{33} *versus log m and log T for* $\eta = 0.1$.

to equalize between different pores. Figure 4d is the profile of the transverse total stress S_{11}. Here the spatial asymmetry about Y = 0 is due to the presence of m. In our formulation it is required that the spatial distribution of S_{11} over the lateral faces is such that there is no net force ($N_x = 0$) and no moment ($M_x = 0$) associated with S_{11}. The transverse total stress S_{11} itself is not zero. This is a major difference between a poroelastic beam and a solid elastic beam, where S_{11} is zero. However, we can see in this case (Figure 4d) that the transverse total stress S_{11} is almost 3 orders of magnitude smaller than the axial total stress S_{33}. The reason for this is that the transverse solid constituent stress $S_{11}^{(s)}$ (Figure 4e) is approximately of the same magnitude but opposite in phase with the fluid constituent stress $S^{(f)}$ (Figure 4c). These two transverse constituent stresses cancel each other to give rise to a very small transverse total stress. For other combinations of the parameters η, m, and T, the ordering relationships among the magnitudes of the total stresses and constituent stresses S_{33}, S_{11}, $S_{33}^{(s)}$, $S_{11}^{(s)}$, and $S^{(f)}$ are the same; $|S_{11}| < |S_{11}^{(s)}|$, $|S^{(f)}| < |S_{33}|$, $|S_{33}^{(s)}|$. These magnitude ordering relationships are also implied by expressions (21) and (22); since $\Psi = 5.3$, $\Lambda = 0.35$, it is obvious that the first term in (22) is the dominating term.

Figure 7 shows a similar loading condition to that of Figure 4, but the leakage condition at Y = ± 1 is placed at the other extreme, free leakage ($\eta = \infty$). S_{33}, $S_{33}^{(s)}$ (Figures 7a,b) appear to be the same as S_{33}, $S_{33}^{(s)}$ for $\eta = 0$ (Figures 4a,b), but $S^{(f)}$, S_{11}, $S_{11}^{(s)}$ differ considerably from the corresponding quantities in Figure 4. Their magnitudes are all at least 1 order lower and they manifest much richer variation in space compared with their counterparts of the no-leakage case. The pore fluid constituent stress $S^{(f)}$ remains at zero at Y = ± 1. The reason for the decreasing of magnitude of $S^{(f)}$ is that, for the free-leakage case, the pore fluid pressure becomes much more difficult to build up, and the axial deformation induces less transverse

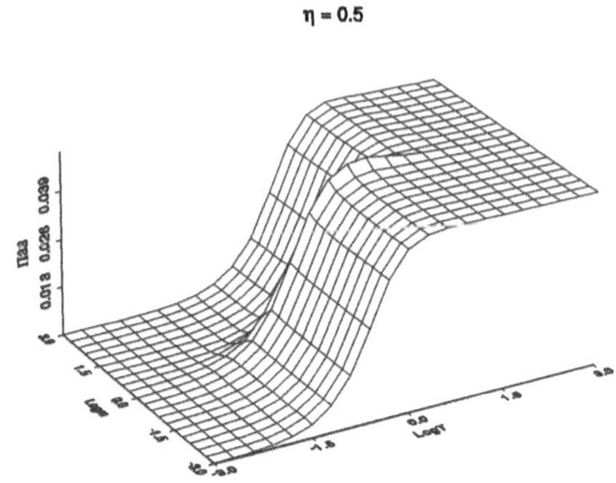

Figure 6c: Π_{33} *versus log m and log T for* $\eta = 0.5$.

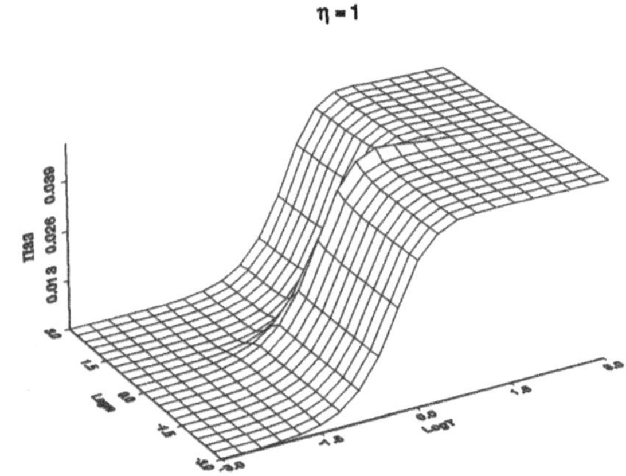

Figure 6d: Π_{33} *versus log m and log T for* $\eta = 1$.

stress through the pore fluid pressure variation.

Figures 8 and 9 are the profiles of the stresses for the pure compression case (m = 0) for the two extreme leakage conditions, $\eta = 0$ (no leakage) and $\eta = \infty$ (free leakage), at $Y = \pm 1$. For the no-leakage case ($\eta = 0$) the pore fluid pressure and thus $S^{(f)}$ is uniform in space, thus S_{11} is uniformly zero (*c.f.*, Eq. (21)) and $S_{33} = -\Psi \sin T\tau$. For the free-leakage case ($\eta = \infty$) $S^{(f)}$ is no longer uniform, thus S_{11} is nonzero though very small. In this case, all profiles are symmetric about Y = 0, suggesting that the spatial asymmetry is caused by the presence of m.

Figure 10 represents the case when the nondimensional frequency is very high, T = 1000, $\eta = \infty$, and m = 2. From the definition of T (Eq. (13)), we know that if T is large, it doesn't necessarily mean that the angular frequency ω is high. A small c (a high fluid viscosity or a low intrinsic permeability) is also able to produce a large value of T for a moderate value of angular frequency ω. From Figure 10c we see that for most part of the thickness $S^{(f)}$ closely follows the variation of S_{33} except near the upper and the lower surfaces, $Y = \pm 1$, where it sharply drops to zero. Near the upper and lower surfaces there are two boundary layers within which the pore fluid pressure experiences a steep gradient. The higher the T, the thinner the boundary layer, and the steeper the gradient. The same boundary layer phenomenon also occurs for S_{11} (Figure 10d), which is zero along the beam thickness except near the boundary surfaces $Y = \pm 1$. The sharp drop of pore fluid pressure near the boundary $Y = \pm 1$ doesn't affect the total axial stress S_{33} much because it constitutes only a minute part of the total stress (below 5%). However, as far as the bone cells are concerned, a sharp pore fluid pressure drop produces significant shear stresses on the cell membranes, stresses that can be sensed by receptors of the bone cells (Weinbaum *et al.*, 1993).

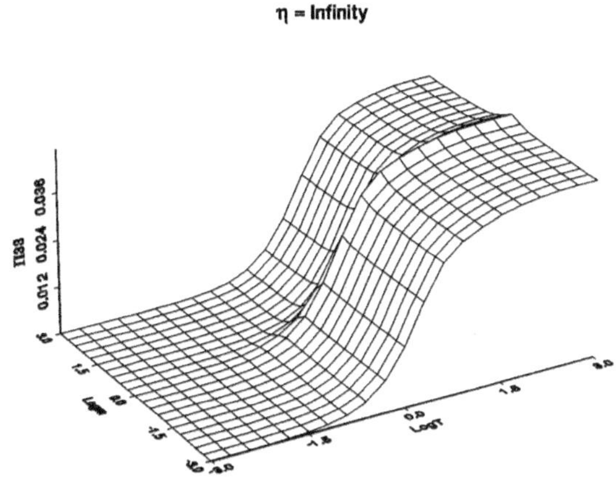

Figure 6e: Π_{33} *versus log m and log T for* $\eta = \infty$.

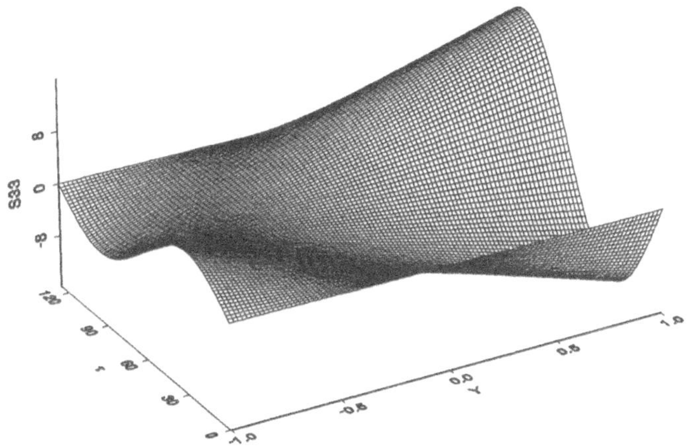

Figure 7a: S_{33} *versus* τ *and Y for* $\eta = \infty$, $m = 2$, $T = 0.05$.

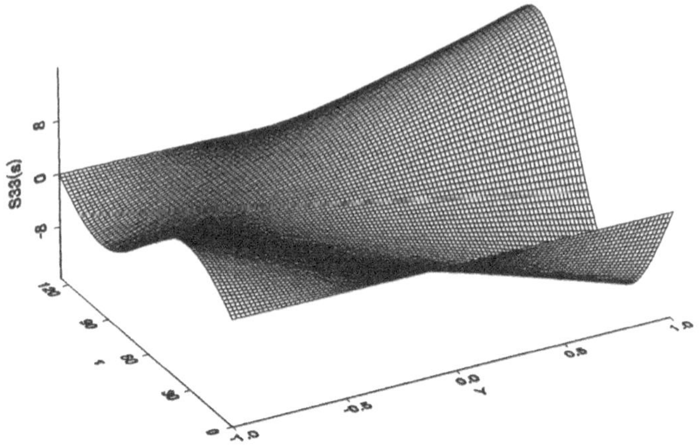

Figure 7b: $S_{33}^{(s)}$ *versus* τ *and* Y *for* $\eta = \infty$, $m = 2$, $T = 0.05$.

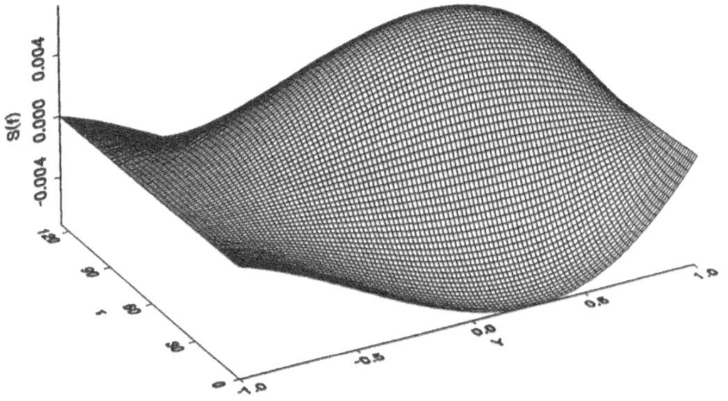

Figure 7c: $S^{(f)}$ *versus* τ *and* Y *for* $\eta = \infty$, $m = 2$, $T = 0.05$.

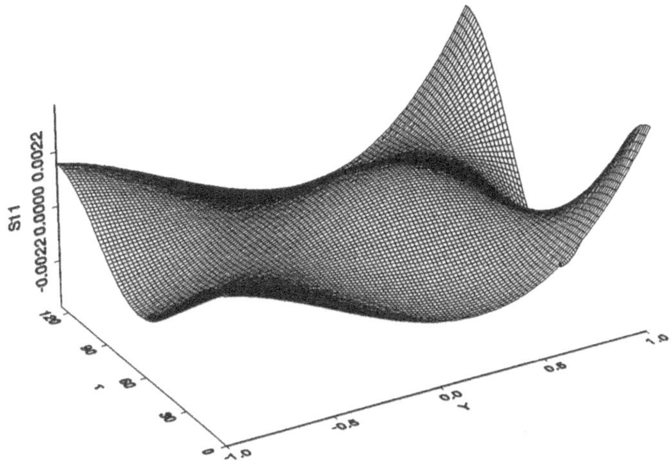

Figure 7d: S_{11} *versus* τ *and* Y *for* $\eta = \infty$, $m = 2$, $T = 0.05$.

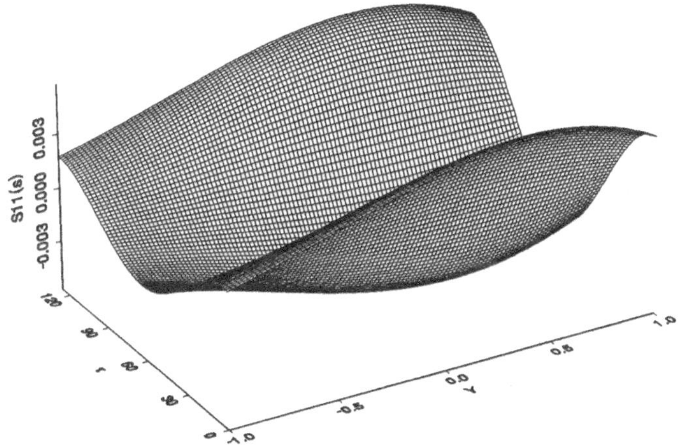

Figure 7e: $S_{11}^{(s)}$ *versus* τ *and* Y *for* $\eta = \infty$, $m = 2$, $T = 0.05$.

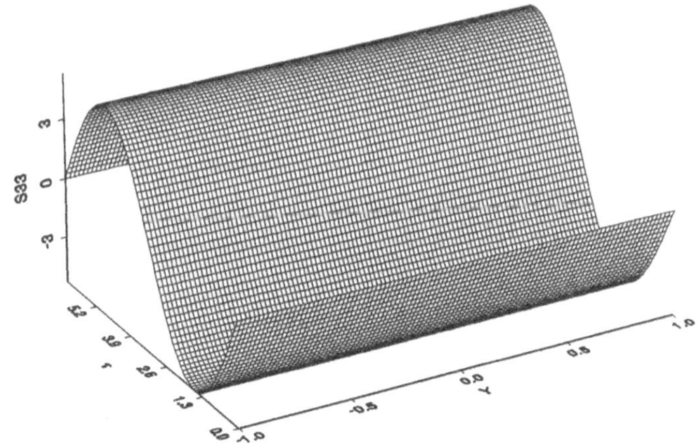

Figure 8a: S_{33} *versus* τ *and* Y *for* $\eta = 0$, $m = 0$, $T = 1$.

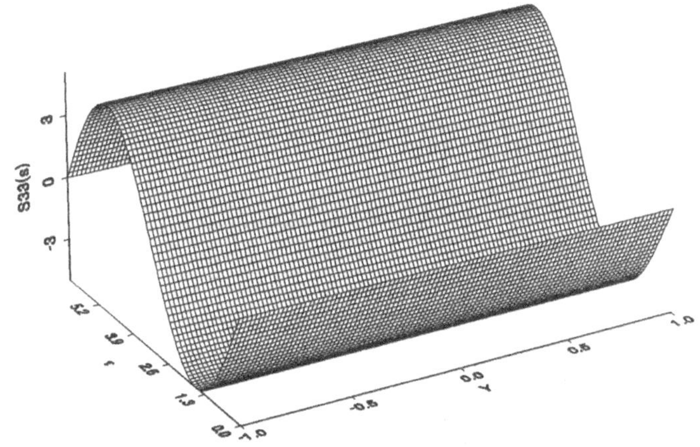

Figure 8b: $S_{33}^{(s)}$ *versus* τ *and* Y *for* $\eta = 0$, $m = 0$, $T = 1$.

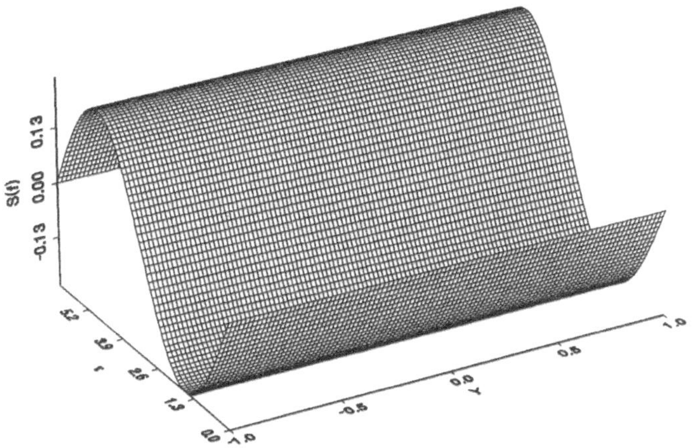

Figure 8c: $S^{(f)}$ *versus* τ *and* Y *for* $\eta = 0$, $m = 0$, $T = 1$.

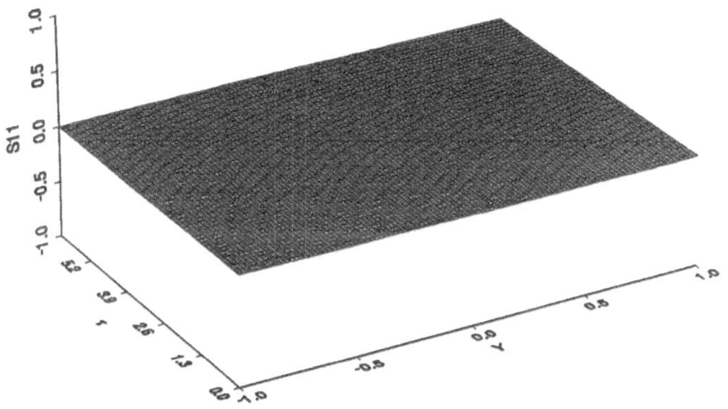

Figure 8d: S_{11} *versus* τ *and* Y *for* $\eta = 0$, $m = 0$, $T = 1$.

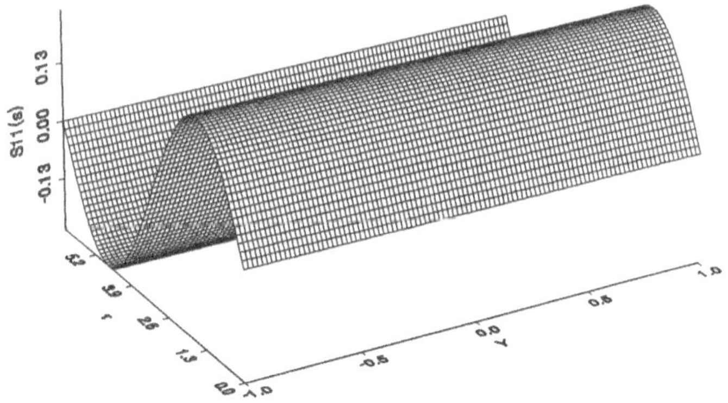

Figure 8e: $S_{11}^{(s)}$ *versus* τ *and* Y *for* $\eta = 0$, $m = 0$, $T = 1$.

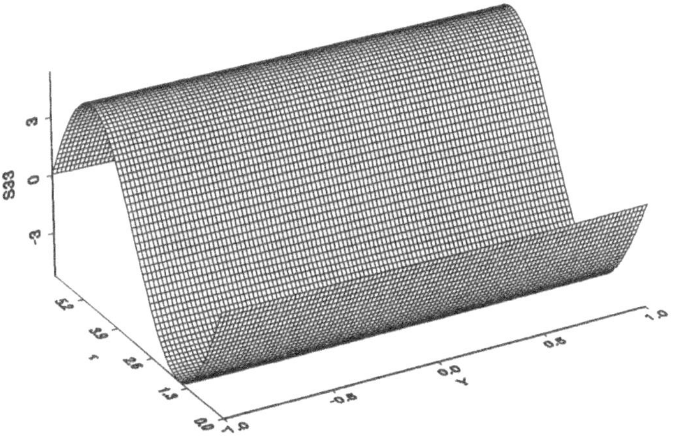

Figure 9a: S_{33} *versus* τ *and* Y *for* $\eta = \infty$, $m = 0$, $T = 1$.

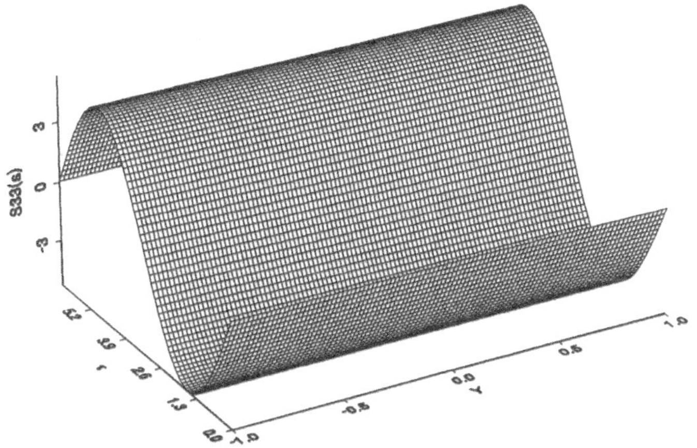

Figure 9b: $S_{33}^{(s)}$ *versus* τ *and* Y *for* $\eta = \infty$, $m = 0$, $T = 1$.

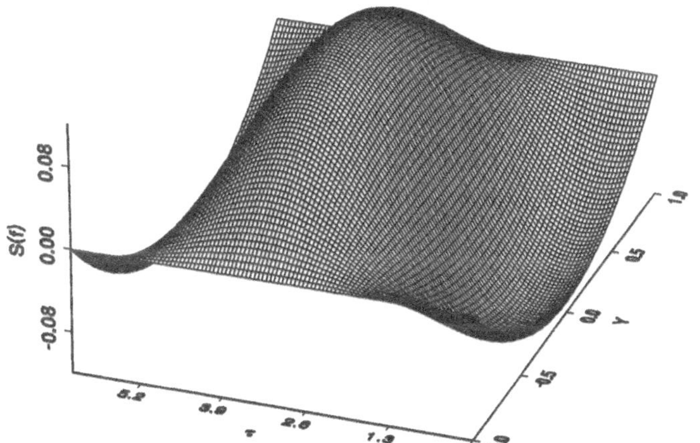

Figure 9c: $S^{(f)}$ *versus* τ *and* Y *for* $\eta = \infty$, $m = 0$, $T = 1$.

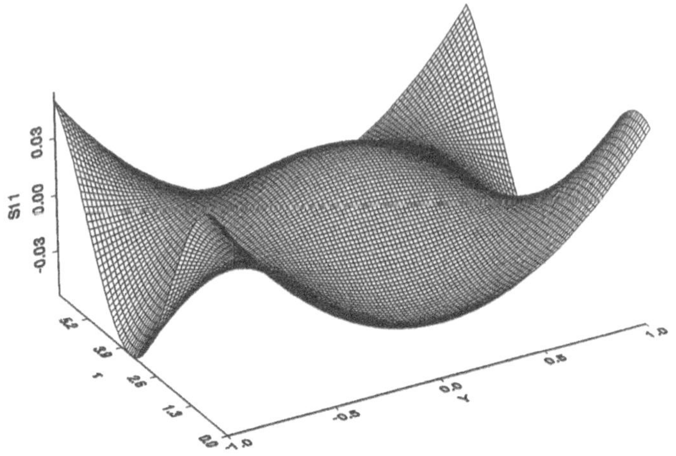

Figure 9d: S_{11} *versus* τ *and* Y *for* $\eta = \infty$, $m = 0$, $T = 1$.

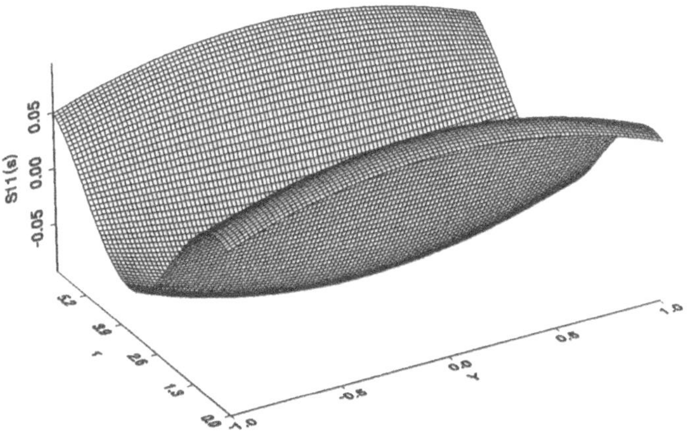

Figure 9e: $S_{11}^{(s)}$ *versus* τ *and* Y *for* $\eta = \infty$, $m = 0$, $T = 1$.

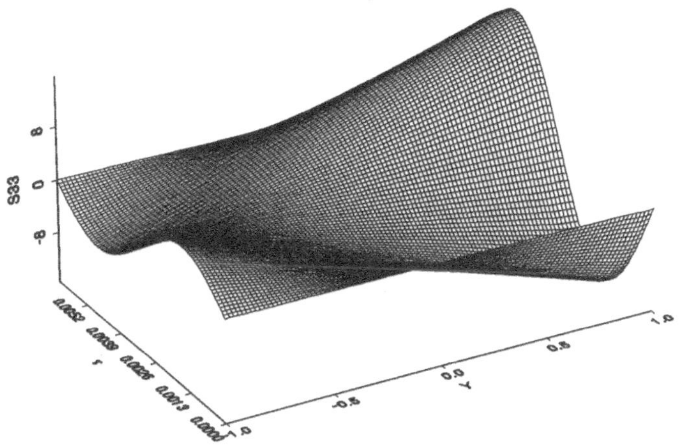

Figure 10a: S_{33} *versus* τ *and* Y *for* $\eta = \infty$, $m = 2$, $T = 1000$.

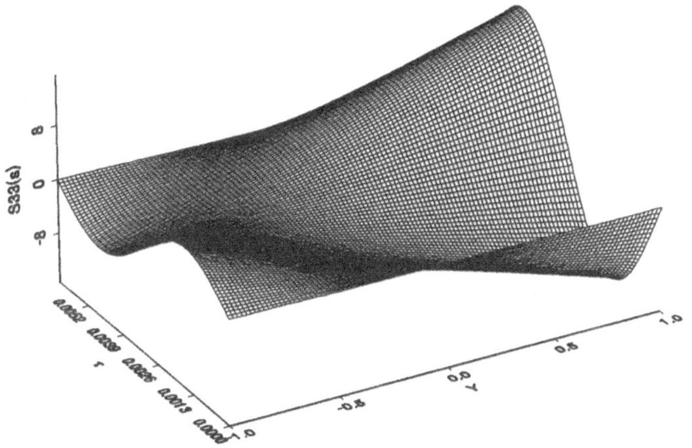

Figure 10b: $S_{33}^{(s)}$ *versus* τ *and* Y *for* $\eta = \infty$, $m = 2$, $T = 1000$.

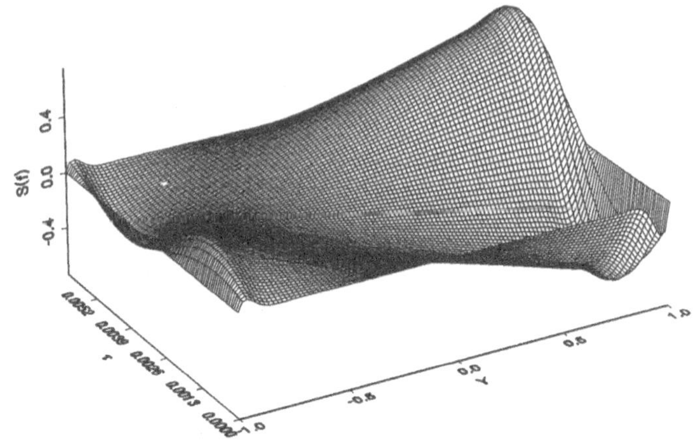

Figure 10c: $S^{(f)}$ *versus* τ *and* Y *for* $\eta = \infty$, $m = 2$, $T = 1000$.

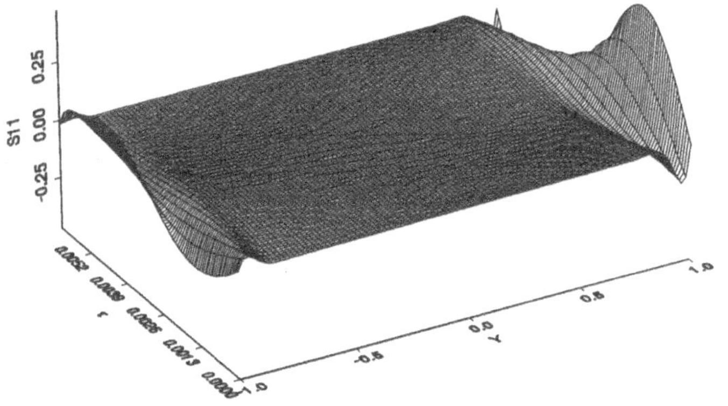

Figure 10d: S_{11} *versus* τ *and* Y *for* $\eta = \infty$, $m = 2$, $T = 1000$.

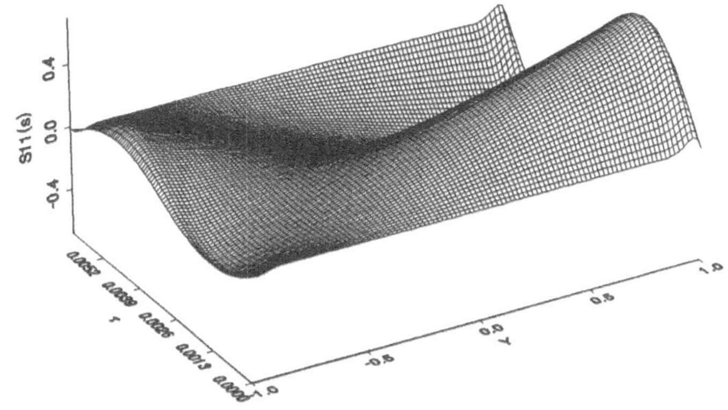

Figure 10e: $S_{11}^{(s)}$ *versus* τ *and* Y *for* $\eta = \infty$, $m = 2$, $T = 1000$.

7 Concluding remarks

The revised definition of the stress contribution factor of the pore fluid pressure in the axial total stress is an improvement over the old definition (Zhang and Cowin, 1994) in that it eliminates a nonphysical numerical artifact. The portion of the axial total stress supported by the pore fluid pressure depends on the ratio of the applied bending moment to the applied force, the loading frequency, and the boundary leakage condition. In the extreme case, the results are consistent with the idea that the total stress is partitioned between the fluid and solid constituents in proportion to their volume fraction.

For the poroelastic beam of bone under combined axial compression and bending, the pore fluid contribution factor in the axial total stress is always below and asymptotic to the value of the porosity. The stress-sharing behavior of the pore fluid pressure changes depending on how the external loading is composed, *i.e.*, whether it is dominated by axial compression or by bending.

The presence of the pore fluid and pore fluid pressure within a beam of bone doesn't have a significant influence on its load-bearing axial total stress. The solid matrix supports most of the axial total stress, which responds almost like a solid elastic beam. The transverse total stress is nonzero but its magnitude is at least one order of magnitude smaller than the axial total stress, and it is closely moderated by the pore fluid pressure. The pore fluid pressure driven by the external loading manifests amazingly rich variations as the loading condition and the leakage condition change, which has interesting implications to the cellular activities mediated by the bone fluid, *e.g.*, the bone remodeling mechanisms.

Acknowledgement: This research was supported by NASA Grant NAGW-3860.

References

Biot, M.A. (1941) General theory of three-dimensional consolidation. *J. Appl. Phys.*, **12**, 155-164.

Biot, M.A. (1956) Theory of propagation of elastic waves in fluid-saturated porous solid. I. Low-frequency range. *J. Acoust. Soc. Am.*, **28**, No.2, 168-178.

Biot, M.A. and Willis, D.G. (1957) The elastic coefficients of the theory of consolidation. *J. Appl. Mech.*, **24**, 594-601.

Nur, A. and Byerlee, J.D. (1971) An exact effective stress law for elastic deformation of rock with fluids, *J. Geophys. Res.*, **76**, No. 26, 6414-6419.

Rice, J.R. and Cleary, M.P. (1976) Some basic stress diffusion solutions for fluid-saturated elastic porous media with compressible constituents, *Reviews of Geophysics and Space physics*, **14**, 227-241.

Rudnicki, J.W. (1985) Effect of pore fluid diffusion on deformation and failure of rock, *Mechanics of Geomaterials*, ed. by Z. Bazant, 315-347, John Wiley & Sons Ltd..

Salzstein, R.A., Pollack, S.R., Mak A.F.T., and Petrov, N. (1987). Electromechanical potentials in cortical bone-I. A continuum approach. *J. Biomechanics*, **20**, No. 3, 261-270.

Weinbaum, S., Cowin, S.C., and Zeng, Y. (1993) Excitation of osteocytes by mechanical loading-induced bone fluid shear stresses, *J. Biomechanics*, **27**, 339-360.

Zhang, D. and Cowin S.C. (1994) Oscillatory bending of a poroelastic beam, *J. Mech. Phys. Solids*, **42**, No. 10, 1575-1599.

Predictions of the Swelling-Induced Pre-Stress in Articular Cartilage

Lori A Setton[1] , Weiyong Gu, W. Michael Lai, and Van C. Mow
Departments of Mechanical Engineering and Orthopaedic Surgery
Columbia University, New York NY 10032

Abstract
Due to the presence of fixed charges in articular cartilage, the tissue *in situ* is in a swollen state with its associated pre-stresses. In this paper, the triphasic theory, which models the tissue as a mixture of three phases (solid, fluid and ions), is used to predict the swelling-induced stresses and strains in a geometrically simplified (spherical layer) model of articular cartilage on the femoral or humoral head. The model predictions indicate that the residual stresses and strains depend strongly on the magnitude and distribution of the fixed charged density, matrix stiffness and the curvature of the layer of the tissue. This study indicates that specific models for each diarthrodial joint and cartilage layer must be developed in order to assess the relative importance of these three factors on influencing the residual-stress states within the tissue.

Key words: Fixed charge density, Pre-stress, Residual stress, Residual strain, Swelling, Triphasic

1. Introduction

The extracellular matrix of articular cartilage consists of a dense fibrillar network of collagen and proteoglycans, with fluid, mobile cations and anions contained within the pore structure. The major functional role of articular cartilage is to provide for load support and a smooth, nearly frictionless interface between the articulating bony surfaces within a diarthrodial joint. The ability of this tissue to provide for these functions *in situ* derives largely from its hydrated state, which in turn depends on the negatively-charged nature of the tissue. These negative charges (the sulfate and carboxyl groups on the proteoglycans) and the mobile counter-ions associated with them, give rise to a large swelling pressure which contributes to the compressive, tensile, electrokinetic and transport properties of the tissue (2,8,23,30). Furthermore, in the absence of externally applied loads, this swelling pressure must be balanced by stresses generated in the solid matrix of articular cartilage at equilibrium (17,22). Therefore, the solid matrix of articular cartilage, when equilibrated in an external bathing solution at normal physiological electrolyte concentrations, will experience a stress from the swelling pressure; this stress is often denoted as a residual stress or a pre-stress. While the functional role of this residual stress in the cartilage matrix is not known, it can be expected that the residual stress will play an important role in ensuring the viability of the

---

[1] Currently, Assistant Professor, Department of Biomedical Engineering, Duke University, Durham NC 27708-0281

P. S. Selvadurai (ed.), Mechanics of Poroelastic Media, 299–320.
1996 *Kluwer Academic Publishers.*

tissue for as many as seven or eight decades of joint functions. Furthermore, the swelling-induced residual stresses will contribute to the local mechanical environment of the chondrocyte, and so may be involved in regulation of the cartilage matrix through mechanical signal transduction mechanisms. In this study, a mechano-electrochemical constitutive formulation for charged-hydrated-soft tissues, the triphasic theory, is used to predict the magnitude and distribution of this residual stress in certain *in situ* models of articular cartilage.

2. Proteoglycans and the swelling pressure in articular cartilage

Glycosaminoglycans (chondroitin sulfates, keratan sulfates and hyaluronan) are the fundamental building block of the proteoglycan molecule in articular cartilage. Chondroitin sulfate has both carboxyl and sulfate groups, keratan sulfate has only the sulfate group, and hyaluronan has only the carboxyl group. Both carboxyl and sulfate groups become negatively charged when they are dissolved in solution, as it is the case within the tissue. Numerous chondroitin sulfates and carboxyl groups are covalentl bonded to a single protein core to form an aggrecan molecule. In normal cartilage, mos of the aggrecans are further linked to a single filament of hyaluronan to form larg proteoglycan aggregates of molecular weights ranging from 100-200 x 10^6 (13,28,34 The extremely large size and complex structure of the proteoglycan aggregate is require to immobilize and restrain the molecule within the extracellular matrix of articul cartilage, not only by entanglement with each other (10,31), but also by interactions wi the network formed by the collagen fibers and other quantitatively minor collagen type and glycoproteins (9,19). Therefore, the negative charges associated with th proteoglycans in articular cartilage are considered to be *fixed* charges (21,23). The densi of these aggrecan-associated charges is known as the fixed negative charge density (c and may be expressed as moles of charges, or "equivalents", per unit tissue volume weight. It may also be expressed as moles per unit tissue water volume or weight, whic may be based on either the quantity of total water or water extrafibrillar to the collag network (24,39).

Both carboxyl and sulfate groups are negatively charged and restrained within t tissue. Each negative charge group requires a positive charge on counter-ions (Na^+ and a lesser degree Ca^{++}) nearby to maintain electroneutrality within the interstitium. Wh bathed in a dilute ionic solution, because of the high c^F, the total ion concentration insi the tissue is much greater than the ion concentration of the external bathing soluti This imbalance of mobile ions gives rise to a pressure in the interstitial fluid which higher than the ambient pressure in the external bath; this is known as a Donnan osmo pressure (7). Furthermore, the very high concentration of proteoglycans in articu cartilage (over 50 mg/ml of interstitial water) implies the occurrence of strong interactions between the charge groups on the glycosaminoglycans (15,28). This electrical interaction causes charge-to-charge repulsive forces to occur between charge groups and between molecules. These ionic interactions are modulated by the counter-ions in the interstitial space of articular cartilage, and thus by the nature of the electrolytes (valency) and their ionic strengths within the external bathing solution. With increasing ion concentration, the equivalent Debye length between the counter-ion "cloud" and the fixed charge is decreased, so that the net charge-to-charge repulsive force is decreased (12). These repulsive forces will therefore, also contribute to the swelling of the extracellular matrix. The absence of both osmotic effects and charge-charge repulsive effects in articular cartilage at hypertonic concentrations is often referred to as a "charge-shielding" effect (8,25,30).

Both the composition and the structure of the proteoglycan molecule have also been found to vary with depth from the articular surface. The proteoglycan content is lowest in the superficial zone, and increases rapidly with depth to a maximum in the middle zone (5,20,23,40). There is also evidence of changes in the structure of the aggregate, as well as the glycosaminoglycan content with depth in the cartilage layer (32,29). For human femoral head cartilage, the variation in the c^F through the depth is shown in Figure 1.

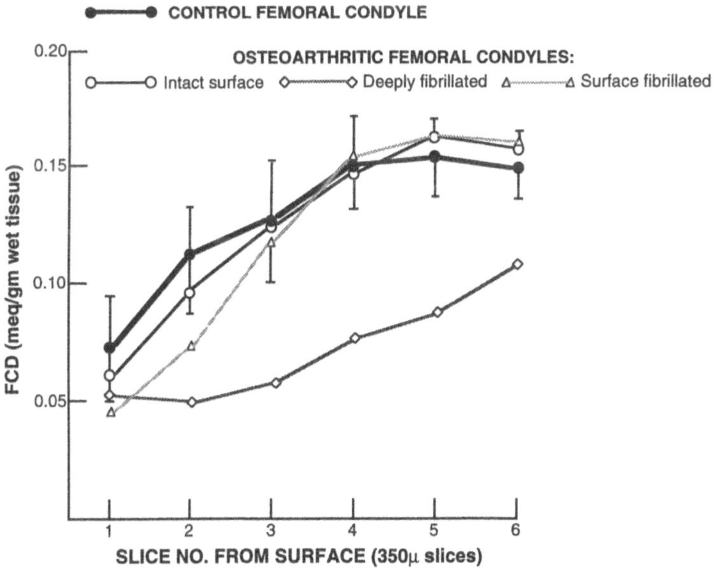

Figure 1. *Variation of the c^F through the thickness of articular cartilage from a normal and a fibrillated femoral head (reproduced with permission from Brocklehurst et al., 1984).*

This variation in the c^F gives rise to an inhomogeneous swelling pressure throughout the cartilage layer, and therefore, a variation in the propensity to swell. Indeed, data on dimensional free-swelling behavior of strips of articular cartilage varies with depth in the cartilage layer, so that the maximum swelling will occur in the deep zones of articular cartilage (30). The extent of swelling, however, is regulated by the intrinsic moduli for the material in addition to the c^F. Experimentally, the tensile modulus of the surface zone of the tissue has been found to be greater than those of the middle zone, which is in turn greater than that of the deep zone (1,35; Figure 2). This finding is related to the higher concentration of collagen at the articular surface (1,2,28), as well as the anisotropic organization of the collagen network (6). While there is clear evidence of inhomogeneities in both the tissue composition and material properties, the physiological significance for such inhomogeneities is not well understood. It may be generally concluded, however, that these spatial variations serve an important role in cartilage

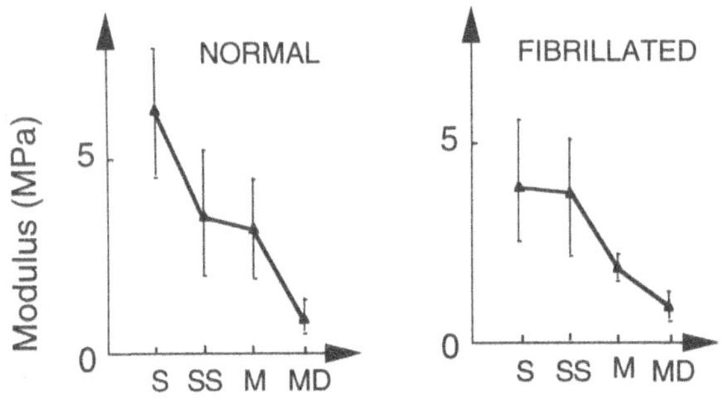

Figure 2. *(Above) Variation of tensile modulus with depth for human femoral head articular cartilage (S=surface zone, SS=sub-surface zone; M=middle zone, MD=middle/deep zone)(data from Akizuki et al., 1986).*

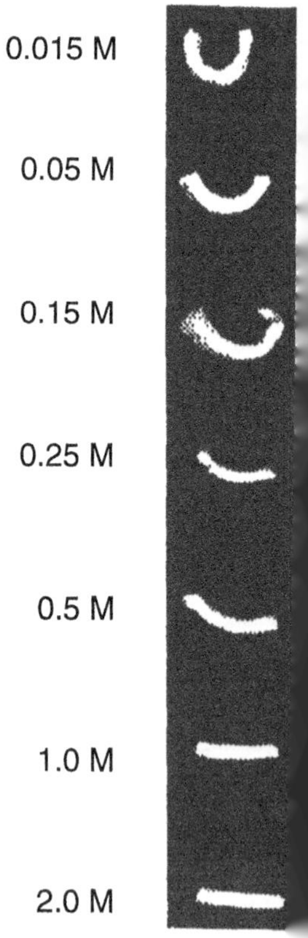

Figure 3. *(right) Strips of bovine articular cartilage from the medial femoral condyle after excision from the bone and soaked in various bathing solutions for thirty minutes.*

mechanics and matrix regulation based on data showing that these distributions will change in response to altered joint loading and osteoarthritis (1,5,16,18,19,23,40).

3. Zero stress-strain reference state for articular cartilage

For articular cartilage, it has been demonstrated that the unloaded state *in situ* does not correspond to the zero initial stress-strain state. For example, it is known that articular cartilage will "curl" and change its shape upon excision from the bone (14,30,37,38). As an example, the free-swelling behavior of strips of bovine articular cartilage after removal from the subchondral bone are shown in Figure 3 for bathing solutions of 0.015M-2M NaCl. Cutting the collagen attachments of cartilage to subchondral bone releases some "initial" stresses existing within the collagen-proteoglycan solid matrix *in situ.* We may efer to these initial stresses as a residual stress or a pre-stress. The "curling" behavior of hese excised specimens observed at physiological saline concentrations (0.15M NaCl) lemonstrates the existence of a state of pre-stress in cartilage *in situ.* Note that at a ypertonic concentration (e.g., 2.0M NaCl), all swelling effects are diminished due to shielding" of the matrix charges from ionic interactions so that there will be no ontribution of either the Donnan osmotic effect or charge-to-charge repulsive forces to he swelling pressure. This behavior is apparent in Figure 3 since the curling effect is iminished with increasing concentration of the bathing solution, so that at hypertonic aline the strip is virtually flat. Since the swelling pressure will be zero at hypertonic oncentrations, the hypertonic state serves as a convenient reference point for the easurement of swelling pressure and matrix deformations (8,17). In other words, only t the hypertonic salt concentration, and in the absence of applied loading, may the tissue e assumed to be stress-free. The existence of such a state of pre-stress under physiologic onditions has long been hypothesized (e.g., 11). In the current study, the non-hysiologic hypertonic state will be denoted as the initial zero stress-strain state; we efine this as the hypertonic reference configuration. All strains and stresses will be etermined relative to this reference configuration.

. The triphasic constitutive model for articular cartilage

order to describe the swelling arising from the charged nature of the matrix, it becomes cessary to utilize a constitutive model which incorporates the effects of *both* fixed arges and mobile charges within articular cartilage. In this study, the triphasic nstitutive model for articular cartilage (17) will be used to predict the residual or pre-ess resulting from the swelling pressure within cartilage *in situ.* The triphasic theory was developed based on the theory of mixtures for one incompressible solid phase and N-1 incompressible fluid phases (4). A brief summary of the triphasic theory is presented below.

4.1 Balance of linear momentum

Articular cartilage is modeled as a mixture of three phases: 1) an incompressible fluid phase (water); 2) an incompressible solid phase (the collagen-proteoglycan matrix); and 3) a ion phase consisting of two species of ions (cations and anions). For the purpose of this study, only univalent ions are considered (e.g., Na^+ and Cl^-). The volume fraction of each phase (dV^α) with respect to the total tissue volume (dV) is given by $\phi^\alpha = dV^\alpha/dV$, so that $\phi^s + \phi^w + \phi^+ + \phi^- = 1$, where s,w,+,- denote the solid, water, cation and anion phases, respectively. The quasi-static momentum equations for each constituent phase and species can be written in the following form:

water $$-\rho^w \nabla \mu^w + f_{ws}(\mathbf{v}^s - \mathbf{v}^w) + f_{w+}(\mathbf{v}^+ - \mathbf{v}^w) + f_{w-}(\mathbf{v}^- - \mathbf{v}^w) = \mathbf{0}, \quad [1$$

cation $$-\rho^+ \nabla \mu^+ + f_{+s}(\mathbf{v}^s - \mathbf{v}^+) + f_{+w}(\mathbf{v}^w - \mathbf{v}^+) + f_{+-}(\mathbf{v}^- - \mathbf{v}^+) + F_c(\rho^+/M_+)\mathbf{E}_e = \mathbf{0}, \quad [2$$

anion $$-\rho^- \nabla \mu^- + f_{-s}(\mathbf{v}^s - \mathbf{v}^-) + f_{-w}(\mathbf{v}^w - \mathbf{v}^-) + f_{-+}(\mathbf{v}^+ - \mathbf{v}^-) - F_c(\rho^-/M_-)\mathbf{E}_e = \mathbf{0}, \quad [3$$

tissue $$\text{div}\,\mathbf{T} = 0, \quad [4$$

where ρ^α and $\mathbf{v}^\alpha$ denote the density and the velocity, respectively, of the α-phas (α=s,w,+,-), μ^w, μ^+ and μ^- are the chemical potentials for the water phase, cation an anion species, respectively, F_c is the Faraday constant, M_+ and M_- are the atomic weigh of the cation and anion respectively, $\mathbf{E}_e$ is the electric field vector, and $\mathbf{T}$ is the stres tensor for the mixture. The terms $f_{\alpha\beta}$ denote the frictional coefficients per unit tissu volume between the interdiffusing α and β components and are assumed to be symmetri i.e.,$f_{\alpha\beta}=f_{\beta\alpha}$. The concentrations of the ionic species must obey the followin electroneutrality condition within articular cartilage, given by:

$$c^+ = c^- + c^F, \quad [$$

where c^F denotes the molar concentration (per unit tissue water volume) of the fix negative charges. Furthermore, each phase and ionic species (α = s,w,+,-) must obey t continuity equation, given by:

$$\partial \rho^\alpha/\partial t + \text{div}\,(\rho^\alpha \mathbf{v}^\alpha) = 0.$$

From equations [5] and [6] and the condition that c^F is conserved at all time during deformation of the tissue, an equation of continuity for the electric current density ($\mathbf{I}_e$) can be obtained:

$$\mathrm{div}\{\, F_c \phi^w [c^+(\mathbf{v}^+ - \mathbf{v}^s) - c^-(\mathbf{v}^- - \mathbf{v}^s)] \,\} = 0, \qquad [7]$$

where the expression within the bracket { } is $\mathbf{I}_e$. When $\mathbf{I}_e$ is zero, $\mathbf{v}^+$ and $\mathbf{v}^-$ are related so that cations and anions may be considered together as a single species of "neutral" salt with density $\rho^1 = \rho^+ + \rho^- - \rho^F$, where ρ^F is given by the relation, $\rho^F = \phi^w c^F M_+$. Under such a condition, and assuming negligible f_{+-}, equations [1], [2] and [3] can be reduced to the following two equations:

$$-\rho^w \nabla\mu^w + K_{sw}(\mathbf{v}^s - \mathbf{v}^w) + K_{w1}(\mathbf{v}^- - \mathbf{v}^w) = \mathbf{0}, \qquad [8]$$

$$-\rho^1 \nabla\mu^1 + K_{s1}(\mathbf{v}^s - \mathbf{v}^-) + K_{w1}(\mathbf{v}^w - \mathbf{v}^-) = \mathbf{0}, \qquad [9]$$

where $\mu^1 = (M_+\mu^+ + M_-\mu^-)/(M_+ + M_-)$ is the chemical potential of the neutral salt, and $K_{\alpha\beta}$ are related to c^α and $f_{\alpha\beta}$ (17). Thus, the system of equations [4], [8] and [9] defines the conservation of linear momentum in a triphasic material in the absence of electric current.

4.2 Constitutive formulation for linear triphasic theory

In the general triphasic theory, the constitutive relationships for the stresses ($\mathbf{T}^\alpha$) and the chemical potentials (μ^α) are derived from the Helmholtz energy density function A^α for each phase with appropriate thermodynamic considerations (17). The energy density, A^α, is assumed to be a function of the temperature (T), the right Cauchy-Green deformation tensor, $\mathbf{C}$ ($\mathbf{C}=\mathbf{F}^T\mathbf{F}$, where $\mathbf{F}$ is the deformation gradient), ρ^w, ρ^+, ρ^- and c^F. General forms of the constitutive relations for $\mathbf{T}^\alpha$ and μ^α in terms of T, $\mathbf{C}$, $\mathbf{F}$, ρ^α and c^F are given by Lai and co-workers (17).

In the linear formulation of the triphasic theory, the solid matrix is assumed to be isotropic and the strains are assumed to be infinitesimal as given by the infinitesimal strain tensor, $\mathbf{E}$; this strain is defined relative to the hypertonic reference configuration. The constitutive relations for the total mixture stress $\mathbf{T}$ and the fluid stress $\mathbf{T}^f$ are given as:

$$\mathbf{T} = -(p + T_c)\mathbf{I} + \sigma, \qquad [10a]$$

where

$$\sigma = \lambda_s \,\mathrm{tr}(\mathbf{E})\mathbf{I} + 2\mu_s \mathbf{E}, \qquad [10b]$$

and

$$\mathbf{T}^f = -\,\phi^f\, p\, \mathbf{I}. \qquad [11]$$

In the above equations, σ is the elastic stress tensor, p is the interstitial fluid pressure, T_c is a pressure term to be described below, and λ_s and μ_s are the two Lamé coefficients for a triphasic material.

This linear formulation also leads to the following forms of the chemical potentials for the interstitial water and the neutral salt:

$$\mu^w = \frac{1}{\rho_T^w}\, p + R_w + \frac{B_w}{\rho_T^w}\, \mathrm{tr}\, \mathbf{E}, \qquad [12$$

$$\mu^1 = R_1 + B_1\, \mathrm{tr}\, \mathbf{E}, \qquad [13$$

where R_w, R_1, B_w, and B_1 depend on ρ^α and c^F and ρ_T^w is the true density of water. In th absence of a solid matrix, the above equations must reduce to those giving the chemica potentials for a charged macromolecular solution. Thus, the chemical potentials for th neutral salt and the interstitial water in a triphasic medium are chosen to be:

$$\mu^1 = \mu_o^1 + \frac{RT}{M_1} \ln\,[\gamma_\pm^2 c(c+c^F)], \qquad [1$$

and

$$\mu^w = \mu_o^w + \frac{1}{\rho_T^w}\,[p - 2RT\phi(2c+c^F) + B_w\, \mathrm{tr}\, \mathbf{E}], \qquad [1$$

where $\gamma_\pm^2$ is known as the mean activity coefficient, ϕ is the osmotic coefficient, and B is an "inter-phase" coupling coefficient, R is the universal gas constant, and T is th absolute temperature. The term B_w tr**E** may be considered to incorporate some of th known "excluded volume effects" in articular cartilage due to the unavailability of portion of water contained in the collagen fibrils to proteoglycans and solute Furthermore, there is an implicit dependence of μ^α on the dilatation tr(**E**) through t dependence of the fixed negative charge density on the tissue water volume given by:

$$c^F = c_o^F\,(1 - \mathrm{tr}(\mathbf{E})/\phi_o^w), \qquad [1$$

where c_0^F and ϕ_0^w are the fixed charge density and tissue porosity defined at the hypertonic reference state. For inhomogeneous materials, both of these quantities may be assumed to vary with position. The variation of the tissue porosity with its deformation is given by:

$$\phi^w = [1 - \phi_0^w] \cdot \mathrm{tr}(\mathbf{E}) + \phi_0^w, \qquad [17]$$

The external NaCl bathing solution can be considered to be a reduced triphasic medium, where there is no solid component and c^F. The chemical potentials for the solute (NaCl) and solvent (H_2O) thus are given by

$$\mu^{1*} = \mu_0^{1*} + \frac{RT}{M_1} \ln(\gamma_\pm^{*2} c^{*2}), \qquad [18]$$

and

$$\mu^{w*} = \mu_0^{w*} + \frac{1}{\rho_T^w}(p^* - 2RT\phi^* c^*). \qquad [19]$$

Finally, a constitutive form for the term, T_c, in equation [10] must be considered. Equation [10] states that the total stress tensor (**T**) is given by a swelling pressure, $P_{sw} \equiv$ +T_c, and an elastic stress ($\sigma = \lambda_s \mathrm{tr}(\mathbf{E}) + 2\mu_s \mathbf{E}$). The term, $T_c\mathbf{I}$, is defined as the chemical-expansion stress in the triphasic theory. A specific constitutive form for T_c was chosen based on the experimental observation that the swelling pressure in articular cartilage has an exponential dependence on external ion concentration of the external bath (3,27), and that the chemical expansion effect must be negligible at the hypertonic reference configuration (17):

$$T_c(c,c^F) = a_0 c^F \exp[-\kappa\,(\gamma_\pm/\gamma_\pm^*)\,(c(c+c^F))^{1/2}] - P_\infty. \qquad [20]$$

We note that at equilibrium, the molar concentration of the neutral salt c in articular cartilage is related to that in the bath c^* by the Donnan equilibrium ion distribution (7):

$$c^{*2} = c(c+c^F)(\gamma_\pm/\gamma_\pm^*)^2. \qquad [21]$$

Thus, at equilibrium, equation [20] describes the exponential decay of T_c with c^*. The constants, a_0 and κ have been determined to be $a_0 = 2.5$ MPa/(Eq/l) and $\kappa = 7.5$ (mole/l)$^{-1}$ for a sample of articular cartilage (17); and P_∞ is usually negligible (see below).

In summary, the constitutive equations for an isotropic triphasic material under infinitesimal deformation are given in equations [10], [11], [14], and [15]. These equations, together with the momentum equations [4], [8] and [9] and the continuit equations [6] defines a well posed problem from which one may obtain the internal stres strain, swelling pressure, and ion concentrations within a charged-hydrated-soft tissue fo given boundary/initial conditions. Note that both c^F and ϕ^w will change with dilatatio as given by equations [16] and [17], since they are related to the tissue water volum contained within the material. As a consequence, although the constitutive equation fo the total stress (**T**) is linear with respect to the strain tensor (**E**), the effects of io concentration on the predicted swelling behavior is highly nonlinear.

5. One-dimensional formulation of the *in vitro* equilibrium swelling in charged-hydrated-soft tissue

With the triphasic theory it is now possible to predict the swelling-induced stresses ar strains in articular cartilage *in situ*, using data for the physicochemical and materi coefficients available in the literature. In 1991, the first of such problems was solved b Lai and co-workers to describe the one-dimensional free-swelling behavior of articul cartilage (17). In that problem, a cylindrical sample is confined radially so that there only one component of strain (ε_{zz}); initially, the sample is equilibrated in a NaCl soluti of concentration c*. At physiological salt concentration c* (e.g., 0.15M NaCl), the tiss swells relative to the hypertonic reference state, inducing a residual stress. T appropriate boundary condition for this problem is that the chemical potential of ea phase must be continuous across the surface; also the total stress acting on each surfa must be zero (i.e., the free swelling condition). Therefore, the following bounda conditions must be satisfied at the free surface:

$$T_{zz}=0, \quad \mu^{1} = \mu^{1*} \text{ and } \mu^{w} = \mu^{w*}. \qquad [$$

The chemical potentials for the internal and external NaCl solutions are given equations [14] and [15], and equations [18] and [19] respectively. With the velocities being zero, the solution of equations [4], [8] and [9] for the above boundary conditi easily yields the following equations for the concentration of NaCl (c), interstitial fl pressure (p) and elastic stress (σ) within the material:

$$c(c+c^F) = (\gamma_{\pm}^{*}/\gamma_{\pm})^2\, c^{*2}, \qquad [$$

$$p = RT[\phi(2c+c^F) - 2\phi^{*}c^{*}] + P_{\infty} - B_w\, \mathrm{tr}(\mathbf{E}), \qquad [$$

$$\sigma = P_{sw}\mathbf{I}, \quad \text{or} \quad \sigma_{zz} = p + T_c.$$

Equation [23] is the well-known Donnan equilibrium ion distribution (i.e., equation [21]), and is recovered from the boundary condition that $\mu^1 = \mu^{1*}$ and the assumption that $\mu_0^1 = \mu_0^{1*}$. In equation [24], $P_\infty = -\rho_T^w(\mu_0^w - \mu_0^{w*})$; it represents the osmotic pressure resulting from differences in the reference chemical potential of water in free solution and in articular cartilage; it represents the colligative effect of the proteoglycans molecules which can generally be neglected because of their relatively large molecular size.

For this free-swelling problem, equation [25] states that the swelling pressure generated in articular cartilage, P_{sw}, must be balanced by the elastic matrix stress, $\sigma_{zz}=(\lambda_s+2\mu_s)\varepsilon_{zz}$. Equation [24] states that there are three sources contributing to the fluid pressure p in the tissue: 1) the Donnan osmotic pressure, $p_D = RT\Delta$ where Δ is the effect of the difference in ion distribution between external and internal environments and is given as $\Delta = \phi(2c+c^F)-2\phi^*c^*$; 2) contribution to osmotic pressure due to the colligative properties of the proteoglycans, denoted as P_∞; and 3) a pressure due to the deformation-induced "excluded volume effect" (23) given by $B_w\varepsilon_{zz}$. Note that the total mixture stress, $T = -P_{sw}\mathbf{I} + \boldsymbol{\sigma}$, is zero in this configuration while the elastic stress tensor $\boldsymbol{\sigma}$, representing the residual stress in the extracellular matrix, is not. In this one-dimensional problem, the zeroth order solution for the only non-zero swelling-induced residual tensile strain is given by the ratio of the swelling pressure P_{sw} to the one-dimensional aggregate modulus, $H_A \equiv \lambda_s + 2\mu_s$. Similarly, in any free-swelling configuration, one can obtain the dilatation due to swelling is given by the ratio of the swelling pressure P_{sw} to the bulk modulus of the solid matrix, $B_s \equiv \lambda_s + (2/3)\mu_s$.

. Swelling-induced residual stress in articular cartilage *in situ*

In this section, the linear triphasic theory will be applied to another one-dimensional equilibrium swelling problem in order to study the effect of geometry, material inhomogeneity as well as the fixed charge density on the swelling-induced residual stresses and strains in articular cartilage *in situ*. Consider the schematic for the proximal end of a humerus or femur shown in Figure 4. The articular cartilage layer may be modeled as a single layer upon a rigid substrate representing the subchondral bone. For cartilage at a distance from the joint margins, it is assumed that the subchondral bone surface may be approximated as a sphere of radius R_1, and the cartilage surface may be approximated by a concentric sphere of radius R_0. A similar model may be applied to nearly cylindrical articulations, such as the femoral condyle or patellofemoral groove (6). In these models for articular cartilage, it may be assumed that the deformation of cartilage at the cartilage-bone interface is zero due to the anchoring of collagen fibers into the tidemark (33). For a joint surface equilibrated in bathing solution of known concentration and in the absence of externally applied loads, the boundary conditions at the free surfaces are the same as those of the one-dimensional problem presented in the previous section.

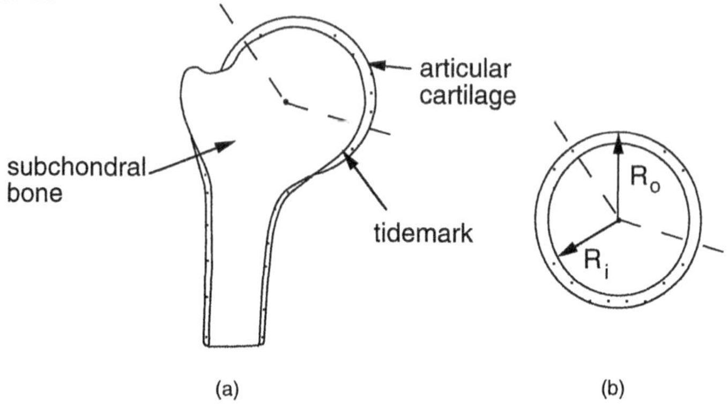

Figure 4. *(a)Schematic of the proximal end of a humerus or femur showing the subchondral bone and cartilage layer. (b) Diagram of the spherical model representing the femoral or humoral head. R_i= radius of subchondral bone; R_o=outer radius of cartilage-bone complex.*

For the spherical symmetry problem shown in Figure 4, all parameters are assumed to vary with radial position (R) only. Let u be the only nonzero radial component of displacement, then, there are only two nonzero components of the strain tensor, **E**: radial strain ε_R = du/dR and a tangential strain, ε_t ($\equiv \varepsilon_{\phi\phi} = \varepsilon_{\theta\theta}$) = u/R . The nonzero elastic stress components are given by $\sigma_R = \lambda_s e + 2\mu_s(du/dR)$, $\sigma_t \equiv (\sigma_{\theta\theta} = \sigma_{\phi\phi})$ = $\lambda_s e + 2\mu_s(u/R)$, where e= du/dR+2u/R.

At equilibrium, there is no relative fluid flux, therefore, the momentum equations [4 [8] and [9] become, in spherical coordinates,

$$\frac{d(H_A e)}{dR} = \frac{dP_{sw}}{dR} \qquad [26a]$$

$$\mu^w = \text{constant} \quad \text{and} \quad \mu^1 = \text{constant}, \qquad [26b,$$

where $H_A = \lambda_s + 2\mu_s$. In arriving at the above equation, we have allowed for the variation of the H_A with depth, whereas the shear modulus μ_s is assumed to be a constant. Using the boundary conditions $\mu^w = \mu^{w*}$ and $\mu^1 = \mu^{1*}$ on the boundary $R=R_o$, equations [26b,c] lead to solutions that are identical to those given in equations [23] and [24]. By combining these two equations, the pressure, p, can be expressed in terms of the fixed charge density (c^F) and external ion concentration (c^*):

$$p = RT\{ \phi[(c^F)^2+(2\gamma^*_{\pm}c^*/\gamma_{\pm})^2]^{1/2} - 2\phi^*c^*\} - B_w \,\mathrm{tr}(\mathbf{E}). \qquad [27]$$

In the following, we shall first present the solution for the case where the material is a layer of homogeneous medium followed by the solution for a non-homogeneous medium.

6.1 Homogenous material

For a compositionally homogeneous material, H_A=constant and $c^F_0(R) = c^F_0$ = constant, so that the governing equation Eq.[26a] reduces to de/dR=0. The boundary conditions are: at R=0, u=0 and at $R=R_0$, $-P_{sw}+\sigma_R = 0$. Thus, the solution of the free-swelling problem of a spherical layer of articular cartilage attached to a rigid substrate is obtained as,

strain $$\varepsilon_R = \frac{e}{3}\{1 + \frac{2R_1^3}{R^3}\} \quad \text{and} \quad \varepsilon_t = \frac{e}{3}\{1 - \frac{R_1^3}{R^3}\}, \qquad [28a]$$

dilatation $$e = P^{(o)}\{B_s - P^{(1)} + \frac{4\mu_s}{3}\frac{R_1^3}{R_o^3}\}^{-1}. \qquad [28b]$$

where

$$P^{(o)} = RT\{ \phi[(c^F_0)^2+(2\gamma^*_{\pm}c^*/\gamma_{\pm})^2]^{1/2} - 2\phi^*c^*\} + a_0c^F_0 \exp(-\kappa c^*). \qquad [28c]$$

$$P^{(1)} = -\frac{RT\phi(c^F_0)^2}{\phi^w_0}\{(c^F_0)^2+(2\gamma^*_{\pm}c^*/\gamma_{\pm})^2\}^{-1/2} - B_w - \frac{a_0c^F_0}{\phi^w_0}\exp(-\kappa c^*). \qquad [28d]$$

n arriving at equations [28] we have expanded P_{sw} as $P_{sw}=P^{(0)}+P^{(1)}e$. This expansion is onsistent with the infinitesimal triphasic theory where e is assumed to be infinitesimal. Note that the dilatation, e, is constant throughout the thickness of the tissue and ependent only on the fixed charge density (c^F), the geometric dimensions (R_0, R_1), the welling parameters (a_0, κ, B_w) and the material parameters (B_s and μ_s), where the bulk nodulus is defined as $B_s = \lambda_s+2\mu_s/3$.

For comparison purposes, a set of material parameters are introduced based on vailable data on normal articular cartilage: c^F_0 = 0.2 Eq/L, c*=0.15 M NaCl (except for igure 6), H_A=1 MPa, ν_s=0.17, B_w=0 MPa, a_0=2.5 MPa/(Eq/L), κ=7.5(mole/L)$^{-1}$, $\phi/\phi^* =$ $_{\pm}/\gamma^*_{\pm}$ = 1, R_1=10 mm, h=1 mm (3,17,21,23,26). The solution for the radial and ngential components of the swelling-induced residual strain (ε_R and ε_t) are presented in igure 5a. Here, the normalized radius is defined by $(R-R_1)/h$, where h is the thickness of e layer R_0-R_1, and the strains are plotted for material of four different values of c^F_0. The maximum value for radial strain occurs at the cartilage-bone interface, while the maximum value for tangential strain occurs at the cartilage surface. In general, the value of c^F_0 is seen to have a significant effect on the magnitude of both strain components, although the dependence of ε_R on R is relatively unchanged (i.e., constant slope). For all values of c^F_0, the value for ε_R is an order of magnitude larger than ε_t.

A plot of the radial and tangential stress components is shown in Figure 5b. The magnitude of c^F_0 significantly affects the magnitude of both radial and tangential stress components, although the distribution through the tissue thickness is relatively unaffected. In Figure 6 are plotted the dilatation and dimensionless swelling pressure against ion concentration of the external bath. The magnitude and decay of the predicted swelling pressure is similar to that seen experimentally (8) and is also sensitive to c^F_0.

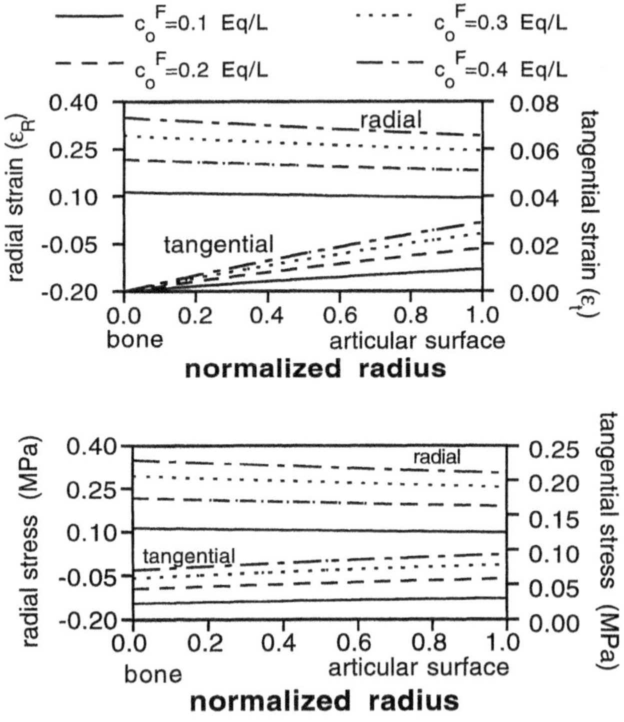

Figure 5. *Distribution of (a)radial and tangential strain (top) and (b)radial and tangential stre (bottom) in a spherical layer of articular cartilage for various reference fixed negative char density (c^F_0). The radial direction is perpendicular to the articular surface. Normalized radius defined as $(R-R_i)/h$, where h is the thickness of the layer.*

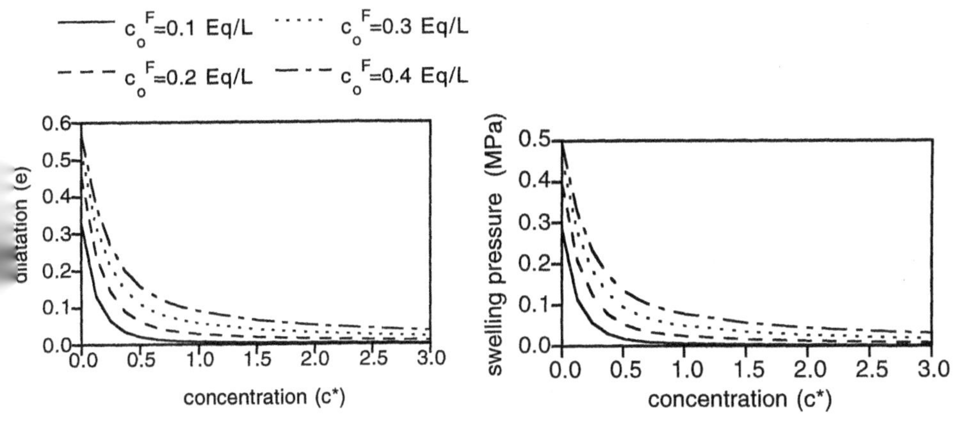

gure 6. *Dilatation at the cartilage-bone interface ($R=R_i$) and swelling pressure in a spherical yer of articular cartilage as a function of external NaCl concentration (c^*) and for various ference fixed negative charged density (c_o^F).*

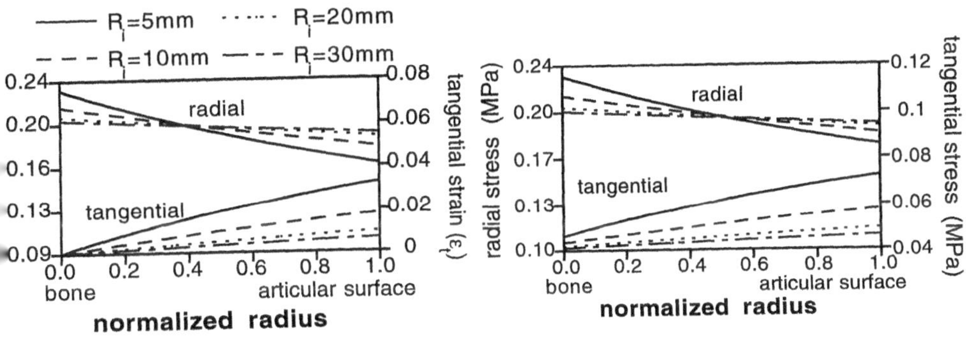

ure 7. *Distribution of radial and tangential strain and stress in a spherical layer of articular tilage for various radius R_i of the subchondral bone. Thickness of layer h = 1 mm.*

To study the curvature effect, values for the radial and tangential strains are plotted in ure 7 for various values of R_1 for a 1 mm thickness of cartilage layer. As the curvature he joint increases (i.e., decreasing R_1), the sensitivity of both strain components to al position increases. In contrast, there is little evidence of a change in either the magnitude or the distribution of either strain component for values of $R_1 > 20$ mm (with h=1 mm). Of interest is the observation that the geometry has relatively little effect on the magnitude of the maximum radial strain (at $R=R_1$), but a very significant effect on the maximum tangential strain (at $R=R_0$). Apparently, the curvature of the joint is a very important determinant of the tangential component of the residual strain. The radial and tangential stress components display similar trends with both radial position and joint geometry.

6.2 Solution for non-homogenous cases

For the non-homogeneous case, either c^F or H_A or both may vary with R.

With

$$P_{sw} = P^{(o)} + P^{(1)}e \quad [29]$$

where $P^{(0)}$ and $P^{(1)}$ are given earlier in equations [28c] and [28d], equation [26a] can be integrated to obtain

$$e = [P^{(o)} + C_1]/[H_A - P^{(1)}] \quad [3$$

and

$$u = \frac{1}{R^2} \int_{R_1}^{R} x^2\{(P^{(o)}+C_1)/(H_A-P^{(1)})\}dx + C_2/R^2 \quad [3$$

where C_1 and C_2 are integration constants. Using the boundary conditions that u=0 R=R_1 and $-P_{sw}+\sigma_R=0$ at R=R_0, these constants are determined to be:

$$C_1 = \mu_s\alpha/(1 - \beta) \text{ and } C_2 = 0 \quad [3$$

where

$$\alpha = \frac{4}{R_0^3} \int_{R_1}^{R_0} P^{(o)}R^2/(H_A-P^{(1)})dR, \quad [3$$

and

$$\beta = \frac{4\mu_s}{R_0^3} \int_{R_1}^{R_0} R^2/(H_A-P^{(1)})dR, \quad [3$$

6.3 Non-homogeneous material properties: Parabolic distribution of C_o^F

In Figure 8 are plotted two parabolic distributions for c_o^F. The equation for the parabolas are given by

$$c_o^F = c_o[1+\varepsilon(bx+cx^2)] \qquad [33]$$

where $x = (R-R_1)/h$, $b= 12.7$, $c= -20$, $c_o = 0.3$ Eq/liter and $\varepsilon= 0.05$ and 0.10. These

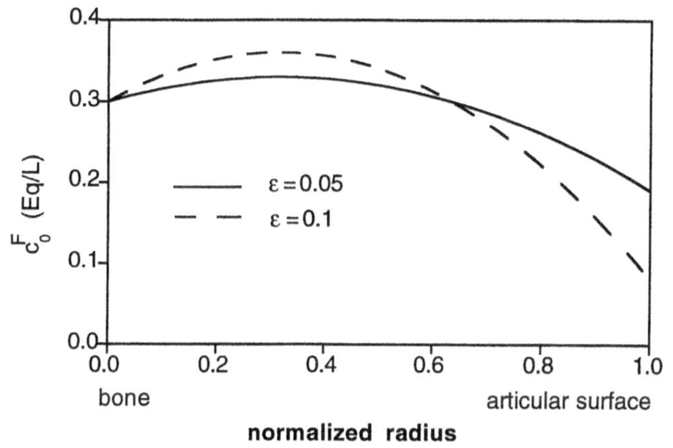

igure 8. *Parabolic distributions for reference fixed charge density c_o^F.*

arameter values are chosen to approximate the fixed charge density distribution bserved in human hip and femoral condylar cartilage (5). We note that the maximum alue for the fixed charge density is located near the bone-cartilage interface and the west value is located at the articular surface.

Using the solution given in the previous section, one can easily calculate the strain d the stress field. The stress distribution is given in Figure 9. We see that the radial ress distribution has more or less the same pattern as that of the c^F distribution, whereas e tangential stress remains fairly constant throughout the thickness. Again, as is in the se of homogeneous material, the radial stresses are much larger than the tangential esses.

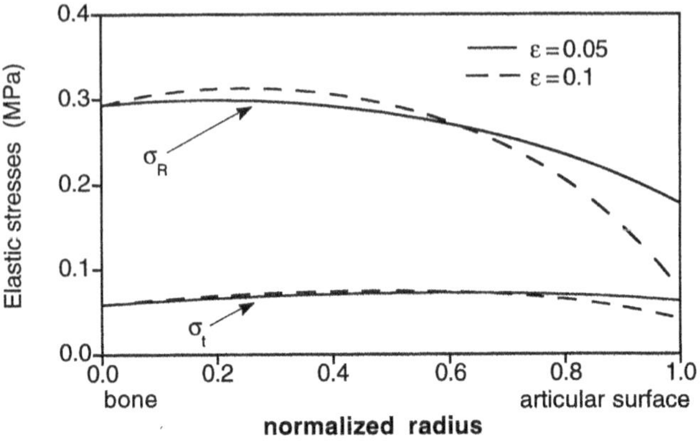

Figure 9. *Radial and tangential stress distributions in a spherical layer of articular cartilage wit parabolic distribution of* c_o^F.

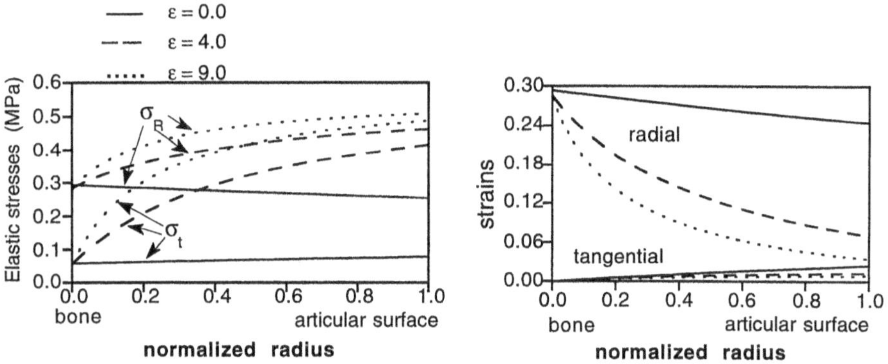

Figure 10. Radial and tangential strain and stress distributions in a spherical layer of articul cartilage with a linear variation of H_A.

6.4 Non-homogeneous material properties: Linear variation in material stiffness

The next case is that of a linear variation in the material stiffness of articular cartila with radial position, i.e., $H_A(x) = H_{Ao}(1 + \varepsilon x)$, where $x = (R-R_1)/h$ and H_{Ao} represents t value for equilibrium aggregate modulus ($H_{Ao} \equiv \lambda_o + 2\mu_o$) at the bone. Two cases fo linear variation in H_A were examined: In one case, H_A varies from 1 MPa to 5 MPa (ε=4) and in the other it varies from 1 MPa to 10 MPa (ε=9). The results for the strain and stress distribution are shown in Figure 10. In this figure, ε=0 corresponds to the case of a uniform H_A. We see that the effect of increasing the magnitude of H_A at the articular surface is an overall decrease in the strain components. Thus, the swelling of the tissue is considerably reduced in these cases. Associated with this decrease of strain components, the corresponding stress components become higher, particularly for the tangential component.

7. Discussion and Conclusions

We note that all components of swelling-induced residual strain and stress predicted in these canonical problems were tensile. It is known that compressive loading of articular joints will produce compressive stresses *in situ*, which may be of the same order of magnitude as the tensile residual stresses presented here. This observation lends support to the hypothesis that the existence of pre-stresses *in situ* may lead to reductions in the magnitude of matrix stresses during joint function, so that the cartilage matrix functions at a nearly zero stress state. The model predictions indicate that the maximum value for radial strain occurs at the cartilage-bone interface, while the maximum value for tangential strain occurs at the articular surface. These predictions are based on an isotropic model of the tissue. For the tissue *in situ*, it is known that at the cartilage-bone junction, most fibers are oriented perpendicular to the interface (i.e., in the radial direction) whereas on the cartilage surface they are predominantly parallel to the articular surface (i.e., in the tangential direction) (19). It is possible that these naturally occurring anisotropic structural arrangements (a feature which is common for all types of animal articular cartilage) will tend to make the strain components more uniform than those predicted here. Therefore, the existence of swelling-induced residual stresses and strains may play a role in determining the collagen ultrastructural organization of articular cartilage. The results of the present study also show a strong dependence of residual strain and stress on joint geometry, thereby suggesting that the variations in surface topography in normal articular cartilage (3) will be associated with different pre-stress environments. This pre-stress pattern will affect the *in vivo* state of stress within articular cartilage during joint motion.

In conclusion, using relatively simple models, we have calculated the swelling induced residual stress and strain distribution in a layer of cartilage. It is seen that the magnitude and the distribution of the swelling-induced stress and strain components depend strongly on the amount and the distribution of the fixed charge density, the material stiffness, and the curvature of the layer of the tissue. Further studies, both theoretical and experimental, directed at the differences in ultrastructural organization, composition and swelling behavior are needed to address the hypothesis that the swelling-induced residual strain may affect the structural organization of the cartilage layer *in situ*, cellular metabolic activities *in vivo*, and states of stresses and strains within the tissue layer during physiologic function.

Acknowledgment This research was supported in part by NIH grant AR41913.

References

1. Akizuki, S., Mow, V.C., Muller, F., Pita, J.C., Howell, D.S., Manicourt, D.H. (1986) Tensile propertie of knee joint cartilage. Part I- Influence of ionic condition, weight bearing, and fibrillation on th tensile modulus. J. Orthop. Res. 4:379-392.

2. Akizuki, S., Mow, V.C., Muller, F., Pita, J.C., Howell, D.S. (1987) The tensile properties of human kne joint cartilage. Part II- The influence of weight bearing, and tissue pathology on the kinetics o swelling. J. Orthop. Res. 5:173-186.

3. Ateshian, G. A., Soslowsky, L.J., Mow, V.C. (1991) Quantitation of articular surface topography an cartilage thickness in knee joints using stereophotogrammetry. J. Biomechanics 24:761-776.

4. Bowen, R.M. (1980) Incompressible porous media models by use of the theory of mixtures. Intl. Engng. Sci. 18:1129-1148.

5. Brocklehurst, R., Bayliss, M.T., Maroudas, A., Coysh, H.L., Freeman, M.A.R., Revell, P.A., Ali, S. (1984) The composition of normal and osteoarthritic articular cartilage from human knee joints. Bone Joint Surg. 66A:95-106.

6. Clarke, I.C. (1971) Articular cartilage: A review and electron microscope study. J. Bone Joint Sur 53B:732-750.

7. Donnan, F.G. (1924) The theory of membrane equilibria. Chem. Rev. 1:73-90.

8. Eisenberg, S.R., Grodzinsky, A.J. (1985) Swelling of articular cartilage and other connective tissue Electromechanochemical forces. J. Orthop. Res. 3:148-159.

9. Eyre, D.R., Apon, S., Wu, J.-J., Ericsson, L.H., Walsh, K.A. (1987) Collagen type IX: evidence f covalent linkages to type II collagen in cartilage. FEBS Letters 320:337-341.

10. Fessler, J.H. (1957) Water and mucopolysaccharide as structural components of connective tiss Nature 179:426-427.

11. Fung, Y.C. (1984) What principle governs the stress distribution in living organs? Biomechanics China, Japan and USA. Y.C. Fung, E. Fukada (eds). Science Press, pp.1-13.

12. Gabler,R. (1978) Electrical interactions in molecular biophysics, an Introduction. Academic Press.

13. Hardingham, T.E., Muir, H. (1975) Structure and stability of proteoglycan aggregates. Ann. Rhe Dis. 34(suppl.): 2,26.

14. Harris, S., Pan, Y., Peterson, R., Stal, S., Spira, M. (1993) Cartilage warping: An experimental model. Plast.Reconstructive Surg. 5:912-915.

15. Hascall, V.C., Hascall, G.K. (1981) Proteoglycans. In: Cell Biology of Extracellular Matrix. E.D. Hay (ed). Plenum: pp.39-60.

16. Helminen, H.J., Jurvelin, J., Kiviranta, I., Paukkonen, K., Saamanen, A.-M., Tammi, M. (1987) Joint loading effects on articular cartilage: A historical review. In: Joint loading. H.J. Helminen, I. Kiviranta, M. Tammi, A.-M. Saamanen, K. Paukkonen, J. Jurvelin (eds). Bristol, England, Wright: pp.1-46.

17. Lai, W.M., Hou, J.S., Mow, V.C. (1991) A triphasic theory for the swelling and deformation behaviors of articular cartilage. J. Biomech. Engng. ASME, 113:245-258.

8. Mankin, H.J., Brandt, K.D. (1992) Biochemistry and metabolism of cartilage in osteoarthritis. In: Osteoarthritis: Diagnosis and Medical/Surgical Management, 2nd edition. R.W. Moskowitz, D.S. Howell, V.M. Goldberg, H.J. Mankin (eds). Philadelphia, Pa., W.B. Saunders Co: pp.109-154.

9. Mankin, H.J., Mow, V.C., Buckwalter, J.A., Ianotti, J.P., Ratcliffe, A. (1994) Structure and function of articular cartilage. In: Orthopaedic Basic Science, S.R. Simon (ed). Rosemont, Il., Amer. Acad. Orthop. Surg. Pub., pp.1-44

0. Maroudas, A., Evans, H., Almeida, L. (1973) Cartilage of the hip joint: Topographical variation of glycosaminoglycan content in normal and fibrillated tissue. Ann. Rheum. Dis. 32:1-9.

1. Maroudas, A. (1975) Biophysical chemistry of cartilaginous tissues with special reference to solute and fluid transport. Biorheol. 12:233-248.

2. Maroudas, A. (1976) Swelling pressure versus collagen tension in normal and degenerate articular cartilage. Nature 260:808.

3. Maroudas, A. (1979) Physicochemical properties of articular cartilage. In: Adult Articular Cartilage. M.A.R. Freeman (ed). Pitman Medical, Kent UK: pp.215-290.

4. Maroudas, A., Wachtel, E., Grushko, G., Katz, E.P., Weinberg, P. (1991) The effect of osmotic and mechanical pressure on water partitioning in articular cartilage. Biochim. Biophys. Acta. 1073:285-294.

. Mow, V.C., Holmes, M.H., Lai, W.M. (1984) Fluid transport and mechanical properties of articular cartilage: A review. J. Biomechanics 17:377-394.

. Mow, V.C., Kuei, S.C., Lai, W.M., Armstrong, C.G. (1980) Biphasic creep and stress relaxation of articular cartilage in compression: Theory and experiments. J.Biomech.Engng. 102:73-84.

Mow, V.C., Schoonbeck, J.M. (1984) Contribution of Donnan osmotic pressure towards the biphasic compressive modulus of articular cartilage. Trans. Orthop. Res. Soc. 9:262.

Muir, H. (1983) Proteoglycans as organizers of the extracellular matrix. Biochem. Soc. Trans. 11:613-622.

Müller, F.J., Pita, J.C., Manicourt, D.H., Malinin, T.I., Schoonbeck, J.M., Mow, V.C. (1989) Centrifugal characterization of proteoglycans from various depth layers and weight-bearing areas of normal and abnormal human articular cartilage. J. Orthop. Res. 7:326-334.

30. Myers, E.R., Lai, W.M., Mow, V.C. (1984) A continuum theory and an experiment for the ion-induce swelling behavior of articular cartilage. J. Biomech. Engng. ASME, 106:151-158.

31. Ogston, A.G. (1970) The biological functions of the glycosaminoglycans. In: Chemistry and Molecul Biology of the Intercellular Matrix. E.A. Balazs (ed). Academic Press, London: pp.1231-1240.

32. Ratcliffe, A., Fryer, P.R., Hardingham, T.E. (1984) The distribution of aggregating proteoglycans i articular cartilage: Comparison of quantitative immunoelectron microscopy with radioimmunoassa and biochemical analysis. J. Histochem. Cytochem. 32:193-201.

33. Redler, I., Zimny, M., Mansell, J., Mow, V.C. (1975) The ultrastructure and biomechanical significan of the tidemark of articular cartilage. Clin. Orthop. Rel. Res. 112:357-362.

34. Rosenberg, L.C., Hellmann, W., Kleinschmidt, A.K. (1975) Electron microscopic studies proteoglycan aggregates from bovine articular cartilage. J. Biol. Chem. 250:1877-1883.

35. Roth, V., Mow, V.C. (1980) The intrinsic tensile behavior of the matrix of bovine articular cartilage a its variation with age. J. Bone Joint Surg. 62A:1102-1117.

36. Setton, L.A., Gu, W.Y., Lai, W.M., Mow, V.C. (1992) Pre-stress in articular cartilage due to intern swelling pressure. Proc Adv. Bioengng. ASME, BED-22:485-488.

37. Setton, L.A., Gu, W.Y., Saed-Nejad, F., Lai, W.M., Mow, V.C. (1993) Swelling-induced pre-stress articular cartilage and its physiological implications. Trans. Orthop. Res. Soc. 18:282.

38. Tobias, D., Ziv, I., Maroudas, A. (1992) Human facet cartilage: Swelling and some physico-chemi characteristics as a function of age: Swelling of human facet joint cartilage. Spine 17:694-700.

39. Torzilli,P.A. (1985) Influence of cartilage conformation on its equilibrium water partition. J. Orth Res. 3:473-483.

40. Venn, M., Maroudas, A. (1977) Chemical composition and swelling of normal and osteoarthr femoral head cartilage. Part I- Chemical composition. Ann. Rheum. Dis. 36:121-129.

SECTION 5

POROELASTICITY: DYNAMICS

POROELASTIC PLATE AND SHELL THEORIES

LARRY A. TABER and ANDREW M. PULEO
Department of Mechanical Engineering
University of Rochester
Rochester, NY 14627

Abstract. Theories are developed for deformation of porous elastic plates and shells saturated with a viscous fluid. The formulations, which are based on consolidation theory, account for detailed variations in fluid pressure and flow across the thickness. First, a linear theory is discussed for general thin shells, with plates and spherical shells examined as special cases. Then, the theory is extended to poroelastic membranes of revolution undergoing large axisymmetric deformation. In both theories, the fluid and solid components are assumed to be incompressible. Results are given for a spherical poroelastic membrane of Blatz-Ko material with a time-varying internal pressure. For high loading rates at large strain, nonlinearities strongly influence the fluid-pressure distribution.

1. Introduction

Recently, interest has intensified concerning the influence of poroelasticity on the mechanical behavior of biological tissues. In general, poroelastic analyses have been based on three-dimensional mixture theory (Bowen, 1976) or three-dimensional consolidation theory (Biot, 1941, 1955, 1962, 1972). Many biological structures, however, are fluid-saturated membranes, beams, plates, or shells. Examples include arteries, hearts, diaphragms, skin, bones, and bladders. One- and two-dimensional poroelasticity theories would be useful for these and other related mechanics problems.

To date, few papers have addressed sub-three-dimensional theories for poroelastic media. Using linear consolidation theory, Biot (1964) examined the problem of cylindrical bending and bucking of a poroelastic plate due to end loads, and Nowinski and Davis (1972) presented a poroelastic beam theory for application to bones. In two-dimensional formulations, Rajagopal et al. (1983) derived nonlinear (spherical) membrane equations from the theory of mixtures, and Taber (1992a; 1992b) and Theodorakopoulos and Beskos (1994) presented linear theories for plates and shells of revolution based on consolidation theory.

This first part of this paper generalizes the linear poroelastic plate and shell theories of Taber (1992a; 1992b) to account for general geometry and for incompressible fluid and solid components. Since a plate is a special case of a shell, we focus primarily on shells. The theory is based on the following assumptions:

1. The shell is "thin," i.e., $R/h >> 1$, where R is the smallest midsurface radius of curvature and h is the thickness.
2. Displacements are small compared to the thickness.
3. Normals to the middle surface of the solid skeleton remain straight and normal during deformation. (Transverse shear deformation is ignored.)

A. P. S. Selvadurai (ed.), Mechanics of Poroelastic Media, 323–337.

4. The shell is in a state of approximately plane stress, i.e., the *total* transverse normal stress is zero.
5. In-plane fluid-velocity *gradients* relative to the solid are small compared to the transverse fluid-velocity gradient.

The first four assumptions are commonly employed in deriving shell theories, and the last assumption is a consequence of the transverse resistance to flow being much smaller than the resistance parallel to the middle surface. In this paper, only isotropic shells are considered.

The second part of this paper extends the linear theory to develop a nonlinear theory for large axisymmetric deformation of poroelastic membranes of revolution. In this theory, assumption 2 is deleted. An asymptotic analysis is presented for the special case of a spherical membrane with an internal pressure, and numerical results are given.

2. Linear Theory

Consider a thin porous elastic shell that is saturated with a viscous fluid. A poin in the shell is located by the coordinates x_i, with x_1 and x_2 being lines-of-curvatur coordinates on the middle surface, and $x_3 \equiv z$ is the transverse coordinate. (In thi paper, the usual summation convention applies, with Latin indices taking the value 1,2,3 and Greek indices taking the values 1,2.) Pressures $p_1(x_\alpha, t)$ and $p_2(x_\alpha, t)$ ac on the inner ($z = -h/2$) and outer ($z = h/2$) surfaces, respectively.

2.1. Governing Equations

2.1.1. *Stress and Strain*

The total (fluid + solid) stress components per unit area of bulk material are denote by $\sigma_{ij}(x_i, t)$. Then, the stress resultants are given by (Fig. 1)

$$N_{\alpha\beta} = \int \sigma_{\alpha\beta}\, dz, \qquad Q_\alpha = \int \sigma_{\alpha 3}\, dz, \qquad (1$$

and the moment resultants are

$$M_{\alpha\beta} = \int \sigma_{\alpha\beta}\, z\, dz \qquad ($$

where $\int \equiv \int_{-h/2}^{h/2}$. (Drilling moments are ignored.) In terms of these quantities, t equations of motion assume the standard forms for a thin shell (Novozhilov, 1970

For a thin shell, assumptions 1–3 imply that the strain components in the sol skeleton can be written in the form

$$e_{\alpha\beta} = \epsilon_{\alpha\beta} + z\kappa_{\alpha\beta} \qquad ($$

where the $\epsilon_{\alpha\beta}$ are middle-surface strains and the $\kappa_{\alpha\beta}$ are curvature change measur (Transverse shear strains are neglected.) In terms of the displacements of the mid surface of the solid skeleton, these quantities have the usual forms for a thin sh (Novozhilov, 1970).

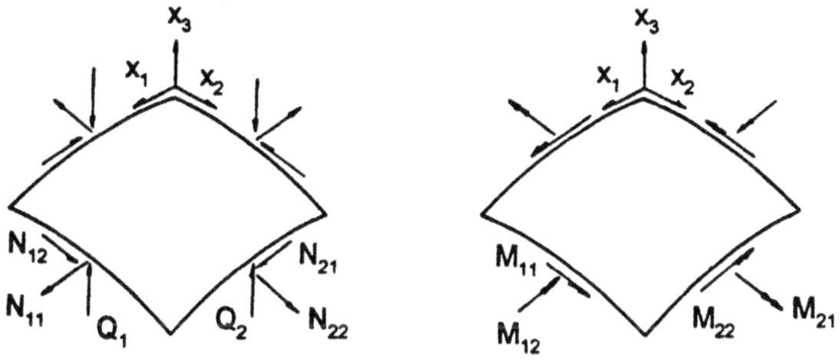

Fig. 1. Force and moment resultants acting on a shell element.

2.1.2. *Constitutive Relations*

For incompressible solid and fluid components, the three-dimensional constitutive elations can be written (Biot and Willis, 1957)

$$\sigma_{ij} = 2\mu e_{ij} + \lambda e \delta_{ij} - p \tag{4}$$

vhere p is the fluid (pore) pressure, δ_{ij} is the Kronecker delta, and λ and μ are the ame constants for the solid skeleton. In terms of the solid displacement $\mathbf{u}(x_i, t)$ nd the fluid displacement $\mathbf{u}^*(x_i, t)$, the solid and fluid dilatations are

$$e = e_{ii} = \nabla \cdot \mathbf{u}, \qquad e^* = \nabla \cdot \mathbf{u}^* \tag{5}$$

ith ∇ being the gradient operator.

For $\sigma_{33} = 0$ (assumption 4), eliminating e_{33} from equations (4) and (5) yields he stresses

$$\begin{aligned} \sigma_{11} &= \frac{E}{1-\nu^2}(e_{11} + \nu e_{22}) - Bp \\ \sigma_{22} &= \frac{E}{1-\nu^2}(e_{22} + \nu e_{11}) - Bp \\ \sigma_{12} &= \sigma_{21} = \frac{E}{1+\nu} e_{12}. \end{aligned} \tag{6}$$

these equations, Young's modulus E, Poisson's ratio ν, and B are material conants for the drained solid skeleton ($p = 0$), as defined by the relations

$$\frac{E}{1-\nu^2} = \frac{4\mu(\lambda+\mu)}{\lambda+2\mu}, \qquad \nu = \frac{\lambda}{2(\lambda+\mu)}, \qquad B = \frac{2\mu}{\lambda+2\mu} = \frac{1-2\nu}{1-\nu}. \tag{7}$$

ote that, although the material of the solid skeleton is incompressible, the bulk aterial is compressible, and so $\nu \neq 0.5$. After substitution of equations (3) and (6) into (1) and (2), integration over the shell thickness gives

$$\begin{aligned} N_{11} &= A(\epsilon_{11} + \nu\epsilon_{22}) + BN \\ N_{22} &= A(\epsilon_{22} + \nu\epsilon_{11}) + BN \\ N_{12} &= N_{21} = A(1-\nu)\epsilon_{12} \end{aligned} \qquad (8$$

and

$$\begin{aligned} M_{11} &= D(\kappa_{11} + \nu\kappa_{22}) + BM \\ M_{22} &= D(\kappa_{22} + \nu\kappa_{11}) + BM \\ M_{12} &= M_{21} = D(1-\nu)\kappa_{12} \end{aligned} \qquad (9$$

where $A = Eh/(1-\nu^2)$ is the extensional rigidity and $D = Eh^3/[12(1-\nu^2)]$ is th flexural rigidity of the shell. In addition,

$$N = -\int p\,dz, \qquad M = -\int p\,z\,dz \qquad (10$$

are the hydrostatic force and moment resultants due to the fluid.

2.1.3. *Fluid Flow*

In three-dimensional consolidation theory, the flow of a viscous fluid through porous elastic solid is governed by Darcy's law (Biot, 1941, 1962)

$$\frac{k}{\mu^*}\nabla p = -\dot{\mathbf{v}}. \qquad (1$$

In this equation, k is the intrinsic permeability, μ^* is the fluid viscosity, dot denot differentiation with respect to time, and

$$\mathbf{v} = \phi(\mathbf{u}^* - \mathbf{u}) \qquad (1$$

is the relative volume flow per unit bulk material area, with ϕ being the porosi of the solid skeleton. Furthermore, the incompressible fluid and solid componer must satisfy the continuity condition

$$(1-\phi)e + \phi e^* = 0. \qquad (1$$

With equations (5) and (12), this expression can be written

$$\boldsymbol{\nabla}\cdot\mathbf{v} + e = 0. \qquad (1$$

Taking the divergence of equation (11) and using equation (14) gives

$$\frac{k}{\mu^*}\nabla^2 p = \dot{e}. \qquad ($$

Next, solving for e_{33} from the constitutive relation (4) for σ_{33} (with $\sigma_{33} = 0$) a substituting the result into $(5)_1$ yields

$$e = B(e_{11} + e_{22} + p/2\mu). \qquad ($$

Finally, assumption 5 and equation (11) imply that the term $p_{,zz}$ dominates $\nabla^2 p$, where comma denotes differentiation with respect to the follower coordinate [see Taber (1992a)]. Thus, putting equation (16) into (15) yields the fluid flow equation

$$Kp_{,zz} = \dot{p} + 2\mu(\dot{e}_{11} + \dot{e}_{22}) \tag{17}$$

where

$$K = \frac{k}{\mu^*}(\lambda + 2\mu) \tag{18}$$

is the effective shell permeability.

As shown in the plate and shell problems considered by Taber (1992a; 1992b), strong transverse gradients in p can occur for high rates of loading. Thus, equation (17) is not integrated across the shell thickness to produce a resultant form.

2.2. Poroelastic Plate

In the present theory, equations (8), (9), and (17) contain fluid-elastic coupling terms. The governing equations are completed by adding the strain-displacement relations, the equilibrium equations, and the boundary conditions.

Consider quasi-static bending of a rectangular poroelastic plate. For $(x_1, x_2, x_3) = (x, y, z)$ and no in-plane stretching, $\epsilon_{\alpha\beta} = 0$ and (Szilard, 1974)

$$\kappa_{xx} = -w_{,xx}\,, \qquad \kappa_{yy} = -w_{,yy}\,, \qquad \kappa_{xy} = -w_{,xy} \tag{19}$$

where $w(x, y, t)$ is the transverse displacement of the middle surface. In addition, the equation of transverse force equilibrium is

$$Q_{x,x} + Q_{y,y} = -p_0 \tag{20}$$

where $p_0 = p_1 - p_2$, and moment equilibrium yields

$$\begin{aligned} Q_x &= M_{xx,x} + M_{xy,y} \\ Q_y &= M_{xy,x} + M_{yy,y}\,. \end{aligned} \tag{21}$$

Substituting equations (9), (19), and (21) into (20) yields

$$D\nabla^4 w = p_0 + B\nabla^2 M, \tag{22}$$

nd inserting equations (3) and (19) into (17) gives

$$Kp_{,zz} = \dot{p} - 2\mu z\nabla^2 w. \tag{23}$$

quations (22) [with $(10)_2$] and (23), which represent two coupled equations for (x, y, t) and $p(x, y, z, t)$, are the governing equations of linear poroelastic plate heory. As written, these equations apply to plates of any geometry.

The boundary conditions on w have the form of the usual conditions for a thin late (Szilard, 1974). The conditions on p depend on the nature of the interfaces at e upper and lower plate surfaces. If the surface is permeable, then p must match e applied pressure; if the surface is sealed, then equation (11) implies that $p_{,z} = 0$. ifferent combinations of fluid boundary conditions were examined by Taber (1992a; 92b).

Equations (22) and (23) can be solved using a variety of analytical techniques. aber (1992a; 1992b) presented Laplace transform and perturbation solutions for iasi-static loading of plates and forced vibration of shells of revolution.

2.3. Spherical Poroelastic Membrane

Consider now a spherical poroelastic membrane of radius R and thickness h that is subjected only to an internal pressure $p_1(t)$ ($p_2 = 0$). In this case, there is stretching but no bending ($\kappa_{\alpha\beta} = 0$). In addition, we assume that, for a thin membrane, the total stresses are approximately constant across the thickness, i.e., the fluid pressure is small compared to σ_{11} and σ_{22}.

Duc to symmetry, $e_{12} - 0$ and $e_{11} = e_{22} \equiv e_0$, and so equations (6) give $\sigma_{12} = 0$ and

$$\sigma_{11} = \sigma_{22} \equiv \sigma_0 = \frac{E}{1-\nu} e_0 - Bp. \tag{24}$$

Substituting this expression into the equilibrium equation

$$\sigma_0 = \frac{p_1 R}{2h}, \tag{25}$$

solving for e_0, and inserting the result into equation (17) yields

$$Kp_{,zz} = \frac{1-\nu}{1+\nu}\left(3\dot{p} + \frac{R}{h}\dot{p}_1\right). \tag{26}$$

Given $p_1(t)$, this equation can be solved for $p(z,t)$ using standard techniques.

3. Nonlinear Membrane Theory

3.1. Governing Equations

The equations governing large deformation of a poroelastic membrane are derived analogously to those of the linear theory. Here, we restrict our attention to axisymmetric (torsionless) deformation of membranes of revolution. Relative to the undeformed reference surface, x_1 and x_2 represent the meridional and circumferential coordinates, respectively, and $x_3 = z$ is the transverse coordinate.

3.1.1. *Stress and Strain*

For a thin membrane, if the fluid pressure is small compared to the total stress components, then the stresses and strains are approximately constant across the thickness. In this case, the stress resultants per unit undeformed length are $N_\alpha \simeq h\sigma_\alpha$, with the $\sigma_\alpha \equiv \sigma_{\alpha\alpha}$ (α not summed) being stresses per unit *undeformed* area and h being the undeformed membrane thickness. For axisymmetric deformation, moreover, there is no shear relative to the x_i coordinates, and the deformation can be described by the stretch ratios λ_i, which can be written in terms of the coordinates of a point in the membrane before and after deformation (Yang and Feng, 1970).

3.1.2. *Constitutive Relations*

For incompressible solid and fluid components, generalizing equation (4) gives the three-dimensional constitutive relations (Biot, 1972; Kenyon, 1976)

$$\sigma_i = \frac{\partial W}{\partial \lambda_i} - \frac{J}{\lambda_i} p \tag{2}$$

where $W(\lambda_i)$ is the strain-energy density function per unit undeformed volume, and

$$J = \lambda_1 \lambda_2 \lambda_3 \tag{28}$$

is the dilatation ratio. Following the previous linear analysis, we plan to set $\sigma_3 = 0$, solve equation (27) for λ_3, and substitute the result into the remaining equations (27) to obtain expressions for σ_1 and σ_2 in terms of λ_1, λ_2, and p [see equations (6)]. For nonlinear materials, however, a closed-form expression for λ_3 is generally not possible, and so we will seek an approximate solution (see Section 3.1.4).

3.1.3. *Fluid Flow.*

For large deformation, the transverse component of Darcy's law (11) can be written (Biot, 1972)

$$\frac{k}{\mu^*} \frac{\lambda_1 \lambda_2}{\lambda_3} \frac{\partial p}{\partial z} = -\frac{\partial v}{\partial t} \tag{29}$$

where $\mathbf{v}$ now represents the volume flow per unit undeformed area ($v = v_3$). In addition, the continuity condition (14) becomes

$$\boldsymbol{\nabla} \cdot \mathbf{v} + J - 1 = 0 \tag{30}$$

where $\boldsymbol{\nabla}$ is defined in terms of the undeformed membrane coordinates x_i. With assumption 5, this equation can be written

$$\frac{\partial v}{\partial z} + J - 1 = 0. \tag{31}$$

3.1.4. *Asymptotic Expansions*

To reduce the governing equations to forms analogous to those of the linear theory [equations (6) and (17)], we take advantage of the small parameter $\epsilon = h/R$, where R is the minimum radius of curvature of the undeformed membrane. First, we nondimensionalize the governing equations by introducing the nondimensional quantities

$$\begin{aligned}
\epsilon &= \frac{h}{R}, & \bar{z} &= \frac{z}{h}, & \bar{t} &= \frac{tkC}{\mu^* h^2} \\
\bar{\sigma}_i &= \frac{\sigma_i}{C}, & \bar{s}_i &= \frac{s_i}{C}, & \bar{p} &= \frac{p}{C} \\
\bar{W} &= \frac{W}{C}, & \bar{p}_1 &= \frac{p_1 R}{Ch}, & \bar{v} &= \frac{v}{h\epsilon}
\end{aligned} \tag{32}$$

where C is a material constant with units of stress. Then,equations (27), (28), (29), and (31) become

$$\begin{aligned}
\bar{\sigma}_i &= \bar{s}_i - J\bar{p}/\lambda_i \\
J &= \lambda_1 \lambda_2 \lambda_3 \\
\frac{\lambda_1 \lambda_2}{\lambda_3} \frac{\partial \bar{p}}{\partial \bar{z}} &= -\epsilon \frac{\partial \bar{v}}{\partial \bar{t}} \\
\epsilon \frac{\partial \bar{v}}{\partial \bar{z}} &= 1 - J.
\end{aligned} \tag{33}$$

where

$$\bar{s}_i = \frac{\partial \bar{W}}{\partial \lambda_i}. \tag{34}$$

Next, expanding the dependent variables in powers of ϵ gives

$$\begin{aligned}
\lambda_i &= \lambda_i^{(0)} + \epsilon \lambda_i^{(1)} + \cdots \\
\bar{\sigma}_i &= \sigma_i^{(0)} + \epsilon \sigma_i^{(1)} + \cdots \\
\bar{s}_i &= s_i^{(0)} + \epsilon s_i^{(1)} + \cdots \\
\bar{p} &= p^{(0)} + \epsilon p^{(1)} + \cdots \\
\bar{v} &= v^{(0)} + \epsilon v^{(1)} + \cdots .
\end{aligned} \tag{35}$$

The terms in the expansion for $\bar{s}_i$ can be expressed in terms of $\bar{W}$ by using the Taylor series

$$\bar{s}_i = \bar{s}_i|_{\epsilon=0} + \epsilon \left.\frac{\partial \bar{s}_i}{\partial \epsilon}\right|_{\epsilon=0} + \cdots \tag{36}$$

where, for $\bar{s}_i = \bar{s}_i(\lambda_j)$,

$$\begin{aligned}
\frac{\partial \bar{s}_i}{\partial \epsilon} &= \frac{\partial \bar{s}_i}{\partial \lambda_j}\frac{\partial \lambda_j}{\partial \epsilon} = \frac{\partial \bar{s}_i}{\partial \lambda_j}(\lambda_j^{(1)} + \cdots) \\
&= \frac{\partial^2 \bar{W}}{\partial \lambda_i \partial \lambda_j}(\lambda_j^{(1)} + \cdots)
\end{aligned} \tag{37}$$

in which equations (34) and $(35)_1$ have been used and the summation convention applies. Equations (34), $(35)_3$, (36), and (37) show that the first two terms in the series for $\bar{s}_i$ are given by

$$\begin{aligned}
s_i^{(0)} &= W_i \\
s_i^{(1)} &= W_{ij}\lambda_j^{(1)}
\end{aligned} \tag{38}$$

where

$$W_i \equiv \left.\frac{\partial \bar{W}}{\partial \lambda_i}\right|_{\epsilon=0}, \qquad W_{ij} \equiv \left.\frac{\partial^2 \bar{W}}{\partial \lambda_i \partial \lambda_j}\right|_{\epsilon=0}. \tag{39}$$

Given $\bar{W}$, these terms can be computed explicitly in terms of the $\lambda_i^{(0)}$.

Now, inserting equations (35) and (38) into equations (33) and equating like powers of ϵ in the usual manner gives the $O(\epsilon^0)$ relations

$$\begin{aligned}
\sigma_1^{(0)} &= W_1 - \lambda_2^{(0)}\lambda_3^{(0)}p^{(0)} \\
\sigma_2^{(0)} &= W_2 - \lambda_1^{(0)}\lambda_3^{(0)}p^{(0)} \\
\sigma_3^{(0)} &= W_3 - \lambda_1^{(0)}\lambda_2^{(0)}p^{(0)} \\
\frac{\partial p^{(0)}}{\partial \bar{z}} &= 0 \\
\lambda_1^{(0)}\lambda_2^{(0)}\lambda_3^{(0)} &= 1
\end{aligned} \tag{40}$$

and the $O(\epsilon^1)$ relations

$$
\begin{aligned}
\sigma_1^{(1)} &= W_{11}\lambda_1^{(1)} + W_{12}\lambda_2^{(1)} + W_{13}\lambda_3^{(1)} - (\lambda_2^{(0)}\lambda_3^{(1)} + \lambda_3^{(0)}\lambda_2^{(1)})p^{(0)} - \lambda_2^{(0)}\lambda_3^{(0)}p^{(1)} \\
\sigma_2^{(1)} &= W_{21}\lambda_1^{(1)} + W_{22}\lambda_2^{(1)} + W_{23}\lambda_3^{(1)} - (\lambda_1^{(0)}\lambda_3^{(1)} + \lambda_3^{(0)}\lambda_1^{(1)})p^{(0)} - \lambda_1^{(0)}\lambda_2^{(0)}p^{(1)} \\
\sigma_3^{(1)} &= W_{31}\lambda_1^{(1)} + W_{32}\lambda_2^{(1)} + W_{33}\lambda_3^{(1)} - (\lambda_1^{(0)}\lambda_2^{(1)} + \lambda_2^{(0)}\lambda_1^{(1)})p^{(0)} - \lambda_1^{(0)}\lambda_2^{(0)}p^{(1)} \\
\frac{\partial p^{(1)}}{\partial \bar{z}} &= -\frac{\lambda_3^{(0)}}{\lambda_1^{(0)}\lambda_2^{(0)}}\frac{\partial v^{(0)}}{\partial \bar{t}} \\
\frac{\partial v^{(0)}}{\partial \bar{z}} &= -(\lambda_1^{(0)}\lambda_2^{(0)}\lambda_3^{(1)} + \lambda_1^{(0)}\lambda_3^{(0)}\lambda_2^{(1)} + \lambda_2^{(0)}\lambda_3^{(0)}\lambda_1^{(1)}).
\end{aligned}
\tag{41}
$$

Equations (40) correspond to those for an incompressible hyperelastic membrane (Yang and Feng, 1970), with $p^{(0)}$ being the so-called hydrostatic pressure, which is constant across the thickness. Once these relations are solved, equations (41) provide a first-order correction for the fluid-elastic effects.

Consider the $O(\epsilon^0)$ equations. On setting $\sigma_3^{(0)} = 0$ (assumption 4), equation $(40)_3$ can be solved for $p^{(0)}$, and $(40)_5$ provides $\lambda_3^{(0)}$. Inserting the results into $(40)_{1,2}$ yields

$$
\begin{aligned}
\sigma_1^{(0)}(\lambda_1^{(0)}, \lambda_2^{(0)}) &= W_1 - W_3/(\lambda_1^{(0)\,2}\lambda_2^{(0)}) \\
\sigma_2^{(0)}(\lambda_1^{(0)}, \lambda_2^{(0)}) &= W_2 - W_3/(\lambda_1^{(0)}\lambda_2^{(0)\,2}) \\
p^{(0)}(\lambda_1^{(0)}, \lambda_2^{(0)}) &= W_3/(\lambda_1^{(0)}\lambda_2^{(0)}).
\end{aligned}
\tag{42}
$$

Supplementing these relations with the geometric and equilibrium equations yields he governing equations for an incompressible membrane. Solutions to these equaions have been presented, for example, by Yang and Feng (Yang and Feng, 1970) nd Tielking and Feng (1974).

The $O(\epsilon^1)$ equations are handled somewhat differently. First, equation $(41)_3$ ith $\sigma_3^{(1)} = 0$ is solved for $\lambda_3^{(1)}$, which is put into $(41)_{1,2}$. Substituting these latter elations into the equilibrium equations then provides two equations to solve for $\lambda_1^{(1)}$ nd $\lambda_2^{(1)}$ in terms of $p^{(1)}$ and the known $O(\epsilon^0)$ terms. After inserting the result to equation $(41)_5$, this equation can be combined with $(41)_4$ to produce a single quation for $p^{(1)}$. The next section details the procedure for a pressurized spherical oroelastic membrane.

2. Asymptotic Analysis for a Spherical Poroelastic Membrane

onsider a spherical membrane with an internal pressure $p_1(t)$ $(p_2 = 0)$. Symmetry emands that

$$
\begin{gathered}
\lambda_1 = \lambda_2 \equiv \lambda = r/R \quad \lambda_1^{(0)} = \lambda_2^{(0)} \equiv \lambda^{(0)} \quad \lambda_1^{(1)} = \lambda_2^{(1)} \equiv \lambda^{(1)} \\
\sigma_1 = \sigma_2 \equiv \sigma \qquad \sigma_1^{(0)} = \sigma_2^{(0)} \equiv \sigma^{(0)} \quad \sigma_1^{(1)} = \sigma_2^{(1)} \equiv \sigma^{(1)} \\
W_1 = W_2 \qquad W_{11} = W_{22} \qquad W_{13} = W_{23}
\end{gathered}
\tag{43}
$$

ere R and r are the undeformed and deformed membrane radii, respectively. addition, summing the forces on a deformed hemisphere yields the equilibrium equation

$$2\pi Rh\sigma = \pi r^2 p_1$$

or, in nondimensional form [see equations (32)],

$$\bar{\sigma} = \tfrac{1}{2}\bar{p}_1\lambda^2. \tag{44}$$

Substituting equations (35) and (43) into this relation yields

$$\begin{aligned}\sigma^{(0)} &= \tfrac{1}{2}\bar{p}_1\lambda^{(0)\,2}\\ \sigma^{(1)} &= \bar{p}_1\lambda^{(0)}\lambda^{(1)}.\end{aligned} \tag{45}$$

3.2.1. *Incompressible Membrane Solution.*
Combining equations $(45)_1$ with $(42)_1$ gives

$$\lambda^{(0)\,3}W_1 - W_3 = \tfrac{1}{2}\bar{p}_1\lambda^{(0)\,5} \tag{46}$$

which, given $\bar{p}_1$ and $W|_{\epsilon=0}$ as a function of $\lambda^{(0)}$, can be solved for $\lambda^{(0)}$. Then, (42) gives the hydrostatic pressure as

$$p^{(0)} = W_3/\lambda^{(0)\,2}. \tag{47}$$

3.2.2. *Fluid-Elastic Coupling.*
With the zero-order terms now known, we examine equations (41). First, on setting $\sigma_3^{(1)} = 0$, equations $(41)_3$ and (43) yield

$$\lambda_3^{(1)} = W_{33}^{-1}[2\lambda^{(0)}\lambda^{(1)}p^{(0)} + \lambda^{(0)\,2}p^{(1)} - 2W_{13}\lambda^{(1)}]. \tag{48}$$

Second, substituting this expression and equations $(41)_1$ and (43) into $(45)_2$ gives

$$\lambda^{(1)} = ap^{(1)} \tag{49}$$

where

$$\begin{aligned}a(\bar{t}) &= \frac{\hat{a}\lambda^{(0)\,2} + (\lambda^{(0)})^{-1}}{W_{11} + W_{12} - 2W_{33}\hat{a}^2 - (\lambda^{(0)})^{-2}p^{(0)} - \bar{p}_1\lambda^{(0)}}\\ \hat{a}(\bar{t}) &= W_{33}^{-1}(\lambda^{(0)}p^{(0)} - W_{13})\end{aligned} \tag{50}$$

are known quantities.

Next, putting equations (43), (48), and (49) into $(41)_5$ yields

$$\frac{\partial v^{(0)}}{\partial \bar{z}} = -bp^{(1)} \tag{51}$$

where

$$b(\bar{t}) = W_{33}^{-1}\lambda^{(0)\,4} + 2a[\lambda^{(0)\,2}\hat{a} + (\lambda^{(0)})^{-1}]. \tag{52}$$

Finally, differentiating equation $(41)_4$ with respect to $\bar{z}$ and (51) with respect to $\bar{t}$ and eliminating $v^{(0)}$ gives, since the $\lambda_i^{(0)}$ are independent of $\bar{z}$,

$$\hat{K}\frac{\partial^2 p^{(1)}}{\partial \bar{z}^2} = \frac{\partial p^{(1)}}{\partial \bar{t}} + \frac{\dot{b}}{b}p^{(1)} \tag{53}$$

where

$$\hat{K}(\bar{t}) = \frac{\lambda_o^4}{b(\bar{t})}. \tag{54}$$

With $\hat{K}$ and b being known functions of time, equation (53) is a *linear* partial differential equation that reduces to equation (26) in the limit of small strains. Thus, after solving the $O(\epsilon^0)$ equations, we can express the fluid pressure as a closed form separation-of-variables solution to equation (53). Of note, a similar expansion using the equations of the linear theory yields the exact linear solution. For internal pressurization only, the appropriate nondimensional boundary conditions are [see equations (32)]

$$\begin{aligned} \bar{z} = -\tfrac{1}{2}: \quad & \bar{p} = \epsilon\bar{p}_1 \quad \longrightarrow \quad p^{(1)} = \bar{p}_1 - p^{(0)}/\epsilon \\ \bar{z} = \tfrac{1}{2}: \quad & \bar{p} = 0 \quad \longrightarrow \quad p^{(1)} = -p^{(0)}/\epsilon. \end{aligned} \tag{55}$$

3.3. Results

Numerical results are given for a poroelastic spherical membrane composed of a material with a strain-energy density function of the form (Blatz and Ko, 1962)

$$\bar{W} = I_1 + \frac{1-2\nu}{\nu} I_3^{-\nu/(1-2\nu)} - \frac{1+\nu}{\nu} \tag{56}$$

where C and ν are material constants and

$$\begin{aligned} I_1 &= \lambda_1^2 + \lambda_2^2 + \lambda_3^2 \\ I_3 &= J^2 = \lambda_1^2\lambda_2^2\lambda_3^2 \end{aligned} \tag{57}$$

are strain invariants. For small strain, ν can be identified as Poisson's ratio and $C = \mu/2$. Note that, since the bulk material is compressible, W must have a compressible form. Blatz and Ko (1962) proposed equation (56) for foam rubber.

The behavior of the membrane was explored for the specified internal pressure

$$\bar{p}_1 = \dot{\bar{p}}\bar{t}$$

where $\dot{\bar{p}}$ is the nondimensional loading rate. For the Blatz-Ko membrane, the pressure reaches a peak ($\bar{p}_1 \cong 2$) near $\lambda_0 = 1.5$ (Fig. 2). Thus, the specified pressure cannot be increased indefinitely in the nonlinear case.

Plots of the pressure distribution across the wall at various times are shown (Figs. 3 and 4). In the linear problem, the normalized fluid pressure p/p_1 is independent of the loading rate (Fig. 3a). Moreover, the quantity $\epsilon p/p_1 = \bar{p}/\bar{p}_1$ is nearly independent of ϵ in the outer layers of the wall (Fig. 3b). Immediately after the load is applied, the membrane deforms elastically, expanding the pores and producing a suction effect. At $\bar{t} = 0.001$, the corresponding negative pressure is nearly uniform

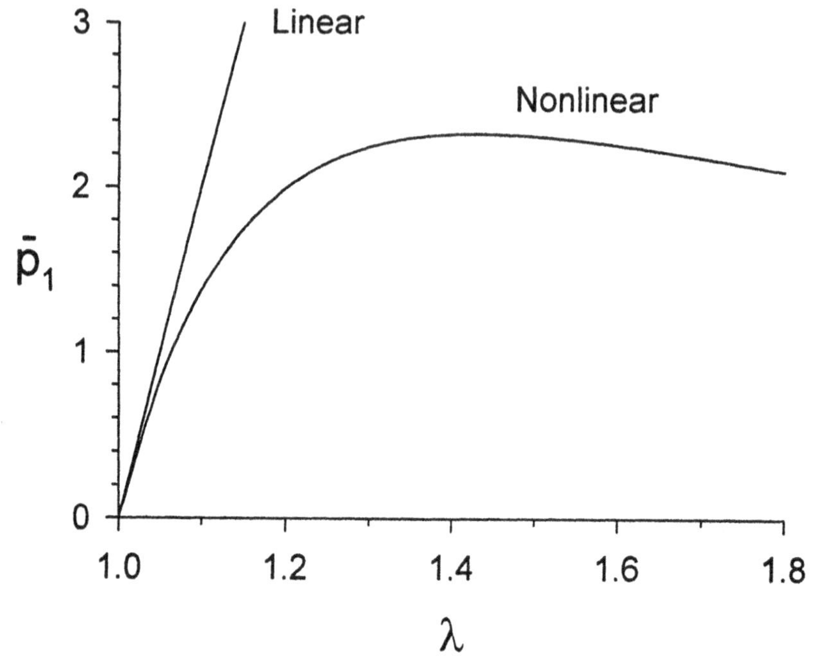

Fig. 2. **Inflation pressure versus stretch ratio for spherical poroelastic membrane.**

over most of the interior of the wall. Near the membrane surfaces, however, bound ary layers form to enable the boundary conditions to be satisfied. In response to th strong pressure gradients, the fluid then flows away from midwall as the boundar layers widen and eventually dissipate. As a steady state is reached for long times the negative pressures disappear, the pressure gradient becomes constant, and constant flow rate is set up from inside to outside.

For short times, the applied pressure and the strains are small, and the linea and nonlinear solutions nearly coincide. Thus, like the linear solution, the nonlinea solution is initially independent of the loading rate (Figs. 3a and 4). As time goe on, however, the strains become large and nonlinear effects become significant. Fc low rates of loading, the large-strain effect is relatively small (Fig. 3a). Howeve for high rates of loading, in contrast to the behavior of the linear solution, the pea fluid pressure increases with time before it begins to decrease toward its steady-sta value (Fig. 4).

Finally, we note that equation (44) gives $\bar{\sigma}/\bar{p}_1 = \frac{1}{2}\lambda^2 > 0.5$, while the comp tations give $|\bar{p}/\bar{p}_1| = |\epsilon p/p_1| \cong 0.3$ (Fig. 3b). Thus, $\bar{p}$ is not much smaller tha $\bar{\sigma}$, as we assumed when deriving the membrane equations. Moreover, the absolu pressure gradients are quite large and grow larger as the membrane becomes thinn (Fig. 3b). We conclude, therefore, that transmural variations in total stress must accounted for even for very thin membranes. Thus, the utility of a true poroelast membrane theory is limited.

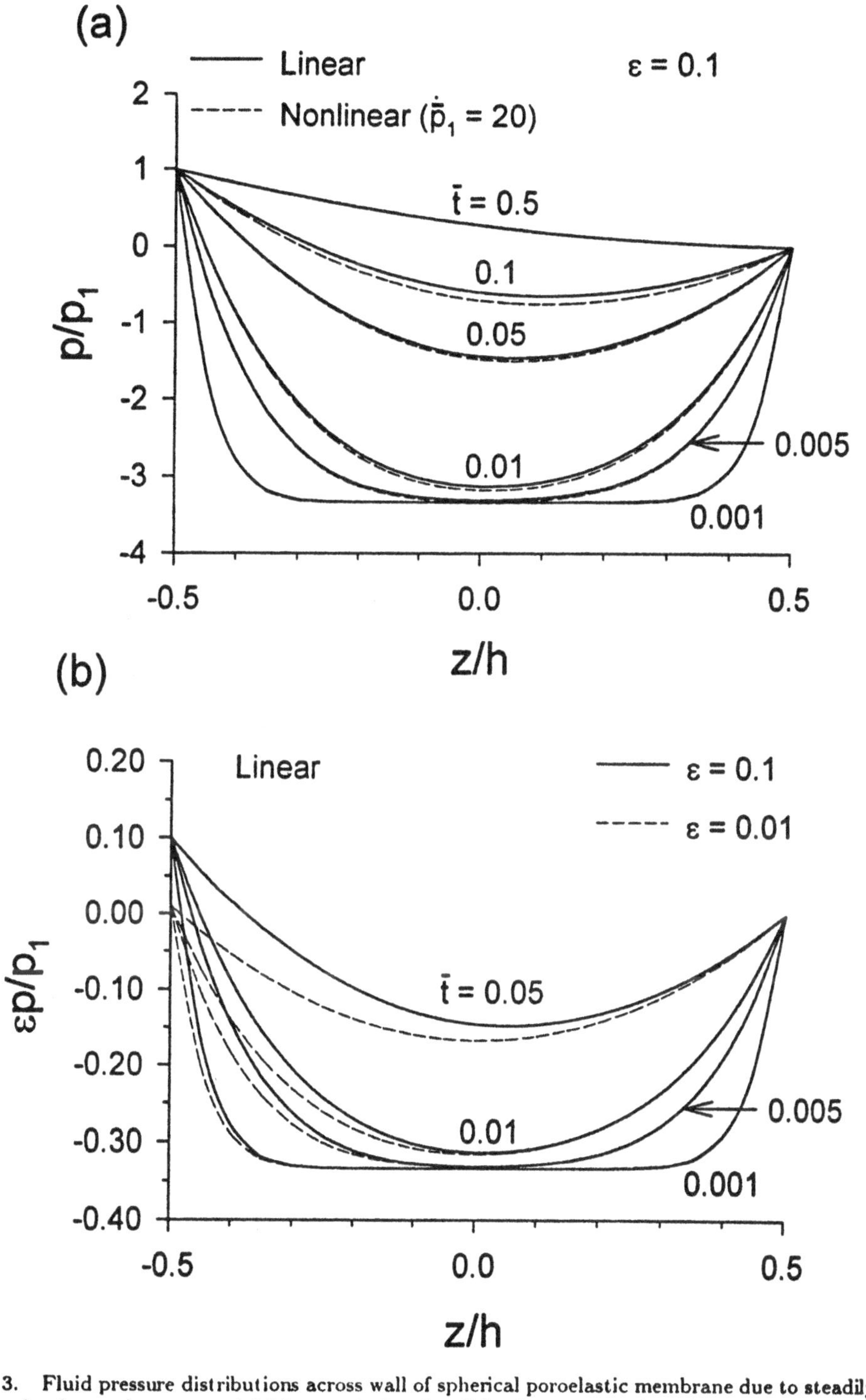

g. 3. Fluid pressure distributions across wall of spherical poroelastic membrane due to steadily creasing internal pressure. (a) Linear solution for all loading rates and nonlinear solution for a atively slow loading rate. (b) Linear solution for two values of membrane thickness ($\epsilon = h/R$).

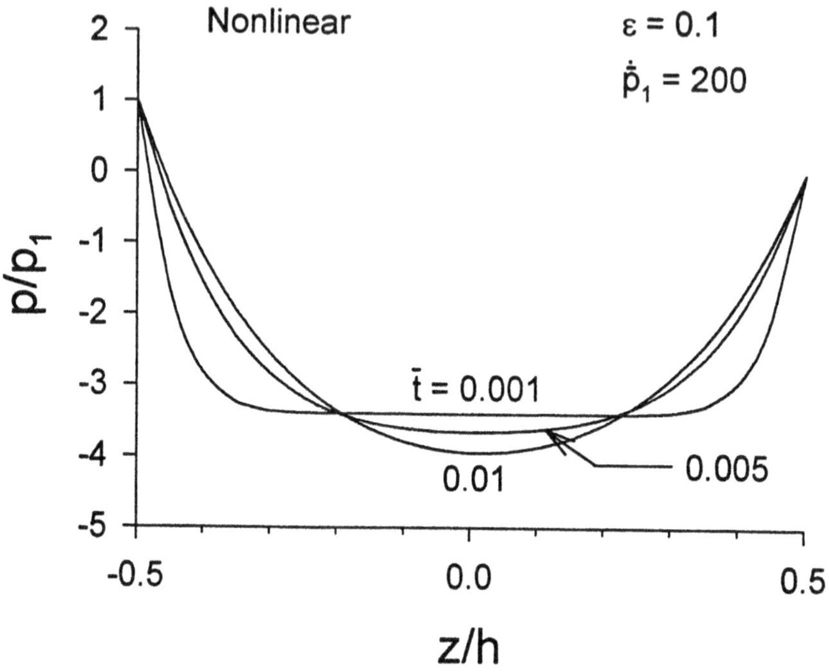

Fig. 4. Fluid pressure distributions across wall of spherical poroelastic membrane due to steadi increasing internal pressure: nonlinear solution for a relatively fast loading rate.

Acknowledgements

This research was supported by an NIH SCOR in Pediatric Cardiovascular Disease (P50 HL51498) at the University of Rochester and by NIH grant R01 HL4636 (LAT).

References

Biot, M.A. (1941) General Theory of Three-Dimensional Consolidation. *J. Appl. Phys.* **12**, 155–16

Biot, M.A. (1955) Theory of Elasticity and Consolidation for a Porous Anisotropic Solid. *J. Ap Phys.* **26**, 182–185.

Biot, M.A. (1962) Mechanics of Deformation and Acoustic Propagation in Porous Media. *J. Ap Phys.* **33**, 1482–1498.

Biot, M.A. (1964) Theory of Buckling of a Porous Slab and Its Thermoelastic Analogy. *J. Ap Mech.* **31**, 194–198.

Biot, M.A. (1972) Theory of Finite Deformations of Porous Solids. *Indiana Univ. Math. J.* **2** 597–620.

Biot, M.A. and Willis, D.G. (1957) The Elastic Coefficients of the Theory of Consolidation. *Appl. Mech.* **24**, 594–601.

Blatz, P.D. and Ko, W.L. (1962) Application of Finite Elasticity to the Deformation of Rubbe Materials. *Trans. Soc. Rheology* **6**, 223–251.

Bowen, R.M. (1976) Theory of Mixtures. In *Continuum Physics. III—Mixtures and EM Field Theories*, (Edited by Eringen, A.C.), pp. 1–127. Academic Press, New York.

Kenyon, D.E. (1976) The Theory for an Incompressible Solid-Fluid Mixture. *Arch. Ration. Mech. Anal.* **62**, 131–147.

Novozhilov, V.V. (1970) *Thin Shell Theory*, Wolters-Noordhoff, Groningen, The Netherlands.

Nowinski, J.L. and Davis, C.F. (1972) The Flexure and Torsion of Bones Viewed as Anisotropic Poroelastic Bodies. *Int. J. Eng. Sci.* **10**, 1063–1079.

Rajagopal, K.R., Wineman, A.S., and Shi, J.J. (1983) The Diffusion of a Fluid Through a Highly Elastic Spherical Membrane. *Int. J. Eng. Sci.* **21**, 1171–1183.

Szilard, R. (1974) *Theory and Analysis of Plates*, Prentice-Hall, Englewood Cliffs, NJ.

Taber, L.A. (1992a) A Theory for Transverse Deflection of Poroelastic Plates. *J. Appl. Mech.* **59**, 628–634.

Taber, L.A. (1992b) Axisymmetric Deformation of Poroelastic Shells of Revolution. *Int. J. Solids Struct.* **29**, 3125–3143.

Theodorakopoulos, D.D. and Beskos, D.E. (1994) Flexural Vibrations of Poroelastic Plates. *Acta Mech.* **103**, 191–203.

Tielking, J.T. and Feng, W.W. (1974) The Application of the Minimum Potential Energy Principle to Nonlinear Axisymmetric Membrane Problems. *J. Appl. Mech.* **41**, 491–496.

Yang, W.H. and Feng, W.W. (1970) On Axisymmetrical Deformations of Nonlinear Membranes. *J. Appl. Mech.* **37**, 1002–1011.

Harmonic dynamics of poroelastic plates with one or two degrees of porosity

D. D. Theodorakopoulos, D. E. Beskos
Department of Civil Engineering
University of Patras
GR-26500 Patras, Greece

Abstract. The derivation of the governing equations of lateral harmonic motion of thin, fluid-saturated poroelastic plates with one and two degrees of porosity is presented. Use is made of Kirchhoff's plate theory and Biot's and Aifantis-Beskos' theories of poroelasticity for the one and two degrees of porosity models, respectively. One degree of porosity models are more suitable for soils, while two degrees of porosity models are more suitable for fissured rocks. The dynamic response of a rectangular, simply supported, poroelastic plate to uniform lateral load harmonically varying with time is determined analytically-numerically for both one and two degrees of porosity models. The effects of porosities and permeabilities on the plate response are studied and a comparison between the results of the two material models is made. The quasi-static problem is analysed as a special case of the dynamic one and an assessment of the inertial effect is made.

. Introduction

The determination of the quasi-static or dynamic response of fluid-saturated, oroelastic, flexural plates is an important problem in geotechnical engineering, eophysics and bioengineering. This work deals with the dynamic analysis of oroelastic plates with one or two degrees of porosity. Biot (1956) was the first to evelop a three-dimensional theory of wave propagation in fluid-saturated porous lastic solids such as soils. An extensive literature survey on wave propagation in orous media can be found in Corapcioglu (1991). Using his theory, Biot (1964) udied the bending of a poroelastic plate. However, he did not consider the general uckling case of poroelastic plates and certainly he did not take into account inertial ffects. Taber (1992) and Theodorakopoulos and Beskos (1994) presented a general eory of poroelastic flexural plates on the basis of Biot's (1956) model under quasi-atic and dynamic conditions, respectively.

It has been observed that fissured, poroelastic rocks have a permeability of fissures uch higher than the permeability of pores and the porosity of porous blocks, separated y fissures is much larger than the porosity of the fissures. This clearly indicates that in

P. S. Selvadurai (ed.), *Mechanics of Poroelastic Media*, 339–351.

rocks there are two degrees of porosity, one due to the fissures and one due to pores. Beskos (1989) derived the governing equations of motion for fully-saturated, poroelastic, fissured rocks on the basis of Aifantis' (1980) model of double porosity and the principles of the continuum theory of mixtures. Using this two degrees of porosity model of Aifantis-Beskos, Theodorakopoulos and Beskos (1993) derived a general theory of poroelastic flexural plates under dynamic conditions and studied their behaviour for particular boundary conditions.

This work, which draws material from the works of Theodorakopoulos and Beskos (1993, 1994), describes the development of the governing equations of flexural harmonic vibrations of thin, fluid-saturated poroelastic plates by using Kirchhoff's theory and the one and two degrees of porosity models of Biot (1956) and Aifantis-Beskos (1980, 1989), respectively. In addition, the dynamic response of a rectangular, simply supported, poroelastic plate to uniformly applied harmonic loading is obtained analytically-numerically for both material models and the effects of porosities and permeabilities on the response are assessed. The quasi-static problem is analysed as a special case of the dynamic one and the importance of the inertial effect is discussed.

2. Biot's governing equations of motion

Consider a system composed of a porous elastic solid saturated by a viscous fluid. According to Biot's theory the governing equations of motion of this system under zero body forces have the form

$$\tau_{ij,j}-b(\dot{u}_i-\dot{U}_i) = \rho_{11}\ddot{u}_i+\rho_{12}\ddot{U}_i \tag{1}$$

$$\tau_{,i}+b(\dot{u}_i-\dot{U}_i) = \rho_{21}\ddot{u}_i+\rho_{22}\ddot{U}_i \tag{2}$$

where Eq. (1) describes the motion of the solid as affected by the motion of the fluid and Eq. (2) Darcy's law of the fluid as affected by the motion of the solid. In the above indices i and j take the values 1,2 and 3 and imply summation when they are repeated; commas and dots indicate differentiation with respect to space and time, respectively; u_i and U_i are the displacement components of the solid and fluid phase, respectively; ρ_{11}, $\rho_{12}=\rho_{21}$ and ρ_{22} are mass density coefficients which are related to the relative solid and fluid mass densities ρ_1 and ρ_2, respectively, by

$$\rho_1 = \rho_{11}+\rho_{12} \quad , \quad \rho_2 = \rho_{21}+\rho_{22} \tag{3}$$

where ρ_1 and ρ_2 are given in terms of the actual mass densities of the solid ρ_s and fluid ρ_f by

$$\rho_1 = (1-n)\rho_s \quad , \quad \rho_2 = n\rho_f \tag{4}$$

with n being the porosity; b is a coefficient related to Darcy's coefficient of permeability K by

$$b = vn^2/K \tag{5}$$

with v being the fluid viscosity and τ_{ij} and $\tau\delta_{ij}$ are the stress tensors acting on the solid and fluid components, respectively, with δ_{ij} being the Kronecker's delta. These stress tensors are connected with the system (total) stress sensor σ_{ij} by

$$\sigma_{ij} = \tau_{ij} + \tau\delta_{ij} \tag{6}$$

Stress tensors σ_{ij} and τ are expressible in terms of the deformation as

$$\sigma_{ij} = 2\mu e_{ij} + (\lambda e + \frac{Q+R}{R}\tau)\delta_{ij} \quad , \tag{7}$$

$$\tau = -np = Qe + Re \tag{8}$$

where λ and μ are the Lamé elastic constants, Q and R are material coefficients given in terms of measurable quantities in Biot and Willis (1957) and e and ε are the solid and fluid dilatations, respectively given by

$$e = e_{\kappa\kappa} = u_{\kappa,\kappa} \qquad \varepsilon = \varepsilon_{\kappa\kappa} = U_{\kappa,\kappa} \qquad (\kappa = 1,2,3) \tag{9}$$

with the strain tensors e_{ij} and ε_{ij} for the solid and fluid, respectively, expressible as

$$e_{ij} = (1/2)\,(u_{i,j} + u_{j,i})$$

$$\varepsilon_{ij} = (1/2)\,(U_{i,j} + U_{j,i}) \tag{10}$$

In view of (6), (7) and (2), (3), Eq. (1) can take the form

$$\sigma_{ij,j} = 2\mu e_{ij,j} + \lambda e_{,i} + \frac{Q+R}{R}\tau_{,i} = \rho_1 \ddot{u}_i + \rho_2 \ddot{U}_i \tag{11}$$

while, a differentiation of (2) with respect to x_i and subsequent elimination of ε with the aid of Eq. (8), yields

$$R\tau_{,ii} - b\dot{\tau} + b(R+Q)\dot{e} = \rho_{22}\ddot{\tau} + (R\rho_{21} - Q\rho_{22})\ddot{e} \tag{12}$$

Thus the system of Eqs (1) and (2) can be replaced by the equivalent one of Eqs (11) and (12), where i=1,2 and 3.

3. Flexural vibrations of a poroelastic soil plate

Consider a thin plate whose material obeys Biot's theory while its flexural motion is governed by Kirchoff's linear flexural theory. Assuming that the middle surface coincides with the (x_1,x_2) plane and that the in-plane fluid flow relative to the solid is negligible compared to the lateral one, one obtains

$$u_1 = U_1, \quad u_2 = U_2, \quad u_3 = w \neq U_3 \tag{13}$$

On the basis of the above assumptions one has

$$\sigma_{33} = 0 \tag{14}$$

where σ_{33} is the normal total stress component along the x_3 direction and

$$u_i = U_i = x_3\, w_{,i} \qquad i=1,2 \tag{15}$$

$$e_{ij} = \varepsilon_{ij} = x_3 w_{,ij} \quad , \qquad i,j=1,2 \tag{16}$$

Assuming a dynamic disturbance harmonic in time, the plate response is also harmonic and hence a typical function $f(x_1,x_2,t)$ of the problem has the form

$$f(x_1,x_2,t)=\tilde{f}(x_1,x_2)e^{i\omega t} \tag{17}$$

where, in what follows, the factor $e^{i\omega t}$ is omitted for simplicity.

Considering Eq. (2) in the frequency domain and solving for $\tilde{U}_i$ one obtains

$$\tilde{U}_i = \frac{\tilde{\tau}_{,i} + (i\omega b + \omega^2 \rho_{12})\tilde{u}_i}{i\omega b - \omega^2 \rho_{22}} \tag{18}$$

Thus $\tilde{U}_i$ is substituted in the frequency domain version of (11) and the result is

$$\tilde{\sigma}_{ij,j} = 2\mu\tilde{e}_{ij,j} + \lambda\tilde{e}_{,i} + \left(\frac{Q+R}{R} - \frac{-\omega^2\rho_2}{i\omega b - \omega^2\rho_{22}}\right)\tilde{\tau}_{,i}$$

$$= -\omega^2 \left(\rho_1 + \rho_2 \frac{i\omega b + \omega^2 \rho_{12}}{i\omega b - \omega^2 \rho_{22}}\right) \tilde{u}_i \; , \tag{19}$$

i.e, an equation involving only four variables ($\tilde{u}_i$ with $i=1,2,3$ and $\tilde{\tau}$) instead of seve[n] ($\tilde{u}_i,\tilde{U}_i$ with $i=1,2,3$ and $\tilde{\tau}$) it involved before. Equation (19) suggests that $\tilde{\sigma}_{ij}$ can b[e] written in the form

$$\tilde{\sigma}_{ij} = 2\mu\tilde{e}_{ij} + \lambda\tilde{e}\delta_{ij} + \left(\frac{Q+R}{R} - \frac{-\omega^2 \rho_2}{i\omega b - \omega^2\rho_{22}}\right) = \tilde{\tau}\delta_{ij} \tag{20}$$

Assuming that the upper surface is subjected to a pressure $q=q(x_1,x_2,t)$ and applying the classical bending theory of the plates one obtains in the frequency domain

$$\tilde{M}_{ij,ij} + \tilde{q} = -\omega^2\left(\rho_1 + \rho_2 \frac{i\omega b + \omega^2\rho_{12}}{i\omega b - \omega^2\rho_{22}}\right)\tilde{w}h \tag{21}$$

where the moment $\tilde{M}_{ij}$ is defined by

$$\tilde{M}_{ij} = \int_{-h/2}^{h/2} \tilde{\sigma}_{ij} x_3 dx_3 \tag{22}$$

It is noted that the stress $\tilde{\sigma}_{ij}$ that appears in (22) should not be taken from (20), which is valid for i,j=1,2,3 but from the equation resulting from (20) after the elimination of $\tilde{e}$ with the help of (14). Thus, use of $\tilde{\sigma}_{ij}$ as above in conjunction with Eq. (16), result in

$$\tilde{M}_{ij} = -D[(1-\nu)\tilde{w}_{,ij} + (\nu\tilde{w}_{,\kappa\kappa} + B_p \tilde{I}_\tau)\delta_{ij}] \tag{23}$$

where

$$D = \frac{Eh^3}{12(1-\nu^2)} , \quad B_p = -\frac{12}{h}\frac{\nu}{\lambda}\left(\frac{Q+R}{R} - \frac{-\omega^2\rho_2}{i\omega b - \omega^2\rho_{22}}\right) \tag{24}$$

$$\tilde{I}_\tau = \int_{-h/2}^{h/2} \tilde{\tau} x_3 dx_3 , \quad \mu = \frac{E}{2(1+\nu)} , \quad \lambda = \frac{E}{(1+\nu)} \cdot \frac{\nu}{(1-2\nu)}$$

Finally, substitution of $\tilde{M}_{ij}$ in (21) by its expression in (23) yields the first governing equation of flexural motion of the poroelastic plate, which reads

$$D(\nabla^4\tilde{w} + B_p \nabla^2\tilde{I}_\tau) + \tilde{q} = (-\omega^2)\left(\rho_1 + \rho_2 \frac{i\omega b + \omega^2\rho_{12}}{i\omega b - \omega^2\rho_{22}}\right)\tilde{w}h \tag{25}$$

Solving now Eq. (20) for e under the condition $\sigma_{33}=0$ of (14) one can obtain

$$\tilde{e}=\frac{-1}{2\mu+\lambda}\left(\frac{Q+R}{R}-\frac{\omega^2\rho_2}{i\omega b-\omega^2\rho_{22}}\right)\tilde{\tau}-\frac{2\mu}{2\mu+\lambda}\,x_3\nabla^2\tilde{w} \qquad (26)$$

Elimination of $\tilde{e}$ between (12) and (25) finally yields the second governing equation o flexural motion of the poroelastic plate in the form

$$R\tilde{\tau}_{,ii}-\{i\omega b[1+\frac{1}{2\mu+\lambda}\;\frac{(Q+R)^2}{R}-\frac{Q+R}{2\mu+\lambda}\;\frac{-\rho_2\,\omega^2}{i\omega b-\omega\,\rho_{22}}]+$$

$$+(-\omega^2)[\rho_{22}-\frac{1}{2\mu+\lambda}\;(R\rho_{21}-Q\rho_{22})\;\frac{Q+R}{R}+\frac{R\rho_{21}-Q\rho_{22}}{2\mu+\lambda}\;.$$

$$\frac{\rho_2(-\omega^2)}{i\omega b-\omega^2\rho_{22}}]\}\tilde{\tau}+[-i\omega b(R+Q)-\omega^2(R\rho_{21}-Q\rho_{22})]\;\frac{2\mu x_3}{2\mu+\lambda}\;\nabla^2\tilde{w}=0 \qquad (27$$

The system of Eqs (25) and (27) for the flexural motion of a poroelastic plate in th frequency domain is easily reducible to the equation

$$-D\nabla^4\tilde{w}+\tilde{q}=-\rho_1\omega^2\tilde{w}h \qquad (28$$

when the plate becomes simply elastic with mass density $\rho_1=\rho_s$

4. Aifantis - Beskos' governing equations of motion

A fissured poroelastic rock medium consists of porous elastic blocks separated fro each other by a system of fissures, i.e., it is a poroelastic medium with two porositi one corresponding to the fissures and the other to the pores of the blocks. By utilisi the rock modelling of Aifantis (1980) and employing well-known principles continuum mechanics, Beskos (1989) succeeded in developing the governing equatio of motion for such a fully saturated rock medium. These are 11 partial different equations with 11 unknowns which, for reasons of completeness, are given he explicity as

$$(\lambda+\mu)u_{\kappa,\kappa i}+\mu u_{i,\kappa\kappa}=v_1Q^1{}_i-v_2Q^2{}_i+\bar{\rho}_s\ddot{u}_i \qquad (2$$

$$-\beta_\alpha p^\alpha{}_{,i}=v_\alpha Q^\alpha{}_i+\rho_f\dot{Q}^\alpha{}_i+\bar{\rho}_\alpha\ddot{u}_i \qquad ($$

$$(\beta_\alpha+n_\alpha)\dot{u}_{i,i}+(\gamma_\alpha+\delta n_\alpha)\dot{p}^\alpha+Q^\alpha{}_{i,i}=-(-1)^\alpha\kappa(p^2-p^1) \qquad ($$

in which indices i and k take the values 1,2, and 3 and imply summation when they are repeated, while the index α takes the values 1 (denoting fissures) and 2 (denoting pores) and does not imply summation when it is repeated; commas and dots indicate differentiation with respect to space and time, respectively; u_i are the displacement components of the solid phase of the medium, n_α are the porosities due to the fissures (α=1) and the pores (α=2); λ and μ are the Lamé elastic constants; $\bar{\rho}_s$ and $\bar{\rho}_\alpha$ are the relative solid and fluid mass densities, respectively, while ρ_f is the actual fluid mass density; the coefficients β_α express the deformability of the solid as this affects the flow in the fissures (α=1) and the pores (α=2); δ is the fluid compressibility; the coefficients γ_α measure the compressibility of the fissures (α=1) and the pores (α=2); the coefficient κ measures the transfer of fluid from the pores to the fissures; and the seepage force factors v_α and the relative specific discharges Q^α_i in the fissures (α=1) and the pores (α=2) are given by

$$v_\alpha = \beta_\alpha v / K_\alpha \tag{32}$$

$$Q^\alpha_i = n_\alpha (\dot{U}^\alpha_i - \dot{u}_i) \tag{33}$$

in which v is the dynamic viscosity of the fluid; K_α are the Muskat permeabilities of the fissures (α=1) and the pores (α=2) and U^α are the displacement components of the fluid in the fissures (α=1) and the pores (α=2). The relative mass densities of the composite medium are given in terms of the actual ones by the relations

$$\bar{\rho}_s = \rho_s [1-(n_1+n_2)] \tag{34}$$

$$\bar{\rho}_\alpha = n_\alpha \rho_f \tag{35}$$

In a fissured poroelastic medium the total stress tensor σ_{ij} (i,j=1,2,3) is given by

$$\sigma_{ij} = \sigma^s_{ij} + \sigma^1_{ij} + \sigma^2_{ij} \tag{36}$$

where σ^s_{ij} and σ^α_{ij} are the stress tensors acting on the solid, fluid in fissures (α=1) and fluid in pores (α=2), respectively. Stress tensors σ^s_{ij} and σ^α_{ij} are expressible in terms of deformation as

$$\sigma^s_{ij} = 2\mu e_{ij} + \lambda e \delta_{ij} \tag{37}$$

$$\sigma^\alpha_{ij} = \tau^\alpha \delta_{ij} = -\beta_\alpha p^\alpha \delta_{ij} \tag{38}$$

where δ_{ij} is the Kronecker's delta and the strain tensor e_{ij} and dilatation e of the solid are expressible as

$$e_{ij} = (1/2)(u_{i,j} + u_{j,i}) \tag{39}$$

$$e = e_{\kappa\kappa} = u_{\kappa,\kappa} \tag{40}$$

In view of (37) and (38), Eq. (36) can take the form

$$\sigma_{ij}=2\mu e_{ij}+\lambda e\delta_{ij}+\tau^1\delta_{ij}+\tau^2\delta_{ij} \tag{41}$$

while using Eqs. (30) for $\alpha=1,2$ in turn and adding the resulting equations together with Eq.(29) with u_i's being replaced by e_{ij}'s one obtains

$$2\mu e_{ij,j}+\lambda e_{,i}-\beta_1 p^1_{,i}-\beta_2 p^2_{,i}=\rho_f(\dot{Q}^1_i+\dot{Q}^2_i)+(\bar{\rho}_s+\bar{\rho}_1+\bar{\rho}_2)\ddot{u}_i \tag{42}$$

Spatial differentiation of (41) and use of (42) yields

$$\sigma_{ij,j}=\rho_f(\dot{Q}^1_i+\dot{Q}^2_i)+(\bar{\rho}_s+\bar{\rho}_1+\bar{\rho}_2)\ddot{u}_i \tag{43}$$

substituting Eqs. (33) into Eqs. (30) and (43) one obtains

$$\tau^\alpha_{,i}=v_\alpha n_\alpha(\dot{U}^\alpha_i-\dot{u}_i)+\bar{\rho}_\alpha\ddot{U}^\alpha_i \quad ; \qquad \alpha=1,2 \tag{44}$$

$$\sigma^s_{ij,j}+\tau^1_{,i}+\tau^2_{,i}=\bar{\rho}_1\ddot{u}^1_i+\bar{\rho}_2\ddot{u}^2_i+\bar{\rho}_s\ddot{u}_i \tag{45}$$

5. Flexural vibrations of a poroelastic rock plate

The case of a thin fissured poroelastic plate which obeys Aifantis-Beskos' (Aifantis 1980, Beskos 1989) material model while its flexural motion is governed by Kirchoff's linear flexural theory is analysed in a similar way as that presented in the previous section. Thereafter the stress $\tilde{\sigma}_{ij}$ in frequency domain and the three governing equations of flexural motion of the fissured poroelastic plate in the frequency domain are in the form

$$\tilde{\sigma}_{ij}=2\mu\tilde{e}_{ij}+\lambda\tilde{e}\delta_{ij}+\frac{i\omega v_1}{i\omega v_1-\omega^2\rho_f}\tilde{\tau}^1\delta_{ij}+\frac{i\omega v_2}{i\omega v_2-\omega^2\rho_f}\tilde{\tau}^2\delta_{ij} \tag{46}$$

$$-D[\nabla^4\tilde{w}+B^1\nabla^2\tilde{I}^1+B^2\nabla^2\tilde{I}^2]+\tilde{q}=$$

$$=-\omega^2(\bar{\rho}_s+\bar{\rho}_1+\bar{\rho}_2+\frac{\omega^2\rho_f\bar{\rho}_1}{i\omega v_1-\omega^2\rho_f}+\frac{\omega^2\rho_f\bar{\rho}_2}{i\omega v_2-\omega^2\rho_f})\,\tilde{w}h \tag{47}$$

$$\tilde{\tau}^1_{,ii}\beta_1-\tilde{\tau}^1[i\omega(\gamma_1+\delta n_1)+\kappa+\frac{i\omega(\beta_1+n_1A_1)}{2\mu+\lambda}\beta_1A_1]\frac{i\omega v_1-\omega^2\rho_f}{i\omega}-$$

$$-\tilde{\tau}^2(-\kappa\frac{\beta_1}{\beta_2}+\frac{i\omega(\beta_1+n_1A_1)}{2\mu+\lambda}\beta_1A_2)\frac{i\omega v_1-\omega^2\rho_f}{i\omega}-$$

$$-\frac{2\mu}{2\mu+\lambda}(\beta_1+n_1A_1)\,\beta_1\,(i\omega v_1-\omega^2\rho_f)\,x_3\nabla^2\tilde{w}=0 \tag{48}$$

$$\tilde{\tau}^2_{,ii}\,\beta_2-\tilde{\tau}^1[-\kappa\frac{\beta_2}{\beta_1}+\frac{i\omega(\beta_2+n_1A_2)}{2\mu+\lambda}\,\beta_2A_1]\,\frac{i\omega v_2-\omega^2\rho_f}{i\omega}-$$

$$\tilde{\tau}^2[i\omega(\gamma_2+\delta n_2)+\kappa+\frac{i\omega(\beta_2+n_2A_2)}{2\mu+\lambda}\,\beta_2A_2]\,\frac{i\omega v_2-\omega^2\rho_f}{i\omega}-$$

$$-\frac{2\mu}{2\mu+\lambda}(\beta_2+n_2A_2)\,\beta_2\,(i\omega v_2-\omega^2\rho_f)\,x_3\nabla^2\tilde{w}=0 \tag{49}$$

where $A_\alpha=\frac{i\omega v_\alpha}{i\omega v_\alpha-\omega^2\rho_f}$, $\tilde{I}^\alpha=\int_{-h/2}^{h/2}\tilde{\tau}^\alpha x_3dx_3$, $B^\alpha=-12v\,A_\alpha/\lambda h^3$

The system of Eqs (47), (48) and (49) is again easily reducible to the equation

$$-D\nabla^4\tilde{w}+\tilde{q}=-\omega^2\rho_s\tilde{w}h \tag{50}$$

when the plate material becomes simply elastic with mass density $\rho=\rho_s$

6. Numerical results

The numerical results of this section have been obtained on the basis of numerical values of the coefficients (Theodorakopoulos and Beskos 1993, 1994) which correspond to water fully-saturated poroelastic sandstone of Fatt (1959) and fissured poroelastic medium, provided in Beskos (1989), respectively.
The physical quantity of interest in this investigation is the frequency-dependent lateral deflection amplitude of the plate $\tilde{w}$. The case of a square plate 4.00 x 4.00 x 0.20 m, with permeable surfaces subjected to a harmonic lateral load of amplitude 1000 N/m^2 was examined. The results show that for the kind of dynamic load used, inertial effects are important only around the resonance area.

Case 1 : One degree of porosity

The response of the plate as a function of the frequency $f=\omega/2\pi$ is shown in Figs 1-2 for various values of porosity n and permeabilities K. From Fig.1 it can be seen that both the first and the second natural frequencies of the elastic plate, (n=0.0001), found to be 18.85 Hz and 94.29 Hz, respectively (Szilard 1974), increase with increasing porosity, especially in the case of the second resonant frequency. The shifting is due to the fact that the frequency is in general proportional to the square root of the stiffness to mass ratio and increasing values of porosity decrease the mass more than the stiffness of the plate. It is also evident that the viscous flow dissipation drastically reduces the amplitude response with increasing porosity for a given permeability, as expected, because of the fact that dissipation forces are proportional to the square of the porosity. Figure 2 shows the variation of $|\tilde{w}|$ in the low frequency range for various values of permeability, with porosity being constant at n=0.30. It is observed that increasing values of permeability result in response increase due to the fact that dissipative forces are inversely proportional to permeability.

Case 2 : Two degrees of porosity

The response of the plate as a function of frequency is shown in Figs 3-4 for various combinations of porosities and permeabilities n_α and K_α (α=1,2), respectively. One can argue on the basis of physical reasons that n and K are the two most important material parameters of the problem (Beskos, 1989). From Fig 3. it can be seen that the dynamic behaviour of the plate was found to be the same with that of a single porosity plate as shown in Fig. 1. It is again shown that the first natural frequency of the elastic plate increases with increasing total porosity for reasons similar to those explained in the previous section. It is also observed from Fig. 3 that the viscous flow dissipation reduces the amplitude response with increasing total porosity, as expected, because of the fact that dissipation forces in both fissures and pores increase with their porosities. Figure 4 shows the variation of $|\tilde{w}|$ versus frequency for various values of the ratio n_1/n where $n=n_1+n_2$ =0.26 and for K_1/K_2=1000 (K_2 =$1\text{x}10^{-8}$ m^2). It is apparent from Fig. 4 that the response of the plate is affected by the porosities only around the resonance area with the amplitude being increased with increasing values of the n_1/n ratio. It is also observed that the first natural frequency of the elastic plate slightly increases with increasing n_1/n ratios for analogous reasons as in the previous case of Fig. 3.

7. Conclusions

The governing equations of flexural vibrations of thin, fluid-saturated poroelastic and fissured poroelastic plates were derived. The dynamic response of a square, simply supported plate with a two degrees of porosity material and subjected to harmonic load was found to be the same with that of a single porosity plate. It was shown that the inertial effects are important only around the resonance area. It was found that in both cases the resonance frequency of the plate increases with total porosity, whereas the viscous flow dissipation reduces the amplitude response with increasing total porosity. It was also shown that increasing values of permeability resulted in response increase whereas the existence of fissures affected the plate response only around the resonance area.

References

Aifantis, E.C. (1980) On the problem of diffusion in solids. Acta Mech. 37 : 256-296

Beskos, D.E. (1989) Dynamics of saturated rocks, I: equations of motion. J. Engng. Mech. ASCE 115 : 982-995

Biot, M.A. (1956) Theory of propagation of elastic waves in a fluid-saturated porous solid, I : low-frequency range. J. Acoust. Soc. Am. 28 : 168-178

Biot, M.A. (1964) Theory of buckling of a porous slab and its thermoelastic analogy. J. Appl. Mech. ASME 31: 194-198

Biot, M.A. and Willis, D.G. (1957) The elastic coefficients of the theory of consolidation. J. Appl. Mech. ASME 24: 594-601

Corapcioglu, M.Y. (1991) Wave propagation in porous media-a review, in J. Bear and M.Y. Corapcioglu (eds.), Transport Processes in Porous Media, Kluwer Academic Publishers, Dordrecht, pp 373-469

Fatt, I. (1959) The Biot-Willis coefficients for a sandstone. J. Appl. Mech. ASME 26 : 296-297

Szilard, R. (1974) Theory and Analysis of Plates : Classical and Numerical Methods. Prentice-Hall Inc., Englewood Cliffs, N. Jersey

Taber, L.A.(1992) A theory for transverse deflection of poroelastic plates. J. Appl. Mech. ASME 59 : 628-634

Theodorakopoulos D.D., and Beskos D.E. (1993). Flexural vibrations of fissured poroelastic plates. Arch. Appl. Mech. 63 : 413-423

Theodorakopoulos, D.D. and Beskos, D.E. (1994) Flexural vibrations of poroelastic plates. Acta Mech. 103 : 191 - 203

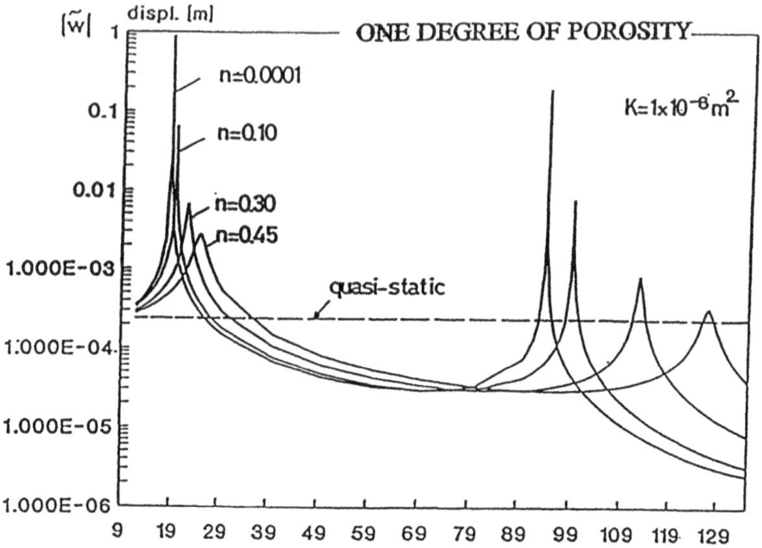

Figure 1. Plate amplitude response versus frequency for various values of porosity : first two resonance areas

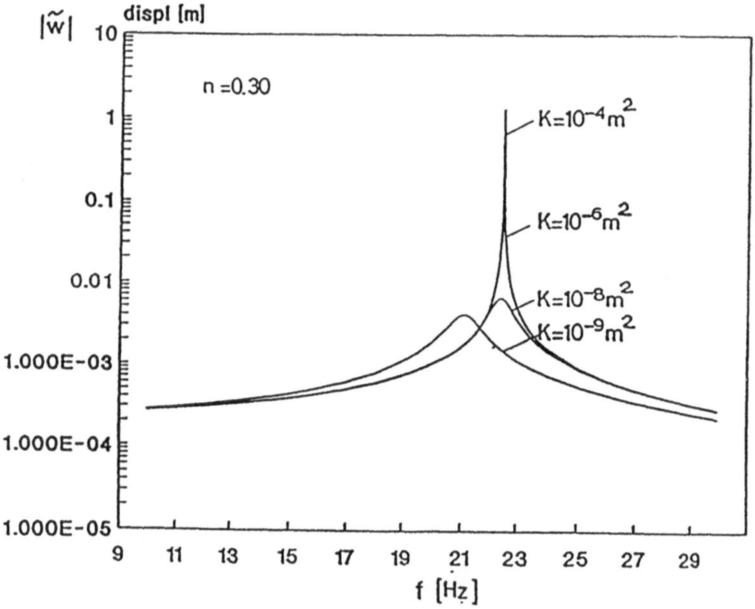

Figure 2. Plate amplitude versus frequency for various values of permeabilities

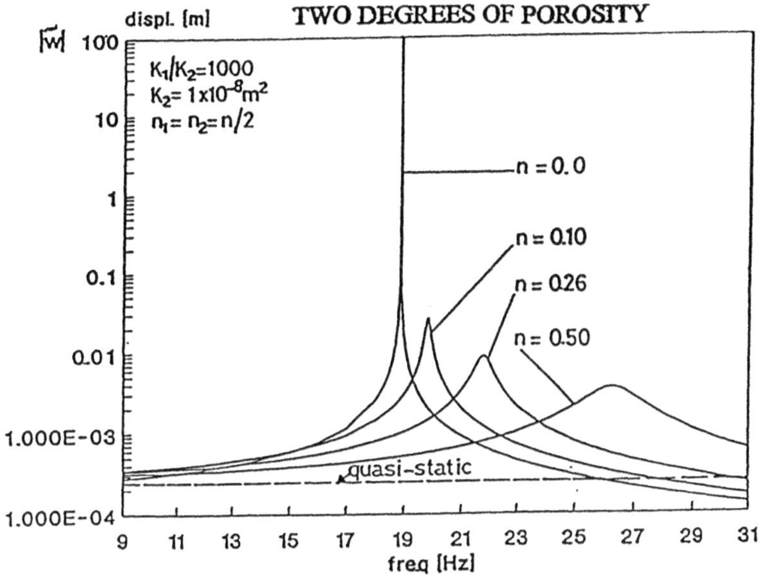

Figure 3. Plate amplitude response versus frequency for various values of total porosity

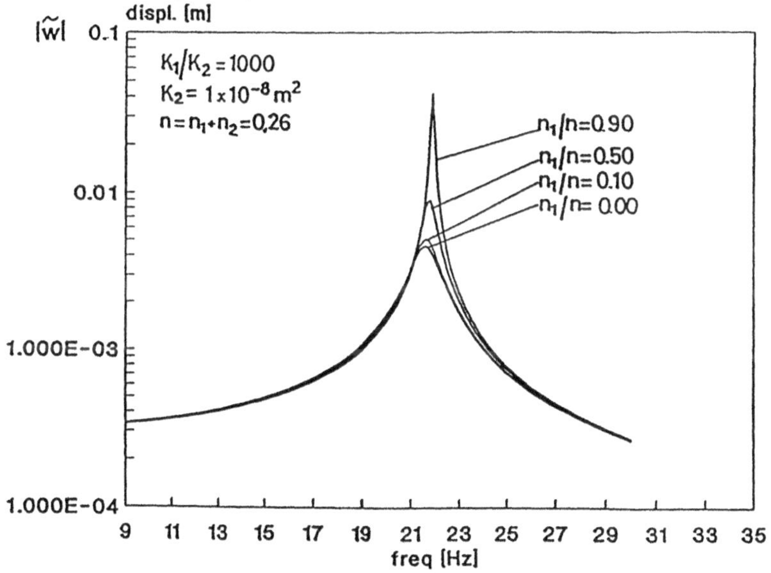

Figure 4. Plate amplitude response versus frequency for various combinations of porosities

Dynamics of a rigid strip bonded to a multilayered poroelastic medium.

T. Senjuntichai and R.K.N.D. Rajapakse
Department of Civil and Geological Engineering
University of Manitoba, Winnipeg, Canada R3T 2N2

Abstract This paper is concerned with the dynamic response of a rigid trip of finite width bonded to the surface of a multilayered poroelastic alf plane. The strip is subjected to time-harmonic vertical, horizontal nd moment loadings. The half plane under consideration consists of number of layers with different thickness and material properties. A iscretization technique based on the rigid body displacement boundary onditions at the strip-half plane contact surface and dynamic Green's unctions of a multilayered poroelastic half plane is used to analyze the nteraction problem. The discretization method used in this paper yields a exibility equation system for determination of contact tractions at strip-alf plane interface. Comparisons with existing studies for a rigid strip n a homogeneous medium are presented to establish the accuracy of the resent scheme. Selected numerical results for impedances of a rigid strip n different poroelastic media are presented to portray the influence of oroelastic effects on the dynamic interaction problem.

Introduction

he study of dynamic response of rigid foundations is very useful to the alysis and design of foundations subjected to dynamic loadings (e.g. brating machines, seismic waves, etc.). A number of problems related vibrations of a rigid foundation on homogeneous and layered ideal e-stic media have been considered in the past by employing a variety of alytical and numerical techniques (Karasudhi *et al.*, 1968; Luco and estmann, 1972; Kausel, 1974; Kausel *et al.*, 1975; Gazetas and Roesset, 79 and Wong and Luco, 1986). Generally, geomaterials are two-phased aterials consisting of a solid skeleton with voids filled with water. Such

. S. Selvadurai (ed.), *Mechanics of Poroelastic Media*, 353–369.

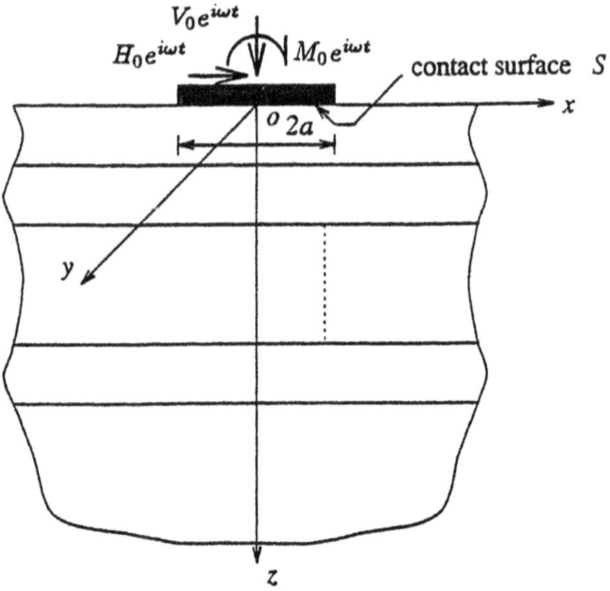

Figure 1: *Geometry of the problem considered in the study*

a material is commonly known as a poroelastic material and is considere as a much more realistic representation for soils and rocks than an ide elastic material.

The first theory of elastic wave propagation in a fluid-saturated poro medium was established by Biot (1956a,b) by adding inertia terms to h quasi-static theory (Biot, 1941). Given the fact that majority of so are naturally poroelastic it is very useful to investigate the influence poroelastic effects in dynamic soil-structure interaction. In recent year some studies have investigated the dynamic response of foundations (poroelastic media. For example, Kassir and Xu (1988) considered t mixed-boundary value problems related to time-harmonic response of rigid strip bonded to a homogeneous poroelastic half plane. Halpern a Christiano (1986) and Philippacopoulos (1989) studied time-harmonic r sponse of a rigid plate in smooth contact with saturated and partia saturated poroelastic half spaces, respectively. Recently, Bougacha *et* (1993) studied problems related to vibrations of a rigid footing on a flu saturated homogeneous layer by extending the finite element concep developed by Lysmer and Waas (1972) and Waas (1972).

In this paper, a methodology is presented to analyze the general case of a rigid permeable strip foundation bonded to a multilayered poroelastic medium. The system under consideration is shown in Fig. 1. The interaction problem is analyzed by dividing the contact surface between a strip foundation and a layered half plane into a number of small strips with uniform traction distribution. Nodal points are selected at the center of each discretized region. A flexibility equation system is established to determine the intensities of contact tractions at different nodal locations by imposing the appropriate rigid body displacement boundary conditions. The influence functions required to establish the flexibility equation system correspond to the displacement of a multilayered half plane due to vertical and horizontal strip loadings applied at the surface level. The numerical solution of the flexibility equation system and the equilibrium of forces result in the impedances of a strip foundation. Selected numerical results are presented to demonstrate the influence of layering and the frequency of excitation on impedances of a rigid strip on different poroelastic media.

Basic Equations

Consider a poroelastic medium with a Cartesian coordinate system (x, y, z) defined as shown in Fig. 1. It is assumed that the deformations are plane strain in the xz–plane, i.e. $\epsilon_{xy} = \epsilon_{yy} = \epsilon_{yz} = 0$. Let $u_i(x, z, t)$ and $w_i(x, z, t)$ denote the average displacement of the solid matrix and the fluid displacement relative to the solid matrix, in the i-direction ($i = x, z$), respectively. Then, the constitutive relation of a homogeneous poroelastic material can be expressed by using standard indicial notations as (Biot, 1941)

$$\sigma_{ij} = 2\mu\epsilon_{ij} + \lambda\delta_{ij}\epsilon_{kk} - \alpha\delta_{ij}p, \qquad i,j = x,z \tag{1}$$

$$p = -\alpha M\epsilon_{kk} + M\zeta \tag{2}$$

where

$$\zeta = -(\frac{\partial w_x}{\partial x} + \frac{\partial w_z}{\partial z}) \tag{3}$$

In the above equations, σ_{ij} is the stress component of the bulk materia ϵ_{ij} and ϵ_{kk} are the strain component and the dilatation of the solid matri respectively, which are related to the displacement u_i as in ideal elasticit μ and λ are Lamé constants of the bulk material; p is the excess po fluid pressure (suction is considered negative); ζ is the variation of th fluid content per unit reference volume and δ_{ij} is the Kronecker delta. I addition, α and M are Biot's parameters accounting for the compressibili of the two-phased material. It is noted that $0 \leq \alpha \leq 1$ and $0 \leq M$ ∞. For a completely dry material $M \rightarrow 0$ whereas for a material wi incompressible constituents $M \rightarrow \infty$ and $\alpha \rightarrow 1$.

The expression for the fluid discharge in the i-direction ($i = x$, denoted by q_i is given by

$$q_i = \frac{\partial w_i}{\partial t} \qquad ($$

The Fourier integral transform of a function $f(x, z)$ with respect to t x-coordinate is defined by (Sneddon, 1951)

$$\bar{f}(\xi, z) = \frac{1}{\sqrt{2\pi}} \int_{-\infty}^{\infty} f(x, z) e^{-i\xi x} dx$$

and the inverse relationship is given by

$$f(x, z) = \frac{1}{\sqrt{2\pi}} \int_{-\infty}^{\infty} \bar{f}(\xi, z) e^{i\xi x} d\xi$$

It is assumed that the motions under consideration are time-harmo of the form $e^{i\omega t}$, where ω is the circular frequency, $i = \sqrt{-1}$ and the te $e^{i\omega t}$ is omitted in the sequel. The equations of motion (Biot, 1962) o poroelastic medium can be expressed in terms of displacements u_i and by using standard indicial notation as

$$\begin{aligned} \mu u_{i,jj} + (\lambda + \alpha^2 M + \mu) u_{j,ji} + \alpha M w_{j,ji} &= \rho \ddot{u}_i + \rho_f \ddot{w}_i \\ \alpha M u_{j,ji} + M w_{j,ji} &= \rho_f \ddot{u}_i + m \ddot{w}_i + b \dot{w}_i \end{aligned}$$

In eqns (7) and (8), ρ and ρ_f are the mass densities of the bulk material the pore fluid, respectively; m is a density-like parameter which depe on ρ_f and the geometry of the pores and b is a parameter accounting the internal friction due to the relative motion between the solid ma and the pore fluid. If internal friction is neglected then $b = 0$. In addition, the superscript dot denotes the derivative with respect to time.

The eqns (7) and (8) can be solved by using Helmholtz representation for a vector field and applying Fourier integral transform with respect to the x-coordinate. The general solutions for poroelastic fields can be expressed in the Fourier-frequency domain in the following matrix form (Senjuntichai, 1994):

$$\mathbf{u}(\xi, z) = \mathbf{R}(\xi, z)\mathbf{C}(\xi) \tag{9}$$

$$\mathbf{f}(\xi, z) = \mathbf{S}(\xi, z)\mathbf{C}(\xi) \tag{10}$$

where

$$\mathbf{u}(\xi, z) = < i\bar{u}_x \quad \bar{u}_z \quad \bar{p} >^T \tag{11}$$

$$\mathbf{f}(\xi, z) = < i\bar{\sigma}_{zx} \quad \bar{\sigma}_{zz} \quad \bar{w}_z >^T \tag{12}$$

$$\mathbf{C}(\xi) = < A \quad B \quad C \quad D \quad E \quad F >^T \tag{13}$$

The arbitrary functions $A(\xi), B(\xi), ..., F(\xi)$ appearing in $\mathbf{C}(\xi)$ are to be determined by employing appropriate boundary and/or continuity conditions and the matrices $\mathbf{R}(\xi, z)$ and $\mathbf{S}(\xi, z)$ are given elsewhere (Senjuntichai, 1994).

Influence functions

The analysis of the problem shown in Fig. 1 by using the proposed discretization technique requires the computation of a set of displacement influence functions of a multilayered half plane due to surface strip loadings of unit intensity (Fig. 2). An exact stiffness matrix method is used to calculate the required influence functions. A brief outline of the stiffness matrix scheme is presented in this section and the details are given elsewhere (Senjuntichai, 1994).

A multilayered half plane with a total of N poroelastic layers overlying a poroelastic half plane is considered. Layers and interfaces are numbered as shown in Fig. 2. A superscript "n" is used to denote quantities associated with the nth layer ($n = 1, 2, ..., N$). For an nth layer, the following

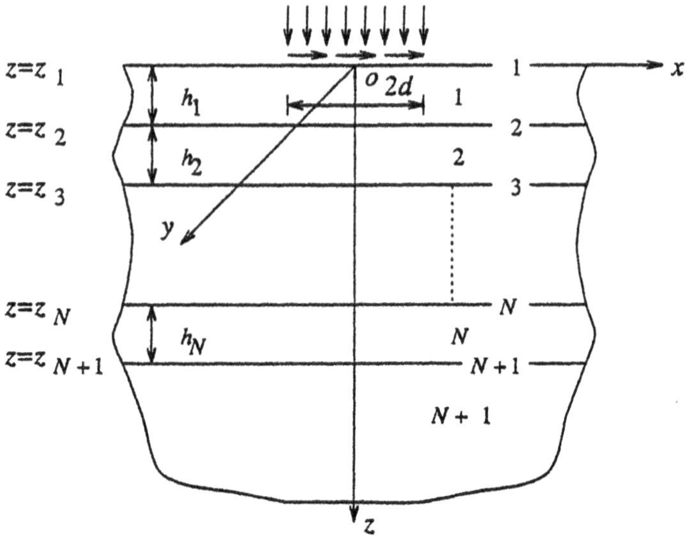

Figure 2: *Multilayered half plane under surface loadings*

relationships can be established by using eqns (9) and (10):

$$\mathbf{U}^{(n)} = \begin{bmatrix} \mathbf{R}^{(n)}(\xi, z_n) \\ \cdots\cdots\cdots \\ \mathbf{R}^{(n)}(\xi, z_{n+1}) \end{bmatrix} \mathbf{C}^{(n)} \quad (14)$$

$$\mathbf{F}^{(n)} = \begin{bmatrix} -\mathbf{S}^{(n)}(\xi, z_n) \\ \cdots\cdots\cdots \\ \mathbf{S}^{(n)}(\xi, z_{n+1}) \end{bmatrix} \mathbf{C}^{(n)} \quad (15)$$

where

$$\mathbf{U}^{(n)} = < \mathbf{u}^{(n)}(\xi, z_n) \quad \mathbf{u}^{(n)}(\xi, z_{n+1}) >^T \quad (16)$$

$$\mathbf{F}^{(n)} = < -\mathbf{f}^{(n)}(\xi, z_n) \quad \mathbf{f}^{(n)}(\xi, z_{n+1}) >^T \quad (17)$$

In eqns (14)-(17), $\mathbf{U}^{(n)}$ denotes a vector of generalized coordinates for the nth layer whose elements are the Fourier transforms of displacements and pore pressure of the top and bottom surfaces of the nth layer. Similarly $\mathbf{F}^{(n)}$ denotes a generalized force vector whose elements are the Fourier transforms of tractions and fluid displacements of the top and bottom faces of the nth layer. The vector $\mathbf{C}^{(n)}$ in eqns (14) and (15) is the bitrary coefficients vector corresponding to the nth layer. The matrices $^{n)}$ and $\mathbf{S}^{(n)}$ in eqns (14) and (15) are identical to $\mathbf{R}$ and $\mathbf{S}$ defined in ns (9) and (10) except that the material properties of the nth layer are ployed in the definition and $z = z_n$ and z_{n+1}. The eqn (14) can be erted to express $\mathbf{C}^{(n)}$ in terms of $\mathbf{U}^{(n)}$ and the substitution in eqn (15) lds

$$\mathbf{F}^{(n)} = \mathbf{K}^{(n)}\mathbf{U}^{(n)}, \qquad n = 1, 2, ..., N \tag{18}$$

ere $\mathbf{K}^{(n)}$ can be considered as an exact stiffness (impedance) matrix in Fourier transform space for the nth layer. Similarly, the stiffness max for the underlying half plane $\mathbf{K}^{(N+1)}$ can be obtained by establishing a ationship between the generalized displacement vector $\mathbf{U}^{(N+1)}$ and the ce vector $\mathbf{F}^{(N+1)}$.

The global stiffness matrix of a multilayered half plane is assembled using the layer and half plane stiffness matrices, and the continuity of ctions and fluid flow at layer interfaces. The number of unknowns in final equation system for the layered half plane shown in Fig. 2 is ial to $3(N+1)$. The final equation system can be expressed as

$$\mathbf{K}^*\mathbf{U}^* = \mathbf{F}^* \tag{19}$$

ere $\mathbf{K}^*$ is the global stiffness matrix, and $\mathbf{U}^*$ is the global vector of eralized coordinates. The global generalized force vector corresponding the loading configurations shown in Fig. 2 can be expressed as

$$= \begin{cases} < -\sqrt{\tfrac{2}{\pi}}\sin(\xi d)/\xi \quad 0 \quad 0 \quad ... \quad 0 >^T & \text{for horizontal strip load of unit intensity} \\ \\ < 0 \quad -\sqrt{\tfrac{2}{\pi}}\sin(\xi d)/\xi \quad 0 \quad ... \quad 0 >^T & \text{for vertical strip load of unit intensity} \end{cases} \tag{20}$$

ere d is the half width of the loading strip. The solution of eqn (19) for crete values of ξ together with the application of numerical quadrature evaluate inverse Fourier transforms defined by eqn (6) yield the solution displacements and pore pressure at layer interfaces. In the analysis of interaction problem shown in Fig. 1, only the displacements at the face level are required.

4 Formulation of interaction problem

Consider a massless rigid strip of width $2a$ bonded to a multilayered po
lastic half plane as shown in Fig. 1. The strip is assumed to be f
permeable and subjected to time-harmonic vertical, horizontal and
ment loadings $V_0e^{i\omega t}$, $H_0e^{i\omega t}$ and $M_0e^{i\omega t}$ per unit length, respectively
the point o. The displacements at an arbitrary point in the strip car
written in terms of vertical displacement $\Delta_V e^{i\omega t}$, horizontal displacen
$\Delta_H e^{i\omega t}$ and rotation $\phi_0 e^{i\omega t}$ about the y-axis at a point o ($x = 0, z = 0$

$$u_x(x,0) = \Delta_H \quad ($$
$$u_z(x,0) = \Delta_V + x\phi_0 \quad ($$

The contact tractions in the x- and z-directions at the strip-half pl
interface due to the applied loadings are denoted by T_x and T_z, respect
ly. Then, the equations of equilibrium of a massless strip can be expres
as

$$V_0 = -\int_{-a}^{a} T_z dx \quad ($$
$$H_0 = -\int_{-a}^{a} T_x dx \quad ($$
$$M_0 = -\int_{-a}^{a} T_z x dx \quad ($$

In the application of the discretization technique used in this pa
the contact surface S between a strip and a layered half plane is discreti
by L equally spaced nodes. It is assumed that the contact stress dis
bution is uniform within the tributary length of each node. Then,
displacement $u_i(x_k, 0), i = x, z$, at the lth node with coordinates $(x_k$
can be expressed as

$$u_i(x_k, 0) = \sum_{l=1}^{L} G_{ij}(x_k; x_l)T_j(x_l)dx, \qquad j = x, z \quad ($$

where $G_{ij}(x_k; x_l)$ denotes the displacement in the i-direction at the
node point on the surface of a layered half plane due to a uniform strip l
of unit intensity applied in the j-direction over the tributary length of

th node point with coordinate $(x_l, 0)$. The influence functions $G_{ij}(x_k; x_l)$ re computed from the stiffness matrix scheme presented in the preceding ection. The eqn (26) can be expressed in the following matrix form

$$\mathbf{GT} = \mathbf{D} \tag{27}$$

here the elements of the vectors **D** and **T** correspond to the surface dis- lacements and the contact stresses, respectively, at nodal locations along he strip-half plane interface. The elements of the matrix **G** correspond to $G_{ij}(x_k; x_l)$. The vector **D** can be expressed in terms of Δ_V, Δ_H and ϕ_0 by sing eqns (21) and (22). Thereafter, eqn (27) can be inverted to express he traction vector **T** in terms of the generalized coordinates Δ_V, Δ_H nd ϕ_0. The substitution of solution for nodal tractions in eqns (23)-(25) ields the following relationship between applied forces and displacements Δ_V, Δ_H and ϕ_0.

$$\begin{Bmatrix} V_0 \\ H_0 \\ M_0 \end{Bmatrix} = \pi\mu \begin{bmatrix} K_V & 0 & 0 \\ 0 & K_H & K_{HM} \\ 0 & K_{MH} & K_M \end{bmatrix} \begin{Bmatrix} \Delta_V \\ \Delta_H \\ a\phi_0 \end{Bmatrix} \tag{28}$$

here $K_V, K_H, K_{HM}(= K_{MH})$ and K_M are the vertical, horizontal, cou- ed and rocking impedances of a rigid strip, respectively.

Numerical results and discussion

ne Figure 3 shows a comparison of horizontal and rocking impedances a rigid strip bonded to an ideal elastic layer of thickness a (Poisson's tio $= 0.3$) with a rigid base. A nondimensional frequency defined as $= \omega a\sqrt{\rho/\mu}$ is used for the comparison. The solutions obtained from the esent scheme shown in Fig. 3 agree very closely with those reported by ang-Liang (1974) using a coupled finite element transmitting boundary ment scheme. The strip-half plane contact surface was discretized by teen [i.e. $L = 16$ in eqn (26)] nodes to obtain the solutions shown in Fig. Additional numerical studies confirm that the solutions for impedances verge for $L = 12 - 20$. All subsequent numerical solutions correspond $L = 16$. The accuracy of the present scheme in analyzing problems

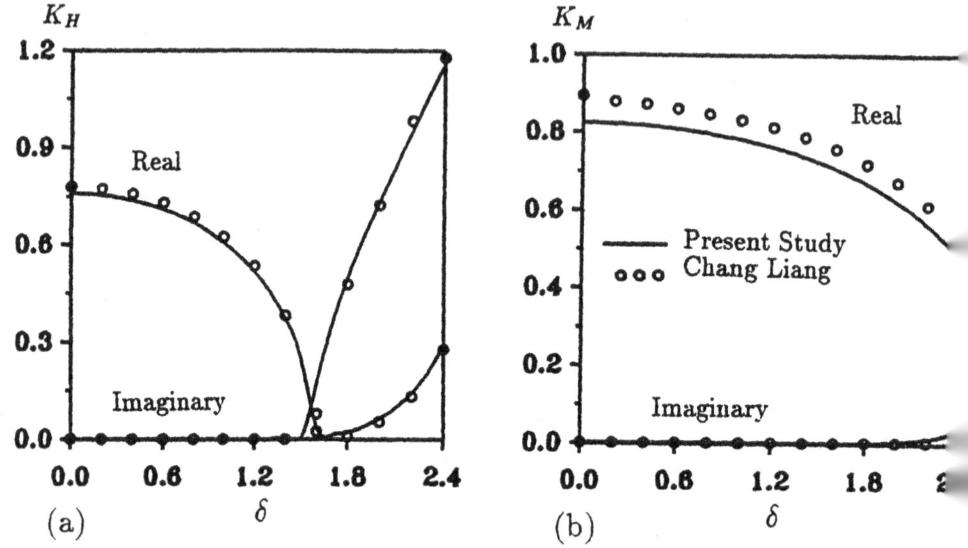

Figure 3: *Impedances of a rigid strip bonded to an ideal elastic layer* (ν 0.3) *with rigid base: (a) horizontal impedance; (b) rocking impedance*

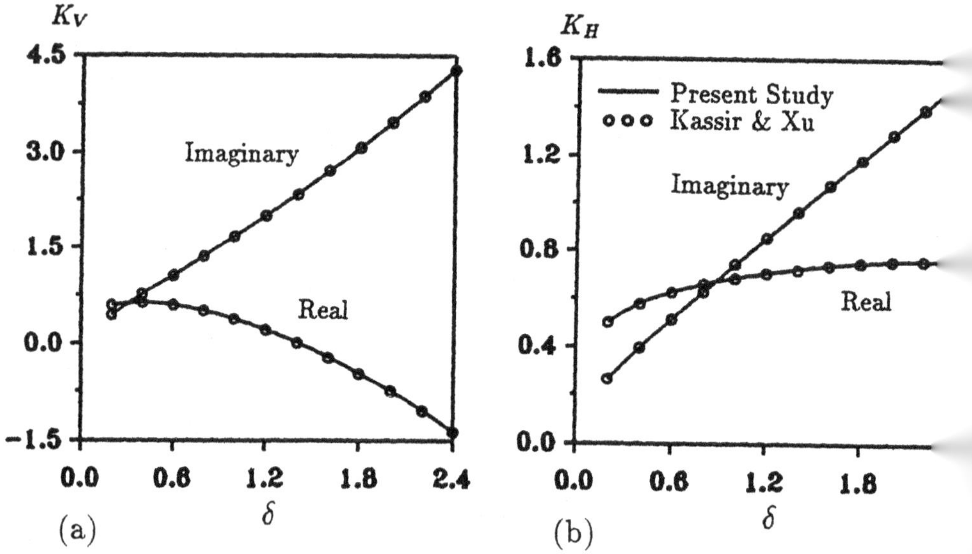

Figure 4: *Impedances of a rigid strip in smooth contact with a hon geneous poroelastic half plane: (a) vertical impedance; (b) horizon impedance*

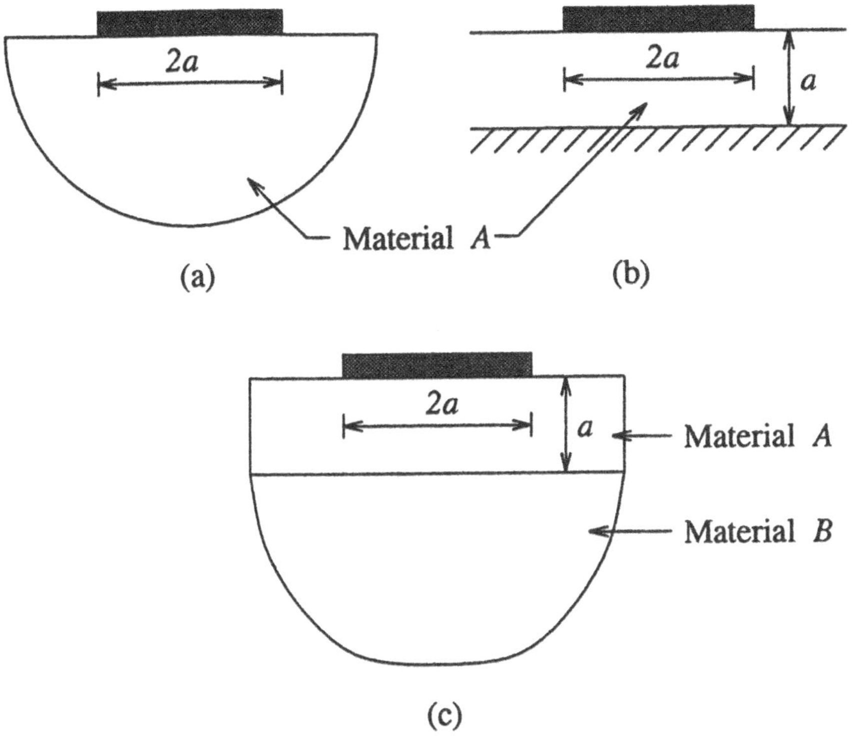

'igure 5: *Rigid strip on different poroelastic systems: (a) homogeneous alf plane; (b) homogeneous layer with impermeable rigid base; (c) layered alf plane*

volving a poroelastic half plane is investigated by considering the case a permeable rigid strip on a homogeneous poroelastic half plane. Kassir d Xu (1988) presented a solution to this problem by numerically solving e governing integral equations for contact tractions. Figure 4 shows the mparison of present solutions for vertical and horizontal impedances th those reported by Kassir and Xu (1988) over the nondimensional quency range $0.2 \leq \delta \leq 2.4$. The two sets of solutions agree very sely. The solutions shown in Fig. 4 correspond to the smooth contact se and are very close to the fully bonded case. The solutions for rocking pedances also agree very well with those given by Kassir and Xu and not presented here for brevity. The accuracy of the present analysis is

Table 1: Properties of material A and B

	$\mu^{\dagger}$	$\lambda^{\dagger}$	$M^{\dagger}$	$\rho^{\ddagger}$	$\rho_f^{\ddagger}$	$m^{\ddagger}$	$b^{\S}$	α
Material A	2.0	3.0	25.0	2.0	1.0	3.0	2.0	0.95
Material B	10.0	10.0	20.0	2.5	1.0	5.0	10.0	0.9

$\dagger$ $\times 10^8$ N/m^2; $\ddagger$ $\times 10^3$ kg/m^3; $\S$ $\times 10^6$ N-s/m^4

clearly established by the comparison shown in Figs. 3 and 4.

In the remainder of this paper, the dynamic response of rigid stri footings shown in Fig. 5 is considered. The study of multilayered systen s involves the consideration of many parameters such as layer thicknes material properties, etc.. Among the many possible choices, the layere systems shown in Fig. 5 are selected as practically useful cases. The prop erties of materials A and B involved in the systems shown in Fig. 5 ar defined in Table 1. A nondimensional frequency defined as $\delta = \omega a\sqrt{\rho_A/\mu}$ is used hereafter in the presentation of the numerical results where ρ_A an μ_A are mass density and shear modulus of material A, respectively. Th Figure 6 shows the variation of vertical impedances of rigid strip sy tems shown in Fig. 5 and that of a strip on an ideal elastic half spa with Lamé constants equal to that of material A. Figures 7-9 show th horizontal (K_H), rocking (K_M) and coupled ($K_{HM} = K_{MH}$) impedanc of the above strip-half plane systems. Comparison of the real parts impedances corresponding to the four different strip-half plane systen shows substantial difference in the values. At low frequencies ($\delta < 1.$ the real part of all components of impedances corresponding to a str on a poroelastic layer with an impermeable rigid base is higher than t corresponding real part of impedances of a strip on other systems. Th behaviour is physically acceptable since the presence of a rigid base mak the system shown in Fig. 5(b) more stiff than the other systems. In a dition, in the low frequency range a strip on the homogeneous poroelas half space of material A show the highest radiation damping except the coupled impedance.

The dependence of impedances on material properties of a poroelas

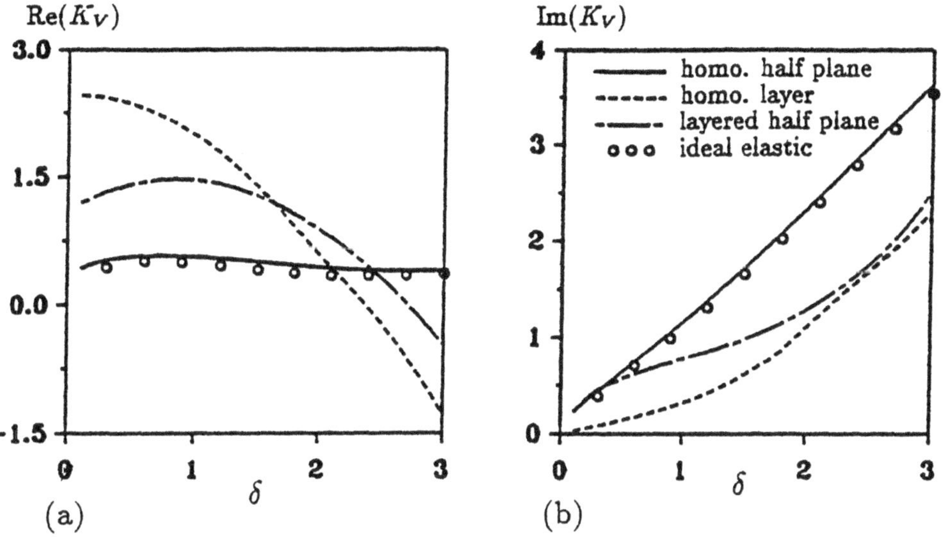

Figure 6: *Vertical impedances of a rigid strip bonded to different poroelastic media: (a) real part; (b) imaginary part*

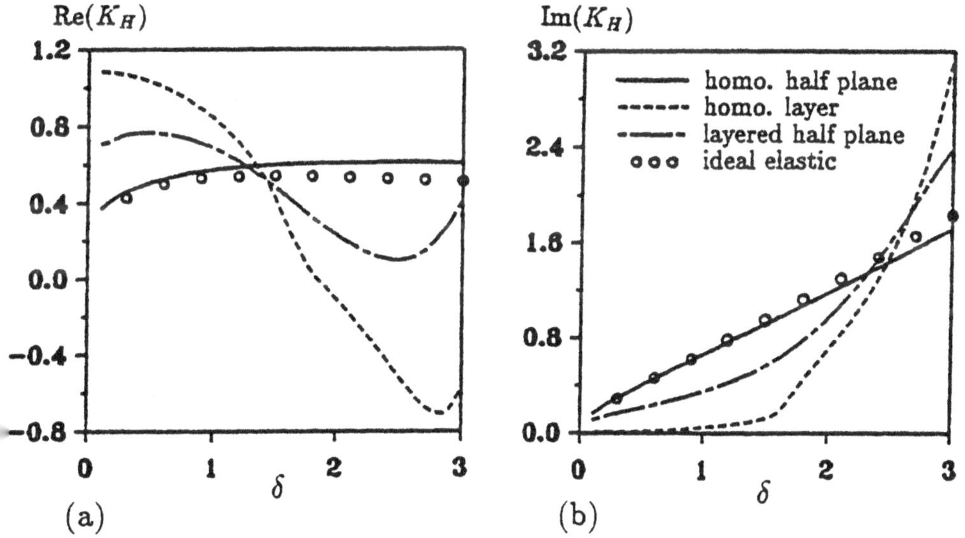

gure 7: *Horizontal impedances of a rigid strip bonded to different poro-stic media: (a) real part; (b) imaginary part*

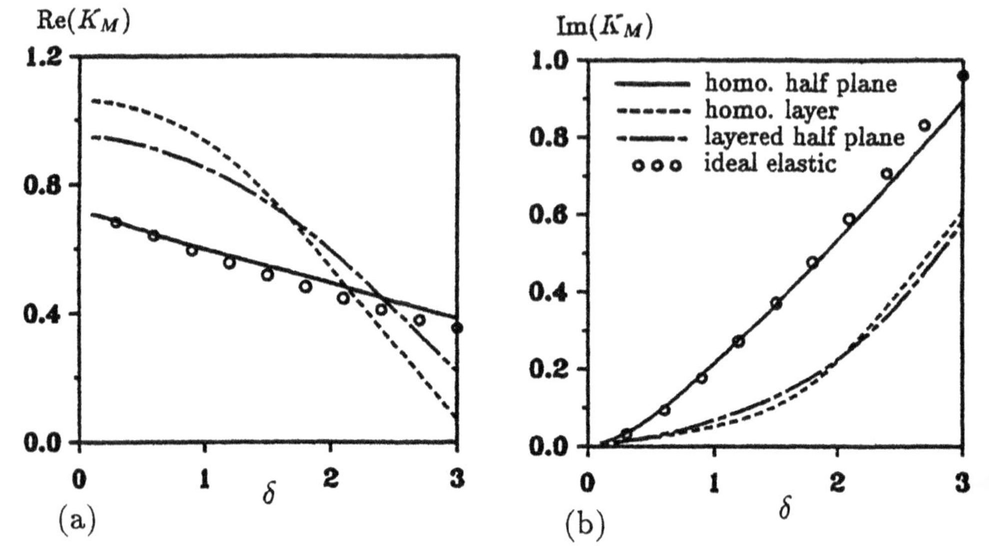

Figure 8: *Rocking impedances of a rigid strip bonded to different poroelasti media: (a) real part; (b) imaginary part*

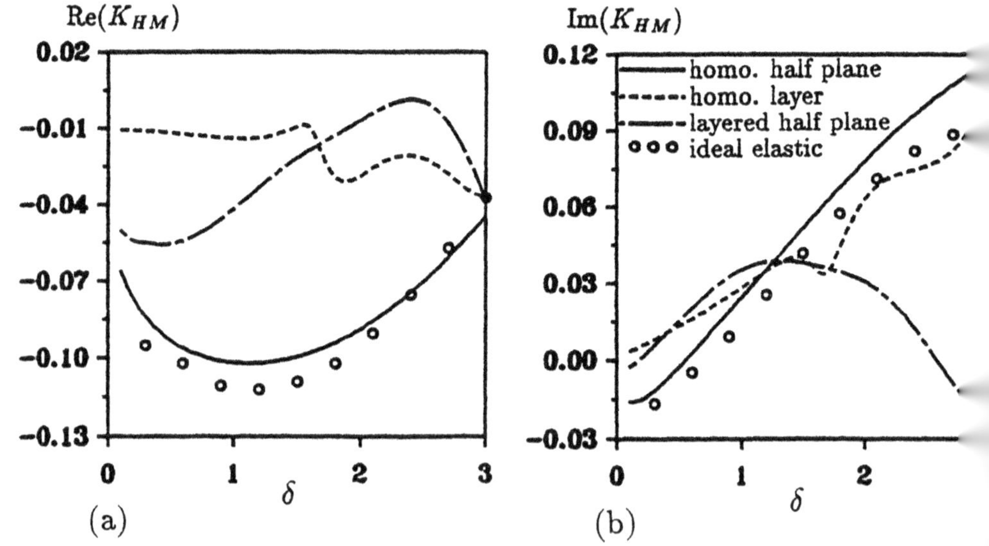

Figure 9: *Coupled impedances of a rigid strip bonded to different poroela tic media: (a) real part; (b) imaginary part*

half plane becomes more complicated for $\delta > 1.5$. In the high frequency range ($\delta > 1.5$) the impedances of systems shown in Fig. 5 show oscillatory variations with the frequency especially for K_H and K_{HM}. The impedances of a strip on a homogeneous half plane vary smoothly with δ over the frequency range considered in this study. The real part of K_V and K_M corresponding to the systems shown in Fig. 5 decreases with δ whereas the imaginary part increases over the frequency range $\delta = 0.0 - 3.0$.

6 Conclusions

A semi-analytical discretization scheme is presented to analyze the time-harmonic response of a rigid strip on a multilayered poroelastic half plane. The present scheme requires the computation of displacement influence functions of a multilayered poroelastic half plane due to surface strip loadings. These influence functions are computed by using an exact stiffness matrix method. The solutions for impedances of a rigid strip obtained from the present scheme agree very closely with existing schemes for a strip on an ideal elastic layer with a rigid base and a strip on a homogeneous poroelastic half plane. The strip-half plane contact surface should be divided to at least twelve nodes to obtain accurate solutions. It is found that the dynamic response of a strip on a poroelastic layered system is substantially different from that on a homogeneous half plane. The frequency of excitation, layer thickness and layer material properties (especially the parameters b and M) significantly influence the impedances of a rigid strip on a layered medium. The solutions presented in this study can be used to evaluate the accuracy of finite element and other numerical methods in the analysis of dynamic poroelastic soil-structure interaction problems.

Acknowledgements: The work presented here was supported by Natural Sciences and Engineering Research Council of Canada Grant A-6507.

References

Biot,M.A. (1941) General theory of three-dimensional consolidation. J. Applied Physics. 12:155-164.

Biot,M.A. (1956a,b) Theory of propagation of elastic waves in a flui saturated porous solid. Part I-Low frequency range, Part II-High fr quency range. J. Acoustical Society of America. 28:168-191.

Biot,M.A. (1962) Mechanics of deformation and acoustic propagation porous media. J. Applied Physics. 33:1482-1498.

Bougacha,S., Roesset,J.M. and Tassoulas,J.L. (1993) Dynamic stiffness foundations on fluid-filled poroelastic stratum. J. Engineering Mecha ics, ASCE. 119:1649-1662.

Chang-Liang,V. (1974) Dynamic response of structure in layered soi Ph.D. Thesis, Massachusetts Institute of Technology.

Gazetas,G. and Roesset,J.M. (1979) Vertical vibration of machine fou dations. J. Geotechnical Engineering Division, ASCE. 105:1435-1454

Halpern,M.R. and Christiano,P. (1986) Steady state harmonic response a rigid plate bearing on a liquid-saturated poroelastic halfspace. Ear quake Engineering and Structural Dynamics. 14:439-454.

Karasudhi,P., Keer,L.M. and Lee,S.L. (1968) Vibratory motion of a bo on an elastic half plane. J. Applied Mechanics, ASME. 35:697-705.

Kassir,M.K. and Xu,J. (1988) Interaction functions of a rigid strip bon to saturated elastic half-space. Int. J. Solid and Structures. 24:915-9

Kausel,E. (1974) Forced vibration of circular foundations on layered r dia. Research Report R74-11, Massachusetts Institute of Technolog

Kausel,E., Roesset,J.M. and Waas,G. (1975) Dynamic analysis of footi on layered media. J. Engineering Mechanics Division, ASCE. 101: 105.

Luco,J.E. and Westmann,R.A. (1972) Dynamic response of a rigid foot bonded to an elastic half space. J. Applied Mechanics, ASME. 39:5 534.

Lysmer,J. and Waas,G. (1972) Shear waves in plane infinite structures. J. Engineering Mechanics Division, ASCE. 98: 85-105.

Philippacopoulos,A.J. (1989) Axisymmetric vibration of disk resting on saturated layered half-space. J. Engineering Mechanics, ASCE. 115: 1740-1759.

Senjuntichai,T. (1994) Green's functions for multi-layered poroelastic media and an indirect boundary element method. Ph.D. Thesis, University of Manitoba.

Sneddon,I.N. (1951) Fourier Transforms. McGraw-Hill, New York.

Waas,G. (1972) Linear two-dimensional analysis of soil-dynamic problems in semi-infinite layered media. Ph.D. Thesis, University of California, Berkeley.

Wong,H.L. and Luco,J.E. (1986) Dynamic response of rigid foundations in a layered half space. Soil Dynamics and Earthquake Engineering. 5:149-158.

Analyses of waves in 3–d poroelastic media

H. Antes
Institute of Applied Mechanics
Technical University Braunschweig
P.O. Box 3329
38023 Braunschweig, Germany

and

T. Wiebe
Computational Structural Dynamics
Ruhr–University Bochum
44780 Bochum, Germany

Abstract. This contribution discusses the application of the boundary element method to wave propagation problems in three–dimensional poroelastic media. Linear elastic porous media, fully saturated with a compressible fluid, have important applications in soil dynamics. Since problems of radiation or scattering of waves often include solutions in infinite domains, the boundary element method is a well suited method for various soil mechanics problems. Different formulations based on Biot's theory are developed and discussed.

1 Introduction

Wave propagation problems in linear elastic porous media can be described by the equations of motion introduced by Biot (1956a; 1956b). Based on the constitutive law and the equations of motion of Biot's theory, the equations of motion, the integral equations and the fundamental solutions for different formulations are examined and derived, respectively. This is done for the time dependent equations as well as for the equations in the Laplace transformed domain.

In Biot's original work, the state of motion is described by the displacement vectors of the solid and the fluid, i.e. by six equations. In order to reduce the numerical effort and to arrive at a more efficient formulation of the boundary conditions, it is advantageous to replace the fluid displacement vector by the scalar pore pressure. This reduction to four state functions is usually used in consolidation theory of porous media.

Finally, one obtains a total of four different formulations: one with the solid and the fluid displacement vectors, and one with the solid displacement vector and the scalar pore pressure in the time domain and two in the Laplace transform domain.

Laplace domain integral equations have already been studied by Manolis and Beskos (1989; 1990) using the fluid displacement and in the context of two–dimensional problems by Cheng et al. (1991) and Dominguez (1991) with the pore pressure. Work on the time domain formulations of the fundamental solutions has been performed by Burridge and Vargas (1979) using the formulation with the fluid displacement vector. A comparison of various integral equations and associ-

A. P. S. Selvadurai (ed.), Mechanics of Poroelastic Media, 371–387.

ated fundamental solutions for wave propagation problems in poroelastic media can be found in (Wiebe, 1993). The formulations using the pore pressure in the time domain can only be developed when the relative motion between the two components of the poroelastic material is assumed to be non–dissipative. Only in this case, it is possible to derive a set of differential equations and to find a closed form fundamental solution.

2 Governing equations of dynamic poroelasticity

As investigated by Biot (1956a), the behavior of a linear elastic, fluid saturated medium can be described by the elastic Lamé constants λ and μ

$$\lambda = \frac{2\nu G}{1-2\nu} \tag{1}$$

$$\mu = G \tag{2}$$

and two special poroelastic constitutive coefficients

$$R = G\Phi^2 B^2 \frac{2(1-2\nu)(1+\nu_u)^2}{9(1-2\nu_u)(\nu_u-\nu)} \tag{3}$$

$$Q = -R + G\Phi B \frac{2(1+\nu_u)}{3(1-2\nu_u)} \tag{4}$$

In these equations, ν_u is the undrained Poisson's ratio, and B is the Skempton pore pressure coefficient (Skempton, 1954). Besides these coefficients, the shear modulus G, the elastic Poisson's ratio ν, and the porosity Φ, are well known material constants in soil mechanics, and can be measured using conventional laboratory techniques; see, e.g., Rice and Cleary (1976) for a discussion.

Using the equations above, the stress–strain relations for a linear elastic fluid saturated porous medium can be given by:

$$\sigma_{ij}^s = 2\mu\varepsilon_{ij}^s + (\lambda + \frac{Q^2}{R})\delta_{ij}\varepsilon_{kk}^s + Q\delta_{ij}\varepsilon_{kk}^f \tag{5}$$

$$\sigma^f = Q\varepsilon_{ii}^s + R\varepsilon_{ii}^f \tag{6}$$

where σ_{ij}^s and σ^f are the stresses, and ε_{ij}^s and ε_{kk}^f are the corresponding elastic strains and fluid dilatation, which act on the elastic and fluid parts, respectively. The total stress is defined as:

$$\sigma_{ij} = \sigma_{ij}^s + \delta_{ij}\sigma^f . \tag{7}$$

The fluid pore pressure p is related to the fluid stress σ^f via the porosity Φ:

$$p = -\frac{\sigma^f}{\Phi} . \tag{8}$$

Using linear strain–displacement relations, the equations of motions

$$\sigma^s_{ij,j} - b(\dot{u}^s_i - \dot{u}^f_i) + F^s_i = \rho_{11}\ddot{u}^s_i + \rho_{12}\ddot{u}^f_i \tag{9}$$

$$\sigma^f_{,i} + b(\dot{u}^s_i - \dot{u}^f_i) + F^f_i = \rho_{21}\ddot{u}^s_i + \rho_{22}\ddot{u}^f_i \tag{10}$$

may be rewritten with the constitutive equations (5) and (6) as

$$(\lambda + \mu + \frac{Q^2}{R})u^s_{j,ji} + \mu u^s_{i,jj} + Qu^f_{j,ji} - b(\dot{u}^s_i - \dot{u}^f_i) + F^s_i = \rho_{11}\ddot{u}^s_i + \rho_{12}\ddot{u}^f_i \tag{11}$$

$$Qu^s_{j,ji} + Ru^f_{j,ji} + b(\dot{u}^s_i - \dot{u}^f_i) + F^f_i = \rho_{21}\ddot{u}^s_i + \rho_{22}\ddot{u}^f_i \tag{12}$$

In these equations, the subscripted commas and overdots indicate space derivatives and time derivatives, respectively. The solid and fluid displacements are denoted by u^s_i and u^f_i. The internal viscous friction, that is caused by the relative movement between fluid and solid, is described by b. F^s_i and F^f_i are the solid and fluid body force vectors, respectively; ρ_{11} and ρ_{22} are the effective mass densities of the solid and the fluid, respectively, and $\rho_{12} = \rho_{21}$ is a dynamic coupling parameter. Details about the material parameters may be found in, e.g., (Cleary, 1977) and (Beskos, 1989).

Instead of directly solving the time dependent equations of motion, it is possible to apply the Laplace transform and solve the resulting equations with the Laplace parameter $s = \delta + i\omega$. The time dependent solution is then obtained by the inverse Laplace transform.

Applying the Laplace transformation to the Eqs. (9) and (10) leads to the equations

$$\begin{aligned} \hat{\sigma}^s_{ij,j} - s^2(\tilde{\rho}_{11}\hat{u}^s_i + \tilde{\rho}_{12}\hat{u}^f_i) + \hat{F}^s_i &= -s(\tilde{\rho}_{11}u^s_{i0} + \tilde{\rho}_{12}u^f_{i0}) \\ &\quad -(\rho_{11}\dot{u}^s_{i0} + \rho_{12}\dot{u}^f_{i0}) \end{aligned} \tag{13}$$

$$\begin{aligned} \hat{\sigma}^f_{,i} - s^2(\tilde{\rho}_{12}\hat{u}^s_i + \tilde{\rho}_{22}\hat{u}^f_i) + \hat{F}^f_i &= -s(\tilde{\rho}_{12}u^s_{i0} + \tilde{\rho}_{22}u^f_{i0}) \\ &\quad -(\rho_{12}\dot{u}^s_{i0} + \rho_{22}\dot{u}^f_{i0}) \, . \end{aligned} \tag{14}$$

in which the transformed variables are denoted by '$(\hat{\ })$'. In addition, the modified density parameters $\tilde{\rho}_{ij}$ defined by

$$\tilde{\rho}_{mn}(s) := \rho_{mn} + (-1)^{m+n}\,\frac{b}{s} \tag{15}$$

are introduced. This definition formally eliminates the dissipation parameter b from the equations of motion. For $s = i\omega$, i.e. $\delta = 0$, these equations also include the case of harmonic excitations with the frequency ω.

In soil dynamics and many other applications it is common to use the pore pressure p as independent variable instead of the fluid displacement vector u^f_i.

The introduction of the pore pressure also reduces the number of equations of motion from six to four.

Starting with the Laplace domain equations (13) and (14), the fluid displacement is replaced by the pore pressure using the relation

$$\hat{u}_i^f = \frac{1}{s^2\tilde{\rho}_{22}}\Big\{ - \phi\hat{p}_{,i} - s^2\tilde{\rho}_{12}\hat{u}_i^s + \hat{F}_i^f + s(\tilde{\rho}_{12}u_{i0}^s + \tilde{\rho}_{22}u_{i0}^f) + (\rho_{12}\dot{u}_{i0}^s + \rho_{22}\dot{u}_{i0}^f)\Big\} \,, \tag{16}$$

which can be derived from Eq. (14). With the additional abbreviations

$$\rho_A = \tilde{\rho}_{11} - \frac{\tilde{\rho}_{12}^2}{\tilde{\rho}_{22}}, \quad \rho_B = \frac{\tilde{\rho}_{12}}{\tilde{\rho}_{22}} - \frac{Q}{R}, \quad \rho_C = \tilde{\rho}_{12} - \tilde{\rho}_{22}\frac{Q}{R}, \quad \rho_D = \frac{\tilde{\rho}_{22}}{R} \tag{17}$$

the resulting system of four equations of motion can be written in matrix notation:

$$\begin{bmatrix} (\lambda+\mu)\partial_{ij} + \delta_{ij}\left(\mu\Delta - s^2\rho_A\right) & \rho_B\partial_i \\ s^2\rho_C\partial_j & \Delta - s^2\rho_D \end{bmatrix} \begin{bmatrix} \hat{u}_j^s \\ \phi\hat{p} \end{bmatrix} + \begin{bmatrix} \delta_{ij} & -\frac{\tilde{\rho}_{12}}{\tilde{\rho}_{22}}\delta_{ij} \\ 0 & -\partial_j \end{bmatrix} \begin{bmatrix} \hat{F}_j^s \\ \hat{F}_j^f \end{bmatrix} = 0 \,, \tag{18}$$

in which Δ denotes the Laplace operator.

A last alternative formulation is a system of time domain equations using the pore pressure instead of the fluid displacements. If the procedure used in the Laplace domain is followed, a direct derivation of the desired equations is not possible. The double time derivatives in the time domain equations (9) and (10) do not explicitly appear in the transformed equations. In the time domain, an additional space derivation of one of these equations is necessary. This leads to a system of only two equations for the pore pressure and the trace of the total stress tensor σ_{ij}. For this system of equations, it is not possible to formulate physically relevant boundary conditions. However, if only the non dissipative case $b = 0$ is considered, Eq. (10) can be solved for the fluid acceleration

$$\ddot{u}_i^f = \frac{1}{\rho_{22}}\left(-\phi p_{,i} + F_i^f - \rho_{12}\ddot{u}_i^s\right) \,. \tag{19}$$

This result can be inserted into Eq. (9). Finally, for $b = 0$, the following time dependent equations of motion are obtained

$$\begin{bmatrix} (\lambda+\mu)\,\partial_{ij} + \delta_{ij}\left(\mu\Delta - \rho_A\partial_{tt}\right) & \rho_B\partial_i \\ \rho_C\partial_j\partial_{tt} & \Delta - \rho_D\partial_{tt} \end{bmatrix} \begin{bmatrix} u_j^s \\ \phi p \end{bmatrix} = \begin{bmatrix} -F_i^s + \frac{\rho_{12}}{\rho_{22}}F_i^f \\ F_{i,i}^f \end{bmatrix} \,. \tag{20}$$

The values ρ_A, ρ_B, ρ_C and ρ_D are defined as in Eq. (17), but the $\tilde{\rho}_{mn}$ have to be replaced by ρ_{mn}.

3 Integral equations

The derivation of the integral equations in elastostatics and Laplace domain elastodynamics is frequently based on Betti's theorem. The transient integral equation are usually derived from Graffi's theorem (Antes and Panagiotopoulos, 1992). Both formulate a relation between two load cases acting independently on a body. An analogous statement for poroelastic material was formulated by Cleary (1977). If two states, denoted by $(\ )^I$ and $(\ )^{II}$, are considered, one obtains:

$$\sigma_{ij}^{sI}\varepsilon_{ij}^{sII} + \sigma^{fI}\varepsilon^{fII} = \sigma_{ij}^{sII}\varepsilon_{ij}^{sI} + \sigma^{fII}\varepsilon^{fI} . \tag{21}$$

In this version, the reciprocity theorem was used by Cheng and Ligget (1984) in consolidation, and by Manolis and Beskos (1989), and Dominguez (1991) in dynamics of poroelastic media. Using the relations

$$\sigma_{ij}^{s} = \sigma_{ij} - \delta_{ij}\sigma^{f} \tag{22}$$

and the relative fluid contents ζ, defined as

$$\zeta = \phi(\varepsilon_{ii}^{s} - \varepsilon_{ii}^{f}) , \tag{23}$$

the reciprocity can be rewritten as

$$\sigma_{ij}^{I}\varepsilon_{ij}^{sII} + p^{I}\zeta^{II} = \sigma_{ij}^{II}\varepsilon_{ij}^{sI} + p^{II}\zeta^{I} , \tag{24}$$

as it was used by Cheng and Predeleanu (1987) and Cheng, Badmus and Beskos (1991). If the reciprocal theorem (21) is written for an actual state (without superscript), and an auxilliary state (denoted by a prime), integration over the domain of interest Ω and the observation time $t^{+} = t + \varepsilon$, $\varepsilon \to 0$, leads to

$$\iint_{0\,\Omega}^{t+} \left(\sigma_{ij}^{s'}\varepsilon_{ij}^{s} + \sigma^{f'}\varepsilon_{ii}^{f}\right) d\Omega d\tau = \iint_{0\,\Omega}^{t+} \left(\sigma_{ij}^{s}\varepsilon_{ij}^{s'} + \sigma^{f}\varepsilon_{ii}^{f'}\right) d\Omega d\tau . \tag{25}$$

In order to obtain integral equations for u_i^s and u_i^f, this expression is integrated by parts and the remaining volume integrals are removed by associating the fundamental solutions with the primed state. Two fundamental solutions are required, one corresponding to a unit impulsive load in the solid ($u_{ik}^{s*(1)}$ and $u_{ik}^{f*(1)}$), the other one describing the responses to the corresponding load in the fluid ($u_{ik}^{s*(2)}$ and $u_{ik}^{f*(2)}$). The boundary tractions $t_{ik}^{s*(l)}$ and $t_{ik}^{f*(l)}$ are obtained by applying the constitutive equations to the fundamental solutions and multiplying the resulting stresses by the local outward unit normal vector.

With the introduction of

$$\mathbf{u} = \begin{bmatrix} \mathbf{u}^s \\ \mathbf{u}^f \end{bmatrix}, \mathbf{t} = \begin{bmatrix} \mathbf{t}^s \\ \mathbf{t}^f \end{bmatrix}, \mathbf{u}^* = \begin{bmatrix} \mathbf{u}^{s*(1)} & \mathbf{u}^{f*(1)} \\ \mathbf{u}^{s*(2)} & \mathbf{u}^{f*(2)} \end{bmatrix}, \mathbf{t}^* = \begin{bmatrix} \mathbf{t}^{s*(1)} & \mathbf{t}^{f*(1)} \\ \mathbf{t}^{s*(2)} & \mathbf{t}^{f*(2)} \end{bmatrix} \tag{26}$$

and the assumption of no dissipation ($b = 0$) the resulting boundary integral equations can be written in the compact form (Γ is the boundary of Ω)

$$c_{ki}u_i(\xi,t) + \int_0^{t+}\int_\Gamma t^*_{ki}(\mathbf{x},\tau;\xi,t)u_i(\mathbf{x},\tau)d\Gamma_x d\tau = \int_0^{t+}\int_\Gamma u^*_{ki}(\mathbf{x},\tau;\xi,t)t_i(\mathbf{x},\tau)d\Gamma_x d\tau \quad i,k = 1,2,\ldots,6\,, \tag{27}$$

in which nonhomogeneous body forces and initial conditions are not taken into account. Equation (27) is formally equivalent to the integral equation formulation in elastodynamics, but with a different range of the indices i and j. If dissipation has to be considered an additional volume integral remains that cannot be removed, see Wiebe and Antes (1991).

The same approach can also be applied to the Laplace domain equations. Similar to Eq. (25), the derivation of the integral equations begins with the reciprocity statement in the form

$$\int_\Omega (\hat{\sigma}^s_{ij}\hat{u}^{s'}_{i,j} + \hat{\sigma}^f\hat{u}^{f'}_{i,i})d\Omega = \int_\Omega (\hat{\sigma}^{s'}_{ij}\hat{u}^s_{i,j} + \hat{\sigma}^{f'}\hat{u}^f_{i,i})d\Omega\,. \tag{28}$$

Following the time domain approach, an integral equation is obtained that again can be written in the form known from harmonic elastodynamics. However, since the dissipation parameter b and the velocity terms $\dot{u}^s_i$ and u^f_i do not explicitly appear, this formulation is valid for arbitrary damping. With the introduction of the normal displacement components

$$\hat{u}^f_n = \hat{u}^f_i n_i \quad \text{and} \quad \hat{u}^{f*(l)}_{nk} = \hat{u}^{f*(l)}_{ik} n_i\,, \tag{29}$$

the equations can be written as

$$\hat{u}^s_k(\xi) + \int_\Gamma \left\{\hat{t}^{s*(1)}_{ik}(x,\xi)\,\hat{u}^s_i(x) + \hat{\sigma}^{f*(1)}_k(x,\xi)\,\hat{u}^f_n(x)\right\} d\Gamma(x) = \int_\Gamma \left\{\hat{t}^s_i(x)\hat{u}^{s*(1)}_{ik}(x,\xi) + \hat{\sigma}^f(x)\hat{u}^{f*(1)}_{nk}(x,\xi)\right\} d\Gamma(x) \tag{30}$$

$$\hat{u}^f_k(\xi) + \int_\Gamma \left\{\hat{t}^{s*(2)}_{ik}(x,\xi)\,\hat{u}^s_i(x) + \hat{\sigma}^{f*(2)}_k(x,\xi)\,\hat{u}^f_n(x)\right\} d\Gamma(x) = \int_\Gamma \left\{\hat{t}^s_i(x)\hat{u}^{s*(2)}_{ik}(x,\xi) + \hat{\sigma}^f(x)\hat{u}^{f*(2)}_{nk}(x,\xi)\right\} d\Gamma(x)\,. \tag{31}$$

This form of the integral equations has also been given by Manolis and Beskos (1989).

The integral equations for the formulations using the pore pressure are derived by applying the method of weighted residuals:

$$\int_0^{t+}\int_\Omega \overset{*}{\mathbf{u}}^T\,(\mathbf{Lu})\,d\Omega d\tau = 0\ . \tag{32}$$

In this expression, $\mathbf{L}$ is the linear differential operator from Eq. (20) and $\overset{*}{\mathbf{u}}^T$ is a weighting function which will be chosen as the time dependent fundamental solution of the adjoint operator of $\mathbf{L}$. u_j $(j = 1,\dots 4)$ is the vector $[u_j^s,\ \phi p], j = 1,2,3$. The usual techniques in deriving boundary integral equations, i.e., integration by parts, introducing the boundary tractions t_i and t_{ki}^*, and taking the limit to the boundary, the integral equation can be written as

$$\begin{aligned} c_{ki}u_i(\xi,t)\ &+ \int_0^{t+}\int_\Gamma t_{ki}^*(x,t;\xi,\tau)u_i(x,\tau)d\Gamma(x)d\tau \\ &= \int_0^{t+}\int_\Gamma u_{ki}^*(x,t;\xi,\tau)t_i(x,\tau)d\Gamma(x)d\tau \quad i,k = 1,2,3,4\ , \end{aligned} \tag{33}$$

which again has the form known from elastodynamics. However, since the differential equations (20) are valid for the non dissipative case only, the same applies to the integral equations.

The integral equations for the Laplace domain using the pore pressure are derived analogously, which finally lead to

$$\begin{aligned} c_{ki}\hat{u}_i(\xi)\ &+ \int_\Gamma \hat{t}_{ki}^*(x;\xi)\hat{u}_i(x)d\Gamma(x) \\ &= \int_\Gamma \hat{u}_{ki}^*(x;\xi)\hat{t}_i(x)d\Gamma(x) \quad i,k = 1,2,3,4\ . \end{aligned} \tag{34}$$

This equation again is valid without restrictions concerning the dissipation.

4 Fundamental solutions

The fundamental solutions required in the integral equations (27),(31), (33), and (34) have to satisfy the equations of motion in the full space with unit impulsive loads, zero initial conditions and radiation conditions at inifinity. As introduced by Burridge and Vargas (1979) for the solid displacement in the time domain and by Manolis and Beskos (1989) for both the solid and the fluid phase in the Laplace domain, the fundamental solutions with six displacement components are derived by using the decomposition of the displacements in the potentials Θ, Φ, Ψ and X. The displacements are then sought in the form:

$$u_{ik}^{s*(l)} = (\Theta/r)_{,ik}^{(l)} + (\Phi/r)_{,jj}^{(l)}\delta_{ik} - (\Phi/r)_{,ik}^{(l)} \quad (35)$$

$$\hat{u}_{ik}^{f*(l)} = (\Psi/r)_{,ik}^{(l)} + (X/r)_{,jj}^{(l)}\delta_{ik} - (X/r)_{,ik}^{(l)} \quad (36)$$

Introducing these expressions into the equations of motion (11) and (12) and exploiting the radial symmetry of the potentials leads to two systems of ordinary differential systems, each for two of the potentials. These systems can be solved by a Laplace transform of the time variable. This leads to the elastic solid fundamental displacements in the Laplace domain:

$$\begin{aligned}
\hat{u}_{ik}^{s(m)} &= \hat{A}_1^{(m)} \exp(\lambda_1 r)\frac{1}{r}\left[r_{,i}r_{,k} + \left(\lambda_1^2 - 3\frac{\lambda_1}{r} + \frac{3}{r^2}\right) - \delta_{ik}\left(\frac{1}{r^2} - \frac{\lambda_1}{r}\right)\right] \\
&+ \hat{A}_2^{(m)} \exp(\lambda_2 r)\frac{1}{r}\left[r_{,i}r_{,k} + \left(\lambda_2^2 - 3\frac{\lambda_2}{r} + \frac{3}{r^2}\right) - \delta_{ik}\left(\frac{1}{r^2} - \frac{\lambda_2}{r}\right)\right] \\
&+ \hat{C}^{(m)} \exp(\lambda_3 r)\frac{1}{r}\left[-r_{,i}r_{,k} + \left(\lambda_3^2 - 3\frac{\lambda_3}{r} + \frac{3}{r^2}\right)\right. \\
&\qquad \left. + \delta_{ik}\left(\frac{1}{r^2} - \frac{\lambda_3}{r} + \lambda_3^2\right)\right]
\end{aligned} \quad (37)$$

and similarly for the fluid displacement:

$$\begin{aligned}
\hat{u}_{ik}^{f(m)} &= \hat{B}_1^{(m)} \exp(\lambda_1 r)\frac{1}{r}\left[r_{,i}r_{,k} + \left(\lambda_1^2 - 3\frac{\lambda_1}{r} + \frac{3}{r^2}\right) - \delta_{ik}\left(\frac{1}{r^2} - \frac{\lambda_1}{r}\right)\right] \\
&+ \hat{B}_2^{(m)} \exp(\lambda_2 r)\frac{1}{r}\left[r_{,i}r_{,k} + \left(\lambda_2^2 - 3\frac{\lambda_2}{r} + \frac{3}{r^2}\right) - \delta_{ik}\left(\frac{1}{r^2} - \frac{\lambda_2}{r}\right)\right] \\
&+ \hat{D}^{(m)} \exp(\lambda_3 r)\frac{1}{r}\left[-r_{,i}r_{,k} + \left(\lambda_3^2 - 3\frac{\lambda_3}{r} + \frac{3}{r^2}\right)\right. \\
&\qquad \left. +\delta_{ik}\left(\frac{1}{r^2} - \frac{\lambda_3}{r} + \lambda_3^2\right)\right] .
\end{aligned} \quad (38)$$

These solutions contain the fundamental solution of elastodynamics as the limiting case of vanishing porosity. The time dependent fundamental solution can then be obtained by the inverse Laplace transform. The constants $\hat{A}_i^{(m)}$, $\hat{B}_i^{(m)}$ are modified by

$$\begin{aligned}
\hat{A}_i^{(m)} &= A_i^{(m)}/s^2 \quad i = 1,2 \\
\hat{B}_i^{(m)} &= B_i^{(m)}/s^2 \quad i = 1,2 \\
\\
\hat{A}_3^{(m)} &= -C^{(m)}/s^2 \\
\hat{B}_3^{(m)} &= -D^{(m)}/s^2
\end{aligned} \quad (39$$

in which the new constants are independent of s. The wave numbers λ_α are decomposed in a similar way:

$$\lambda_\alpha = -\frac{s}{c_\alpha} \qquad \text{with } c_\alpha > 0 \tag{40}$$

in which c_α ,$\alpha = 1, 2, 3$ are the velocities of the fast compressional wave, the shear wave and the slow compressional wave. A closed form inverse Laplace transformation becomes possible only if $b = 0$, i.e., no dissipation is assumed. In this case

$$\tilde{\rho}_{ij} = \rho_{ij} = \text{const.} \tag{41}$$

holds, and one finally obtains (Wiebe and Antes, 1991):

$$\begin{aligned} u_{ik}^{s(m)} &= \sum_{\alpha=1}^{3} A_\alpha^{(m)} \frac{1}{r} \left\{ r_{,i} r_{,k} \left[\frac{1}{c_\alpha^2} \delta\left(t' - \frac{r}{c_\alpha}\right) + \frac{3t'}{r^2} H\left(t' - \frac{r}{c_\alpha}\right) \right] \right. \\ &\qquad \left. - \delta_{ik} \left[\frac{t'}{r^2} H \left(t' - \frac{r}{c_\alpha}\right) \right] \right\} \\ &\quad - A_3^{(m)} \frac{1}{r c_3^2} \delta\left(t' - \frac{r}{c_3}\right) \end{aligned} \tag{42}$$

$$\begin{aligned} u_{ik}^{f(m)} &= \sum_{\alpha=1}^{3} B_\alpha^{(m)} \frac{1}{r} \left\{ r_{,i} r_{,k} \left[\frac{1}{c_\alpha^2} \delta\left(t' - \frac{r}{c_\alpha}\right) + \frac{3t'}{r^2} H\left(t' - \frac{r}{c_\alpha}\right) \right] \right. \\ &\qquad \left. - \delta_{ik} \left[\frac{t'}{r^2} H \left(t' - \frac{r}{c_\alpha}\right) \right] \right\} \\ &\quad - B_3^{(m)} \frac{1}{r c_3^2} \delta\left(t' - \frac{r}{c_3}\right) . \end{aligned} \tag{43}$$

The corresponding singular tractions $t_{ik}^{s*(l)}$ and $t_{ik}^{f*(l)}$ can be calculated using the constitutive equations. It should be mentioned that the fluid traction does contain fast and slow compressional wave terms only, but no shear terms.

The integral equations with the pore pressure require knowledge of either the fundamental solution of the operator that is adjoint to $\mathbf{L}$, see Eq. (32), or to $\tilde{\mathbf{L}}$, the corresponding operator in the Laplace domain. A general method for finding fundamental solutions for static or harmonic problems is the one proposed by Hörmander (1963). If the homogeneous differential equations is written as

$$\hat{B}_{ij}\, \hat{u}_j \;=\; 0, \tag{44}$$

hen the fundamental solution is defined as the solution of

$$\hat{B}_{ij}\, \hat{G}_{jk} \;+\; \delta_{ik}\delta(x - \xi) \;=\; 0\,. \tag{45}$$

This equation has to be solved for $\hat{G}_{jk}$, which is the transposed of $\hat{u}^*_{jk}$, required in the integral equations. Hörmander's method finds the solution by applying the transposed of the cofactors of $\hat{B}_{ij}$ to a scalar function φ. This finally leads to a higher order Helmholtz equation for φ (Cheng and Antes, 1992). Substitution and application of the differentiations to φ finally gives the sought function. This procedure is described, e.g., by Dominguez (1991) for the two–dimensional case and by Wiebe (1993) for three dimensions.

The fundamental solution for the harmonic case can also be obtained by the analogy to thermoelasticity. The equations of motions of poroelasticity and thermoelasticity are similar, and so the fundamental solutions from thermoelasticity that can be found in the literature (Kupradze, 1979). This approach was used by Bonnet (1987), Dominguez (1992) and Cheng et al. (1991). The displacement vector $\hat{u}^s_i$ and the Lamé constants λ and μ are common to both types of problems. The other expressions can be associated to the corresponding thermoelastic terms by using Table I. It has to be noted that the material parameters in poroelasticity

TABLE I
Analogy of thermo– and poroelasticity

Thermoelasticity	Poroelasticity
Θ	$\phi\hat{p}$
γ	$-\tilde{\rho}_B$
κ	$-1/(i\omega\tilde{\rho}_D)$
η	$-i\omega\tilde{\rho}_C$
ρ	$\tilde{\rho}_A$

explicitly depend on the frequency, whereas in thermoelasticity the correspondin values are constant.

The time dependent fundamental solution with the pore pressure is finall obtained by the inverse Laplace transformation. Again as in the formulation wit the fluid displacement, the assumption of no dissipation is required to obtain close form solutions. The details of this derivations are described by Wiebe (1993). Th resulting expressions are again composed of contributions of the three differen wave types.

The terms of the fundamental solution that describe the response of the flui phase to unit impulsive forces in the solid, and the ones, that describe the soli displacement vector caused by a source in the fluid, have no contributions fro the shear wave. The response of an impulsive fluid point source on the fluid itse contains only two impulses propagating with the fast and the slow pressure wav respectively.

It can be concluded, that all developed time dependent formulations lack generality because they are valid for no dissipation only. Best suited for practical applications in soil dynamics is therefore the formulation in the Laplace domain with, in addition to the solid displacement vector, the pore pressure as independent variable.

Some properties of the Laplace domain fundamental solution are now outlined. The properties of the considered material are given in Table II. Concerning the

TABLE II
Material data for Berea sandstone

Material parameter			value
Shear modulus	G	=	6×10^9 N/m^2
Poisson's ratio	ν	=	0.2
Porosity	ϕ	=	0.19
	Q	=	1.399×10^9 N/m^2
	R	=	0.444×10^9 N/m^2
Density of the solid	ρ^s	=	2800 kg/m^3
Density of the fluid	ρ^f	=	1000 kg/m^3
Apparent density	ρ^a	=	150 kg/m^3
Damping coefficient	b	=	$0.19 \times 10^9 Ns/m^4$

wave velocities, it can be observed, that the fast pressure wave travelling with c_1 and the shear wave with the velocity c_2 propagate almost undamped. Both velocities are almost independent of the excitation frequency. The low compressional wave velocity c_3 shows a very different behavior. This velocity is much smaller and the slow wave is strongly damped. This causes the diffusive character of the third wave.

The basic behavior of the poroelastic fundamental solution is similar to the one of elastodynamics, as can be seen in Fig. 1, showing the real part of the displacement in x_1-direction due to a harmonic point force in x_1-direction. The function decays with $1/r$, and the wave length decreases with increasing excitation frequency. Figure 2 shows a comparison of one component of the poroelastic fundamental solution with the equivalent term from elastodynamics and elastostatics. For low frequencies all functions behave similarly, whereas differences increase with the frequency.

5 Examples

The formulation using the pore pressure in the transformed domain is implemented in a boundary element procedure for the analysis of three dimensional problems. It

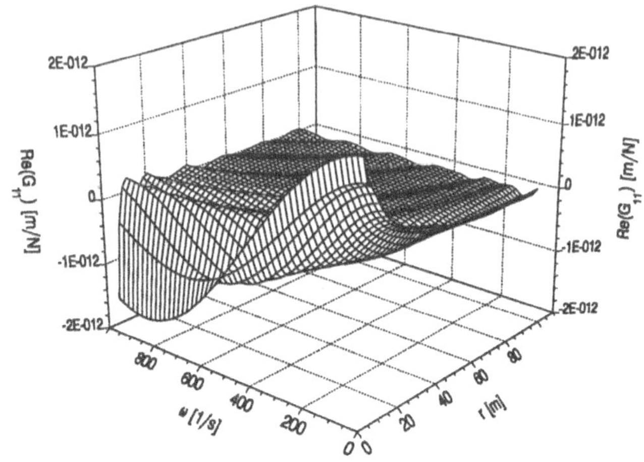

Figure 1. Real part of $\hat{G}_{11}$

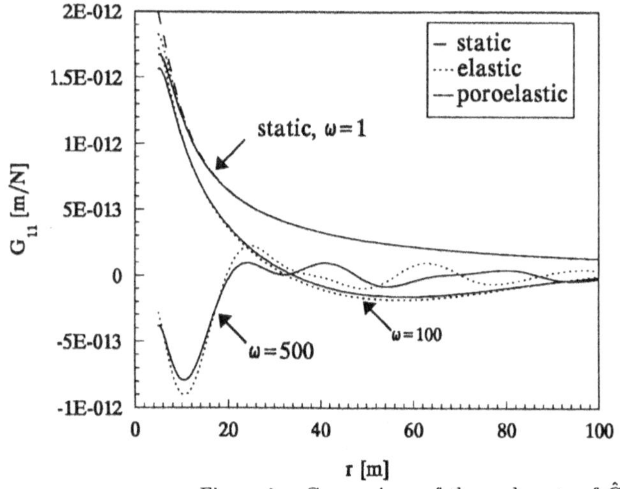

Figure 2. Comparison of the real parts of $\hat{G}_{11}$

is used to study the dynamic behavior of a cube with unit length under harmoni loading. The responses of this poroelastic column are compared to the correspond ing elastodynamic solution, and the influence of the dissipation parameter b i studied. Figure 3 shows the problem under consideration. The cube of length is loaded by a constant harmonic traction t_3. The material data are the same a

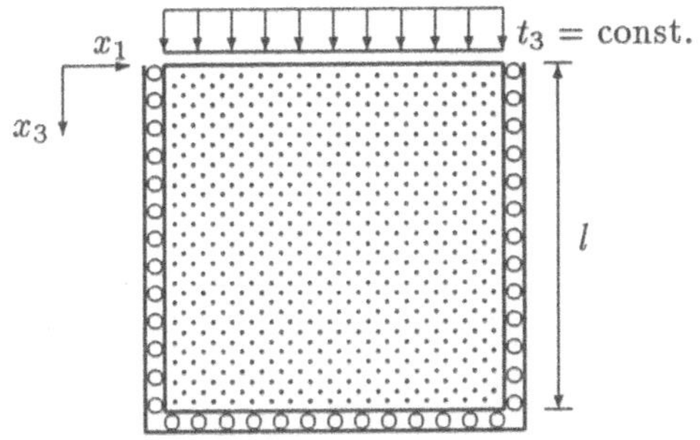

Figure 3. One–dimensional column problem

used in the previous sections. Under the prescribed conditions, the wave propagation is one dimensional. The analytic solution of this problem is given by Cheng et.al. (1991). This problem has also been studied numerically by Cheng et.al. and by Dominguez (1992), but both worked with two dimensional boundary element procedures. In conformance with their representation of the results, the frequency response is non–dimensionalized by the first natural frequency of the corresponding elastic poblem. This is given by:

$$\omega_1 = \frac{\pi}{2l}\sqrt{\frac{E_u}{\rho}}, \tag{46}$$

in which

$$E_u = \frac{2G(1-\nu_u)}{(1-2\nu_u)} \tag{47}$$

is the Young's modulus of the undrained elastic material and

$$\rho = (1-\phi)\rho^s - \phi\rho^f \tag{48}$$

is the density. The following natural frequencies are given by

$$\omega_n = (2n-1)\,\omega_1 \quad n = 1,2,3,\ldots \tag{49}$$

The displacement at the surface $x_3 = 0$ is non–dimensionalized by the undrained elastostatic solution

$$u_0 = \frac{t_3\,l}{E_u} \tag{50}$$

The used discretizations are shown in Fig. 4. The coarse mesh with 190 constant quadrilateral elements is applied for the computations in the frequency range up

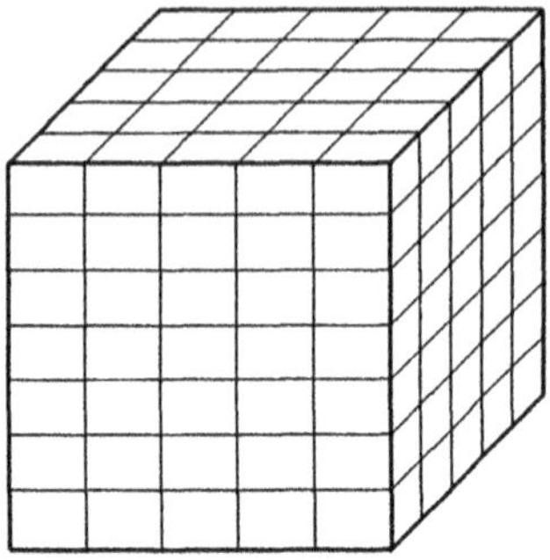

a) 190 elements

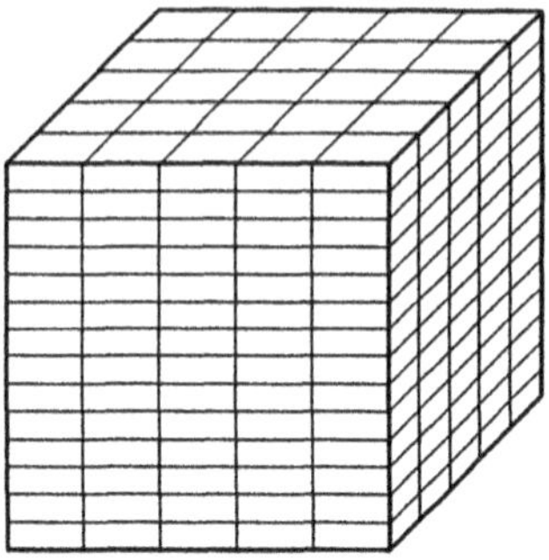

b) 330 elements

Figure 4. 3D discretizations

to $2\omega_1$. The results for frequencies up to $4\omega_1$ are obtained with the fine mesh (330 Elements). Figure 5 shows the analytic and the BEM solution of the column problem. One can observe the excellent agreement of both curves for frequencies up to about $\omega = 2.5\omega_1$. The displacement modes for higher frequencies require a finer discretization in the x_3 direction. In the low frequency range, i.e., below the first natural frequency, the non–dimensional displacement approaches the elastostatic displacement value of the undrained material. Poroelastic and elastic material behave in a similar way in this frequency range. For extremely low frequencies, when ω approaches zero, the non–dimensional displacement rises to a value of *1.33*. This is the non-dimensional static displacement of the drained elastic material. That means, that for almost static processes, the poroelastic problem approaches the drained elastic solution. This behavior can also be observed in Fig. 6 in which the poroelastic solution is compared to both drained and undrained elastic solution. In the one dimensional problem under consideration, the poroelastic material behaves very similarly to the undrained elastic material except for very small frequency values. The natural frequencies of the drained elastic column problem are shifted to lower values and the displacements in the range below the first natural frequency are larger than for the undrained elastic and poroelastic material. In view of the applicability of the formulations that require the assumption of a non dissipative material, i.e. $b = 0$, the influence of the dissipation parameter b is shown in Fig. 7. This figure shows large differences in the behavior of the material with a realistic damping value and the non dissipative case. The latter has additiona resonances, which are caused by the movement of the pore fluid, that is no longe damped by the solid. These resonances do not occur in real soil materials. In addi tion, the displacement at the top of the column for the undamped material is large than in the realistic case. It approaches the solution of the drained elastic mode

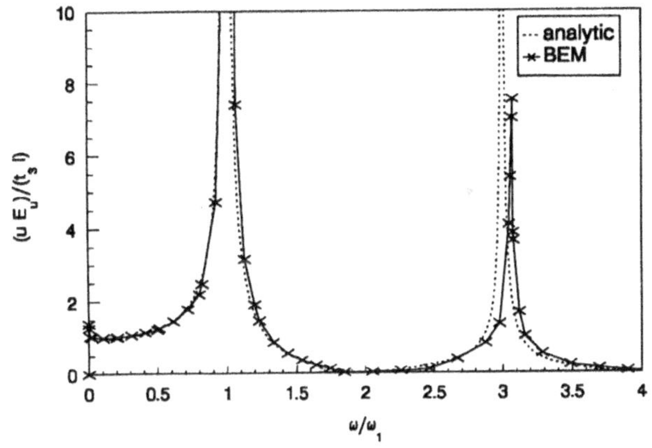

Figure 5. Analytic vs. BEM solution

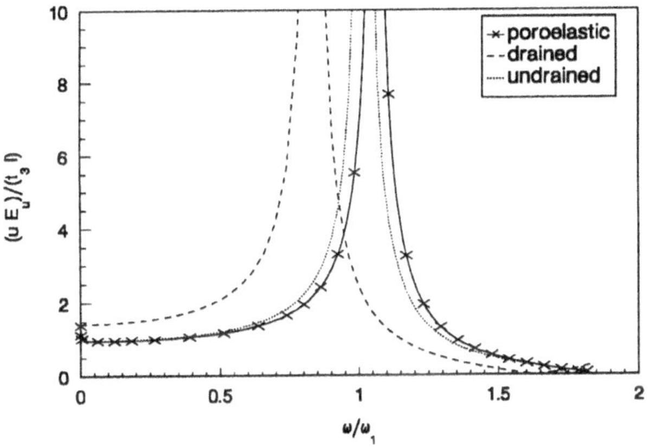

Figure 6. Poroelastic solution vs. elastic solutions

'or low frequencies, whereas the poroelastic material is known to behave similar to he undrained elastic model. This result shows that the influence of the dissipation parameter b is of large importance, and that dissipation may not be neglected in oil dynamics applications. This is the major drawback of the formulations in the ime domain that allow a closed form formulation only for the assumption of no lissipation.

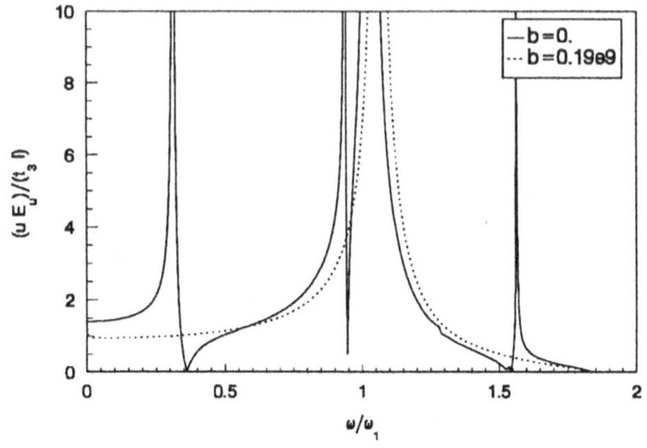

Figure 7. Influence of dissipation

6 Conclusions

The investigation of wave propagation problems in poroelastic media can be done by applying integral equations and the boundary element method. Based on the linear theory of Biot, the theoretical foundations of the boundary element method were developed and studied: the equations of motion, the integral equations, and the fundamental solutions. The problem can be formulated directly in the time domain or in the Laplace transform domain. Formulations using six field variables i.e., the displacement vectors of both solid and fluid, and using four independent variables with the pore pressure instead of the fluid displacements are possible. In the Laplace domain, it can be shown, that the problem can best be described by the formulation with four independent variables. The time domain formulation cannot be applied in the general case, since for their closed form formulation dissipation has to be neglected. However, the dissipation strongly influences the slow compressional wave. As can be seen from the numerical example, neglecting the dissipation leads to different and wrong results.

If a time domain analysis is required, suitable one phase models with properly chosen material parameters should be applied. For problems with a short observation time, dissipation may completely be neglected. The wave propagation is then described by the fast pressure wave and the shear wave only, and can be analyzed by standard time domain elastodynamic boundary element methods. Alternatively, a viscoelastic time domain formulation, which has recently been found by Gaul and Schanz (1992), may be applied.

Acknowledgements

This research was supported by the German Research Foundation (DFG) under grant no. An140/1-1/3.

References

H. Antes and P. D. Panagiotopoulos. (1992) *The Boundary Integral Approach to Static and Dynamic Contact Problems*, volume 108 of *ISNM*. Birkhäuser, Basel.

D. E. Beskos. (1989) Dynamics of saturated rocks. I: Equations of motion. *Journal of Engineering Mechanics* **115**(5):982–995.

M. A. Biot. (1956a) Theory of propagation of elastic waves in a fluid–saturated porous solid. I. low–frequency range. *Journal of the Acoustical Society of America* **28**(2):168–178.

M. A. Biot. (1956b) Theory of propagation of elastic waves in a fluid–saturated porous solid. II. higher–frequency range. *Journal of the Acoustical Society of America* **28**(2):179–191.

G. Bonnet. (1987) Basic singular solutions for a poroelastic medium in the dynamic range. *Journal of the Acoustical Society of America* **82**(5):1758–1762.

R. Burridge and C. A. Vargas. (1979) The fundamental solution in dynamic poroelasticity. *Geophys. J. R. astr. Soc.* **58**:61–90.

A. H.-D. Cheng and J. A. Liggett. (1984) Boundary integral equation method for linear porous–elasticity with applications to soil consolidation. *Int. J. num. Meth. Eng.* **20**:255–278.

A. H.-D. Cheng and M. Predeleanu. (1987) Transient boundary element formulation for linear poroelasticity. *Appl. Math. Modelling* **11**:285–290.

A. H.-D. Cheng, T. Badmus, and D. E. Beskos. (1991) Integral equation for dynamic poroelasticity in frequency domain with bem solution. *Journal of Engineering Mechanics* **117**(5):1136–1157.

A. H.-D. Cheng and H. Antes. (1992) Free space Green's function for high order Helmholtz equations. In S. Kobayashi and N. Nishimura, editors, *Boundary Element Methods–Fundamentals and Applications; Proc. IABEM '91, Kyoto, Japan*, pages 67–71. Springer–Verlag, Berlin.

M. P. Cleary. (1977) Fundamental solutions for a fluid–saturated porous solid. *Int. J. Solids Structures* **13**:785–806.

J. Dominguez. (1991) An integral formulation for dynamic poroelasticity. *J. Appl. Mech.* **58**:588–591.

J. Dominguez. (1992) Boundary element approach for dynamic poroelastic problems. *Int. J. num. Meth. Eng.* **35**:307–324.

L. Gaul and M. Schanz. (1992) BEM formulation in time domain for viscoelastic media based on analytical time integration. In C.A. Brebbia, J. Dominguez and F. Paris, editors, *Proc. Boundary Elements XIV, volume 2, Stress Analysis and Computational Aspects; Seville, Spain*, pages 223–234. CMP Publlications, Southampton.

L. Hörmander. (1963) *Linear Partial Differential Operators*. Springer–Verlag, Berlin, New York.

V. D. Kupradze, editor. (1979) *Three Dimensional Problems of the Mathematical Theory of Elasticity and Thermoelasticity*, volume 25 of *Applied Mathematics and Mechanics*. North–Holland.

G. D. Manolis and D. E. Beskos. (1989) Integral formulation and fundamental solutions of dynamic poroelasticity and thermoelasticity. *Acta Mechanica* **76**:89–104.

G. D. Manolis and D. E. Beskos. (1990) Errata in ... *Acta Mechanica* **83**:223–226.

J. R. Rice and M. P. Cleary. (1976) Some basic stress–diffusion solutions for fluid–saturated elastic porous media with compressible constituents. *Reviews of Geophysics and Space Physics* **14**(2):227–241.

A. W. Skempton. (1954) The pore pressure coefficients A and B. *Geotechnique* **4**:143–147.

T. Wiebe and H. Antes. (1991) A time domain integral formulation of dynamic poroelasticity. *Acta Mechanica* **90**:125–137.

T. Wiebe. (1993) *Wave propagation in poroelastic media: investigation by boundary integral equations (in german)*. volume 11-1993 of Braunschweig Series on Mechanics. Mechanik Zentrum, TU Braunschweig.

AUTHOR INDEX

SUBJECT INDEX

This Index was compiled using keywords supplied by the Authors and supplemented by the Editor: the numbers refer to the first page of the paper concerned.

Destabilization of volcano flows, 215

Mechanics

SOLID MECHANICS AND ITS APPLICATIONS

Series Editor: G.M.L. Gladwell

Aims and Scope of the Series

The fundamental questions arising in mechanics are: *Why?*, *How?*, and *How much?* The aim of this series is to provide lucid accounts written by authoritative researchers giving vision and insight in answering these questions on the subject of mechanics as it relates to solids. The scope of the series covers the entire spectrum of solid mechanics. Thus it includes the foundation of mechanics; variational formulations; computational mechanics; statics, kinematics and dynamics of rigid and elastic bodies; vibrations of solids and structures; dynamical systems and chaos; the theories of elasticity, plasticity and viscoelasticity; composite materials; rods, beams, shells and membranes; structural control and stability; soils, rocks and geomechanics; fracture; tribology; experimental mechanics; biomechanics and machine design.

1. R.T. Haftka, Z. Gürdal and M.P. Kamat: *Elements of Structural Optimization.* 2nd rev.ed., 1990 ISBN 0-7923-0608-2
2. J.J. Kalker: *Three-Dimensional Elastic Bodies in Rolling Contact.* 1990 ISBN 0-7923-0712-7
3. P. Karasudhi: *Foundations of Solid Mechanics.* 1991 ISBN 0-7923-0772-0
4. *Not published*
5. *Not published.*
6. J.F. Doyle: *Static and Dynamic Analysis of Structures.* With an Emphasis on Mechanics and Computer Matrix Methods. 1991 ISBN 0-7923-1124-8; Pb 0-7923-1208-2
7. O.O. Ochoa and J.N. Reddy: *Finite Element Analysis of Composite Laminates.* ISBN 0-7923-1125-6
8. M.H. Aliabadi and D.P. Rooke: *Numerical Fracture Mechanics.* ISBN 0-7923-1175-2
9. J. Angeles and C.S. López-Cajún: *Optimization of Cam Mechanisms.* 1991 ISBN 0-7923-1355-0
10. D.E. Grierson, A. Franchi and P. Riva (eds.): *Progress in Structural Engineering.* 1991 ISBN 0-7923-1396-8
11. R.T. Haftka and Z. Gürdal: *Elements of Structural Optimization.* 3rd rev. and exp. ed. 1992 ISBN 0-7923-1504-9; Pb 0-7923-1505-7
12. J.R. Barber: *Elasticity.* 1992 ISBN 0-7923-1609-6; Pb 0-7923-1610-X
13. H.S. Tzou and G.L. Anderson (eds.): *Intelligent Structural Systems.* 1992 ISBN 0-7923-1920-6
14. E.E. Gdoutos: *Fracture Mechanics.* An Introduction. 1993 ISBN 0-7923-1932-X
15. J.P. Ward: *Solid Mechanics.* An Introduction. 1992 ISBN 0-7923-1949-4
16. M. Farshad: *Design and Analysis of Shell Structures.* 1992 ISBN 0-7923-1950-8
17. H.S. Tzou and T. Fukuda (eds.): *Precision Sensors, Actuators and Systems.* 1992 ISBN 0-7923-2015-8
18. J.R. Vinson: *The Behavior of Shells Composed of Isotropic and Composite Materials.* 1993 ISBN 0-7923-2113-8
19. H.S. Tzou: *Piezoelectric Shells.* Distributed Sensing and Control of Continua. 1993 ISBN 0-7923-2186-3

Kluwer Academic Publishers – Dordrecht / Boston / London

Mechanics

SOLID MECHANICS AND ITS APPLICATIONS

Series Editor: G.M.L. Gladwell

20. W. Schiehlen (ed.): *Advanced Multibody System Dynamics.* Simulation and Software Tools. 1993 ISBN 0-7923-2192-8
21. C.-W. Lee: *Vibration Analysis of Rotors.* 1993 ISBN 0-7923-2300-9
22. D.R. Smith: *An Introduction to Continuum Mechanics.* 1993 ISBN 0-7923-2454-4
23. G.M.L. Gladwell: *Inverse Problems in Scattering.* An Introduction. 1993 ISBN 0-7923-2478-1
24. G. Prathap: *The Finite Element Method in Structural Mechanics.* 1993 ISBN 0-7923-2492-7
25. J. Herskovits (ed.): *Advances in Structural Optimization.* 1995 ISBN 0-7923-2510-9
26. M.A. González-Palacios and J. Angeles: *Cam Synthesis.* 1993 ISBN 0-7923-2536-2
27. W.S. Hall: *The Boundary Element Method.* 1993 ISBN 0-7923-2580-X
28. J. Angeles, G. Hommel and P. Kovács (eds.): *Computational Kinematics.* 1993 ISBN 0-7923-2585-0
29. A. Curnier: *Computational Methods in Solid Mechanics.* 1994 ISBN 0-7923-2761-6
30. D.A. Hills and D. Nowell: *Mechanics of Fretting Fatigue.* 1994 ISBN 0-7923-2866-3
31. B. Tabarrok and F.P.J. Rimrott: *Variational Methods and Complementary Formulations in Dynamics.* 1994 ISBN 0-7923-2923-6
32. E.H. Dowell (ed.), E.F. Crawley, H.C. Curtiss Jr., D.A. Peters, R. H. Scanlan and F. Sisto: *A Modern Course in Aeroelasticity.* Third Revised and Enlarged Edition. 1995 ISBN 0-7923-2788-8; Pb: 0-7923-2789-6
33. A. Preumont: *Random Vibration and Spectral Analysis.* 1994 ISBN 0-7923-3036-6
34. J.N. Reddy (ed.): *Mechanics of Composite Materials.* Selected works of Nicholas J. Pagano. 1994 ISBN 0-7923-3041-
35. A.P.S. Selvadurai (ed.): *Mechanics of Poroelastic Media.* 1996 ISBN 0-7923-3329-
36. Z. Mróz, D. Weichert, S. Dorosz (eds.): *Inelastic Behaviour of Structures under Variable Loads.* 1995 ISBN 0-7923-3397-
37. R. Pyrz (ed.): *IUTAM Symposium on Microstructure-Property Interactions in Composite Materials.* Proceedings of the IUTAM Symposium held in Aalborg, Denmark. 1995 ISBN 0-7923-3427-
38. M.I. Friswell and J.E. Mottershead: *Finite Element Model Updating in Structural Dynamics.* 1995 ISBN 0-7923-3431-
39. D.F. Parker and A.H. England (eds.): *IUTAM Symposium on Anisotropy, Inhomogeneity and Nonlinearity in Solid Mechanics.* Proceedings of the IUTAM Symposium held in Nottingham, U.K. 1995 ISBN 0-7923-3594-
40. J.-P. Merlet and B. Ravani (eds.): *Computational Kinematics '95.* 1995 ISBN 0-7923-3673-
41. L.P. Lebedev, I.I. Vorovich and G.M.L. Gladwell: *Functional Analysis.* Applications in Mechanics and Inverse Problems. 1996 ISBN 0-7923-3849-
42. J. Menčik: *Mechanics of Components with Treated or Coated Surfaces.* 1996 ISBN 0-7923-3700-
43. D. Bestle and W. Schiehlen (eds.): *IUTAM Symposium on Optimization of Mechanical Systems.* Proceedings of the IUTAM Symposium held in Stuttgart, Germany. 1996 ISBN 0-7923-3830

Kluwer Academic Publishers – Dordrecht / Boston / London

Mechanics

FLUID MECHANICS AND ITS APPLICATIONS

Series Editor: R. Moreau

Aims and Scope of the Series

The purpose of this series is to focus on subjects in which fluid mechanics plays a fundamental role. As well as the more traditional applications of aeronautics, hydraulics, heat and mass transfer etc., books will be published dealing with topics which are currently in a state of rapid development, such as turbulence, suspensions and multiphase fluids, super and hypersonic flows and numerical modelling techniques. It is a widely held view that it is the interdisciplinary subjects that will receive intense scientific attention, bringing them to the forefront of technological advancement. Fluids have the ability to transport matter and its properties as well as transmit force, therefore fluid mechanics is a subject that is particularly open to cross fertilisation with other sciences and disciplines of engineering. The subject of fluid mechanics will be highly relevant in domains such as chemical, metallurgical, biological and ecological engineering. This series is particularly open to such new multidisciplinary domains.

1. M. Lesieur: *Turbulence in Fluids.* 2nd rev. ed., 1990 ISBN 0-7923-0645-7
2. O. Métais and M. Lesieur (eds.): *Turbulence and Coherent Structures.* 1991 ISBN 0-7923-0646-5
3. R. Moreau: *Magnetohydrodynamics.* 1990 ISBN 0-7923-0937-5
4. E. Coustols (ed.): *Turbulence Control by Passive Means.* 1990 ISBN 0-7923-1020-9
5. A.A. Borissov (ed.): *Dynamic Structure of Detonation in Gaseous and Dispersed Media.* 1991 ISBN 0-7923-1340-2
6. K.-S. Choi (ed.): *Recent Developments in Turbulence Management.* 1991 ISBN 0-7923-1477-8
7. E.P. Evans and B. Coulbeck (eds.): *Pipeline Systems.* 1992 ISBN 0-7923-1668-1
8. B. Nau (ed.): *Fluid Sealing.* 1992 ISBN 0-7923-1669-X
9. T.K.S. Murthy (ed.): *Computational Methods in Hypersonic Aerodynamics.* 1992 ISBN 0-7923-1673-8
10. R. King (ed.): *Fluid Mechanics of Mixing.* Modelling, Operations and Experimental Techniques. 1992 ISBN 0-7923-1720-3
11. Z. Han and X. Yin: *Shock Dynamics.* 1993 ISBN 0-7923-1746-7
12. L. Svarovsky and M.T. Thew (eds.): *Hydroclones.* Analysis and Applications. 1992 ISBN 0-7923-1876-5
13. A. Lichtarowicz (ed.): *Jet Cutting Technology.* 1992 ISBN 0-7923-1979-6
14. F.T.M. Nieuwstadt (ed.): *Flow Visualization and Image Analysis.* 1993 ISBN 0-7923-1994-X
15. A.J. Saul (ed.): *Floods and Flood Management.* 1992 ISBN 0-7923-2078-6
16. D.E. Ashpis, T.B. Gatski and R. Hirsh (eds.): *Instabilities and Turbulence in Engineering Flows.* 1993 ISBN 0-7923-2161-8
17. R.S. Azad: *The Atmospheric Boundary Layer for Engineers.* 1993 ISBN 0-7923-2187-1
18. F.T.M. Nieuwstadt (ed.): *Advances in Turbulence IV.* 1993 ISBN 0-7923-2282-7
19. K.K. Prasad (ed.): *Further Developments in Turbulence Management.* 1993 ISBN 0-7923-2291-6
20. Y.A. Tatarchenko: *Shaped Crystal Growth.* 1993 ISBN 0-7923-2419-6

luwer Academic Publishers – Dordrecht / Boston / London

Mechanics

FLUID MECHANICS AND ITS APPLICATIONS

Series Editor: R. Moreau

21. J.P. Bonnet and M.N. Glauser (eds.): *Eddy Structure Identification in Free Turbulent Shear Flows.* 1993 ISBN 0-7923-2449-
22. R.S. Srivastava: *Interaction of Shock Waves.* 1994 ISBN 0-7923-2920-
23. J.R. Blake, J.M. Boulton-Stone and N.H. Thomas (eds.): *Bubble Dynamics and Interface Phenomena.* 1994 ISBN 0-7923-3008-
24. R. Benzi (ed.): *Advances in Turbulence V.* 1995 ISBN 0-7923-3032-
25. B.I. Rabinovich, V.G. Lebedev and A.I. Mytarev: *Vortex Processes and Solid Body Dynamics.* The Dynamic Problems of Spacecrafts and Magnetic Levitation Systems. 1994 ISBN 0-7923-3092-
26. P.R. Voke, L. Kleiser and J.-P. Chollet (eds.): *Direct and Large-Eddy Simulation I.* Selected papers from the First ERCOFTAC Workshop on Direct and Large-Eddy Simulation. 1994 ISBN 0-7923-3106-
27. J.A. Sparenberg: *Hydrodynamic Propulsion and its Optimization.* Analytic Theory. 1995 ISBN 0-7923-3201-
28. J.F. Dijksman and G.D.C. Kuiken (eds.): *IUTAM Symposium on Numerical Simulation of Non-Isothermal Flow of Viscoelastic Liquids.* Proceedings of an IUTAM Symposium held in Kerkrade, The Netherlands. 1995 ISBN 0-7923-3262-
29. B.M. Boubnov and G.S. Golitsyn: *Convection in Rotating Fluids.* 1995 ISBN 0-7923-3371-
30. S.I. Green (ed.): *Fluid Vortices.* 1995 ISBN 0-7923-3376-
31. S. Morioka and L. van Wijngaarden (eds.): *IUTAM Symposium on Waves in Liquid/Gas and Liquid/Vapour Two-Phase Systems.* 1995 ISBN 0-7923-3424
32. A. Gyr and H.-W. Bewersdorff: *Drag Reduction of Turbulent Flows by Additives.* 1995 ISBN 0-7923-3485-
33. Y.P. Golovachov: *Numerical Simulation of Viscous Shock Layer Flows.* 1995 ISBN 0-7923-3626-

Kluwer Academic Publishers – Dordrecht / Boston / London